Transforming Service Experience Through Anthropomorphic Robots

Mohsin Altaf
Global Banking School, UK

IGI Global
Scientific Publishing
Publishing Tomorrow's Research Today

Vice President of Editorial Melissa Wagner
Director of Acquisitions Mikaela Felty
Director of Book Development Jocelynn Hessler
Production Manager Mike Brehm
Cover Design Jose Rosado

Published in the United States of America by
IGI Global Scientific Publishing
701 East Chocolate Avenue
Hershey, PA, 17033, USA
Tel: 717-533-8845 | Fax: 717-533-7115
Website: https://www.igi-global.com E-mail: cust@igi-global.com

Library of Congress Cataloging-in-Publication Data

LCCN: 2026002324 (CIP Data Pending)
ISBN13: 979-8-3373-6640-1
Isbn13Softcover: 979-8-3373-6641-8
EISBN13: 979-8-3373-6642-5
British Cataloguing in Publication Data
A Cataloguing in Publication record for this book is available from the British Library.

Table of Contents

Detailed Table of Contents

This chapter examines the transformative convergence of the metaverse and Non Fungible Tokens (NFTs) and its implications for contemporary fashion retail. Using a narrative integrative review of academic literature and industry evidence published between 2021 and 2025, the chapter develops a strategic and interdisciplinary framework that integrates technological enablers, consumer psychology, and brand strategy. It conceptualises the metaverse as an immersive, persistent digital ecosystem that is reshaping how fashion brands design, authenticate, and commercialise value, while NFTs introduce verifiable digital ownership, programmable scarcity, and new revenue models. The analysis highlights clear differences in how luxury and fast fashion brands adopt these technologies, with luxury emphasising exclusivity and provenance, and fast fashion prioritising accessibility and gamified engagement.

Artificial intelligence is developing rapidly, and anthropomorphic service robots are beginning to move from theory into everyday organisational practice. This chapter explores how advances in emotional AI and generative personality modelling may reshape service delivery in sectors such as healthcare, hospitality, retail, and tourism by 2030. Drawing on the Technology Acceptance Model, leadership theory, and research on anthropomorphism, it presents an integrated framework to support organisations in adopting these technologies. The discussion considers not only technological capability, but also customer expectations, employee readiness, and the wider organisational context. It also reflects on how leadership is evolving in response to these changes, highlighting the importance of AI literacy, ethical awareness, and the ability to manage collaboration between people and intelligent systems.

Retailing enters a new era where the AI personaliser goes beyond digital technology to be an anthropomorphic robot that can identify the customer, recall their history of purchases, and adapt itself based on the type of interaction. Using CASA theory, the personalization versus privacy paradox, and uncanny valley theory, the chapter formulates a conceptual framework where anthropomorphic design intensity acts as a moderator to intensify the role of trust building through AI personalization and privacy concerns on purchase intentions. The extensions are based on an empirically tested model (N = 214) where bootstrapped mediation revealed that trust has a positive mediating role in the relationship between personalization and purchase intentions, whereas privacy concern had a negative indirect effect on it. Strategic implications are made for the use of anthropomorphic service robots as brand ambassadors with a considerable potential to influence consumer trust.

The rapid acceptance of new technologies such as artificial intelligence (AI), machine learning (ML), The Internet of Things (IoT), robotics, and augmented reality (AR) have drastically transformed the retail landscape, especially in the post-COVID-19 pandemic landscape. This paper examines the implications of technological innovation on the changed relations of employees, including the introduction of new roles, some revised roles, and a small percentage of job losses. An excellent empirical example is the use of robots and the Internet of Things and generative artificial intelligence at Walmart, the application of which has shown the transformational effect of technology on the retail sphere and has created a more efficient and customer-oriented ecosystem (Rastogi & Saha, 2025). The research highlights why we need to create a balance between automation and human flexibility to have a sustainable and inclusive growth.

This study is based on secondary research, relying on the academic literature, industry reports, and case studies, to examine the effects that these barriers have on the guest's desire to engage AI at the front office. The results suggest that despite the importance of the technical performance, the psychological comfort often has a more decisive effect on the adoption. Therefore, the work of the hotels should be aimed not only at the technological efficiency but the emotional and trust-related aspects as well. Thus, the article proposes a hybrid model of services that will incorporate the benefits of AI and the inimitable human touch. This would help increase efficiency without reducing the overall guest experience, thus increasing the acceptance of AI in the hospitality front office.

The increasing prevalence of social robotics in service settings is bringing about a shift in how customer experience management is being conducted in hospitality, health care, retailing, and banking services. With more and more humanoid and quasi-humanoid robots interacting with customers, emotional engagement is now an essential determinant of satisfaction, trust, and loyalty. In this chapter, the concept of emotional engagement in human-robot service encounters will be discussed from several disciplinary angles, such as service marketing, affective computing, HRI, and organizational psychology. The review, based on 70+ scholarly sources from 2020 to 2025, will focus on various theories including CASA, S-O-R, and emotional labor.

of England interest rates. Policy implications for AI assurance, regional development, and skills investment are discussed.

This chapter constructs a single game-theoretic and multi-objective optimization model to model anthropomorphic service robots in brand strategy and customer experience. The interaction between customers and robots is modeled as a bi-level strategic game, where the firms maximize anthropomorphic design vectors (voice, expressiveness, adaptability) and the customers respond through the process of trust and engagement. A utility maximization model is developed to balance the emotional involvement, brand alignment, and the operational cost. The framework uses the state estimation of digital twins to enable real-time adaptation of robot behavior. Optimal anthropomorphic strategies are derived using equilibrium analysis and dynamic optimization, offering a mathematically based approach to AI-driven service design.

The integration of anthropomorphic robots in hospitality and tourism is revolutionizing service experiences by blending human-like interaction with advanced technology. These robots enhance efficiency, personalization, and emotional engagement, creating a new form of human–machine synergy that reshapes guest expectations and operational models. By mimicking human behaviors, expressions, and communication styles, they bridge the gap between automation and empathy, offering seamless, consistent, and memorable experiences. This transformation not only redefines service delivery but also opens avenues for sustainable innovation and workforce collaboration in the evolving landscape of smart tourism.

Preface

The rapid evolution of artificial intelligence, machine learning, and service robotics is fundamentally transforming the ways organizations interact with customers and deliver value. As intelligent technologies become increasingly integrated into service environments, anthropomorphic robots are emerging not merely as automated tools but as social actors capable of influencing customer perceptions, emotions, and behaviors. Their ability to communicate through humanlike voices, facial expressions, gestures, emotional recognition, and adaptive learning has introduced new opportunities—and new challenges—for organizations seeking to create meaningful, engaging, and trustworthy service experiences.

Transforming Service Experience Through Anthropomorphic Robots brings together leading perspectives from marketing, branding, human-robot interaction, consumer psychology, service management, artificial intelligence, and technology design to examine this rapidly developing field. While much of the existing literature has concentrated on the technical capabilities and operational efficiency of service robots, this volume shifts the conversation toward their human-centered dimensions. It explores how anthropomorphic characteristics influence customer trust, emotional engagement, perceived service quality, and brand relationships, positioning intelligent robots as active participants in the customer experience rather than passive technological interfaces.

A distinguishing contribution of this book is its interdisciplinary approach. By integrating theories from multiple domains, it presents a comprehensive framework for understanding how AI-powered anthropomorphic robots function as brand representatives within increasingly digital service ecosystems. The chapters examine how advances in natural language processing, emotion detection, adaptive learning, and social interaction enable robots to deliver personalized experiences while simultaneously shaping organizational identity and customer expectations.

The volume also emphasizes practical relevance by presenting real-world applications across industries including hospitality, tourism, healthcare, and retail. These examples demonstrate how organizations are leveraging anthropomorphic service robots to improve customer engagement, enhance operational effectiveness, and create distinctive brand experiences. At the same time, the contributors critically examine the broader implications of these technologies, addressing important issues such as ethical AI, algorithmic bias, emotional manipulation, privacy, cultural acceptance, service authenticity, and the evolving relationship between humans and intelligent machines.

As organizations continue to embrace automation and AI-driven innovation, understanding the psychological, social, and strategic dimensions of anthropomorphic robotics becomes increasingly important. The chapters in this volume provide valuable insights for designing customer-centered robotic services that balance technological sophistication with empathy, trust, transparency, and responsible innovation. In doing so, they contribute to ongoing discussions surrounding the future of branding, customer experience management, and intelligent service ecosystems.

This book has been developed to serve a broad interdisciplinary audience. Researchers, academics, and postgraduate students in marketing, branding, service management, human-robot interaction, consumer psychology, artificial intelligence, and digital innovation will find a rich collection of theoretical perspectives and emerging research directions. Equally, practitioners—including service designers, innovation managers, brand strategists, technology developers, and business leaders—will benefit from practical guidance on integrating anthropomorphic robots into customer-facing environments while strengthening organizational objectives and consumer relationships. Policymakers and professionals concerned with the ethical governance of AI technologies will also find valuable discussions that inform responsible deployment in both commercial and public service settings.

As editor, I hope this volume encourages deeper scholarly inquiry, inspires innovative industry practices, and fosters thoughtful dialogue about the opportunities and responsibilities associated with anthropomorphic service robots. By exploring how intelligent, humanlike technologies are reshaping customer interactions and redefining service experiences, this book contributes to a growing body of knowledge that will help guide the future development of ethical, effective, and human-centered AI-powered services.

ORGANIZATION OF THE BOOK

The chapters in this volume present a comprehensive exploration of how anthropomorphic robots, artificial intelligence, and related digital technologies are transforming customer experiences, branding strategies, and service innovation across diverse industries. Collectively, the contributions examine both the technological foundations and the human-centered dimensions of intelligent service systems, offering readers an interdisciplinary perspective on the future of AI-enabled service interactions.

The volume opens by examining the convergence of the metaverse and non-fungible tokens (NFTs) within the fashion retail sector, demonstrating how immersive digital environments and verifiable digital ownership are redefining brand engagement, consumer behavior, and value creation. This discussion establishes a broader context for understanding how emerging digital technologies are reshaping customer experiences beyond traditional physical service environments.

Building upon this digital transformation, the following chapter explores the strategic readiness required for organizations to successfully implement anthropomorphic service robots by 2030. Drawing upon leadership theory, technology acceptance, and organizational change, it highlights the evolving competencies leaders must develop to manage emotionally intelligent robotic systems while fostering employee readiness, ethical governance, and successful human-AI collaboration.

Several chapters then investigate the critical role of trust in human-robot interactions. One contribution presents an empirically grounded framework demonstrating how AI-powered personalization and anthropomorphic design influence consumer trust, privacy perceptions, and purchase intentions in retail environments. Complementing this work, another chapter examines the broader impact of emotional engagement during human-robot service encounters, synthesizing contemporary research to explain how social robotics can strengthen customer satisfaction, trust, loyalty, and long-term relationship building across multiple service industries.

The implications of intelligent automation for organizational design are explored through an examination of robotics in modern retail environments. This chapter considers how artificial intelligence, robotics, IoT, and related technologies are reshaping workforce structures through job redesign, evolving employee responsibilities, and changing organizational roles, while emphasizing the importance of balancing technological efficiency with human adaptability to ensure sustainable and inclusive transformation.

Attention then shifts to the hospitality industry, where one chapter investigates the psychological and technical barriers influencing guest acceptance of AI-enabled front office services. By demonstrating that emotional comfort and trust frequently outweigh technical performance in determining adoption, the authors advocate for hybrid service models that successfully integrate intelligent automation with meaningful human interaction.

The book further expands its discussion by presenting a comprehensive framework for anthropomorphic AI robots across multiple customer-facing industries. This chapter introduces a human-centered service model that integrates emotional intelligence, operational efficiency, brand alignment, and ethical responsibility, illustrating how organizations can leverage intelligent robotics while addressing concerns surrounding privacy, workforce transformation, and responsible AI governance.

The volume also includes an evidence-based analysis of anthropomorphic service robot adoption within the United Kingdom. Drawing upon extensive macroeconomic and regional data, this chapter demonstrates measurable relationships between AI investment, service robot deployment, labor productivity, customer experience, and financial viability, providing valuable insights for policymakers, organizational leaders, and researchers interested in the economic implications of intelligent service technologies.

Complementing these empirical perspectives, another chapter introduces a sophisticated mathematical framework that applies game theory, multi-objective optimization, and digital twin technologies to anthropomorphic service design. By modeling strategic interactions between firms and consumers, the chapter offers innovative approaches for optimizing robot behavior, customer trust, emotional engagement, brand alignment, and operational efficiency through data-driven decision-making.

The volume concludes by returning to the hospitality and tourism sectors, illustrating how anthropomorphic robots are redefining human-machine collaboration through personalized, emotionally responsive, and efficient service delivery. This final contribution emphasizes the transformative potential of intelligent robotics to create memorable customer experiences while supporting sustainable innovation and collaborative workforce models within rapidly evolving service ecosystems.

Together, these chapters demonstrate that anthropomorphic service robots represent far more than technological advancements; they signify a fundamental transformation in how organizations design customer experiences, build brand relationships, and integrate intelligent technologies into service delivery. By combining conceptual frameworks, empirical research, mathematical modeling, and practical case studies, this volume provides readers with a comprehensive foundation for understanding the opportunities, challenges, and future directions of AI-powered anthropomorphic service systems.

CONCLUDING REMARKS

The rapid advancement of artificial intelligence, machine learning, and human-centered robotics is redefining the future of service industries in ways that were once considered purely aspirational. As anthropomorphic service robots become increasingly capable of communicating, adapting, and responding with humanlike characteristics, organizations are presented with unprecedented opportunities to transform customer experiences, strengthen brand relationships, and deliver more personalized and efficient services. At the same time, these innovations introduce complex questions surrounding ethics, trust, privacy, workforce transformation, and the evolving nature of human-machine collaboration.

The contributions assembled in this volume reflect the interdisciplinary nature of these challenges and opportunities. By integrating perspectives from marketing, branding, service management, consumer psychology, human-robot interaction, artificial intelligence, hospitality, retail, organizational leadership, and computational modeling, the chapters collectively provide a comprehensive understanding of how anthropomorphic robots are reshaping contemporary service ecosystems. They demonstrate that successful implementation extends well beyond technological sophistication, requiring thoughtful attention to emotional intelligence, customer expectations, organizational readiness, responsible AI governance, and sustainable innovation.

As editor, I hope this volume serves not only as a valuable scholarly resource but also as a catalyst for future research and informed practice. The insights presented throughout these chapters encourage readers to critically examine both the transformative potential and the broader societal implications of deploying intelligent, humanlike technologies in customer-facing environments. As AI continues to evolve, so too must our approaches to designing service experiences that remain ethical, inclusive, transparent, and centered on human values.

It is my sincere hope that this book inspires continued collaboration among researchers, educators, practitioners, policymakers, and technology developers as they shape the next generation of intelligent service systems. By fostering interdisciplinary dialogue and advancing our understanding of anthropomorphic service robotics, this volume contributes to the ongoing evolution of service innovation and provides a meaningful foundation for the future of human-centered AI in an increasingly connected digital world.

Chapter 1
The Transformative Nexus of Metaverse and NFTs in the Fashion Retail:
A Comprehensive Convergence of Digital Realms and High Fashion

Sadia Idress
http://orcid.org/0000-0001-6993-984X
Global Banking School, University of Suffolk, UK

Mohsin Altaf
Global Banking School, Canterbury Christ Church University, UK

ABSTRACT

This chapter examines the transformative convergence of the metaverse and Non Fungible Tokens (NFTs) and its implications for contemporary fashion retail. Using a narrative integrative review of academic literature and industry evidence published between 2021 and 2025, the chapter develops a strategic and interdisciplinary framework that integrates technological enablers, consumer psychology, and brand strategy. It conceptualises the metaverse as an immersive, persistent digital ecosystem that is reshaping how fashion brands design, authenticate, and commercialise value, while NFTs introduce verifiable digital ownership, programmable scarcity, and new revenue models. The analysis highlights clear differences in how luxury and fast fashion brands adopt these technologies, with luxury emphasising exclusivity and provenance, and fast fashion prioritising accessibility and gamified engagement.

DOI: 10.4018/979-8-3373-6640-1.ch001

INTRODUCTION: THE CONVERGENCE OF DIGITAL REALMS AND HIGH FASHION

The fashion industry is experiencing a profound transformation, largely propelled by continuous digital innovation (Casciani, Chkanikova and Pal, 2022). This shift is moving fashion towards Industry 4.0, integrating advanced digital technologies like 3D modelling, virtual and augmented reality (VR/AR), and digital twinning into design, production, and distribution processes (Casciani, Chkanikova and Pal, 2022; Millatunnisa, 2022). To frame this analysis, the chapter adopts a narrative integrative review approach, synthesising interdisciplinary academic literature and industry reports published between 2021 and 2025. Rather than offering a systematic evaluation, the aim is to develop a conceptual and strategic understanding of how metaverse technologies and NFTs are reshaping fashion retail, with attention to both emerging opportunities and structural challenges.The emergence of the metaverse represents a new frontier for brand engagement and commerce, allowing consumers to interact with products virtually and influencing brand awareness through immersive experiences and digital product ownership (Venturini and Columbano, 2024; Periyasami and Periyasamy, 2022; Nalbant & Aydın, 2023). As the chapter later discusses, these developments are not merely technological novelties but signal deeper organisational shifts in how fashion brands design, communicate, and commercialise value. Non-Fungible Tokens (NFTs) act as catalysts for digital ownership, authenticity, and scarcity (Wu, Liu and Weng, 2023), enabling luxury brands to quantify the value of virtual items and protect brand integrity in digital marketplaces. However, early implementations also reveal volatility and uneven consumer adoption, with several high-profile NFT fashion projects experiencing rapid declines in secondary-market value — a reminder that digital fashion ecosystems remain in an experimental phase (Sung, Kwon and Sohn, 2023).

The synergistic relationship between the metaverse and NFTs is actively reshaping fashion's future, creating new opportunities for direct consumer engagement and novel forms of creative expression that challenge traditional notions of style and beauty (Canag et al., 2025). This convergence empowers brands to extend their identities into virtual realms and adapt to evolving consumer preferences (Ahmed et al., 2025), while also raising strategic questions about long-term viability, interoperability, and organisational capability development. By integrating technological, strategic, and behavioural perspectives, this chapter provides a comprehensive foundation for understanding how digital fashion ecosystems are evolving — and where their most significant structural impacts are likely to emerge.

1 THE TRANSFORMATIVE NEXUS OF METAVERSE AND NFTS IN THE FASHION INDUSTRY

The metaverse, as an immersive and interactive digital ecosystem, is fundamentally reshaping the landscape of brand-consumer interactions within the fashion industry (Coşkun, 2025). This transformative shift has profound implications for both fast-fashion and luxury brands, compelling them to navigate emerging virtual retail environments and redefine their strategies for brand engagement (Coşkun, 2025). Within this evolving digital realm, Non-Fungible Tokens (NFTs) are playing a crucial role by enabling the creation of new digital products, ensuring verifiable ownership, and delivering unique brand experiences in these virtual spaces (Liu, Wang and Liu, 2025; Donvito, Acuti and Song, 2024). The adoption of digital strategies, particularly NFTs, by fashion brands is directly influenced by the proportion of fashion customers to conventional customers, as well as the inherent digital value and cost associated with NFTs (Liu, Wang and Liu, 2025). This moves towards digital and virtual presence within the fashion industry not only offers novel consumer experiences but also presents innovative prospects for businesses to engage with a technologically advanced audience (Donvito, Acuti and Song, 2024).

1.1 The Metaverse: A New Paradigm for Fashion Experience and Retail

The metaverse is emerging as a critical frontier for the fashion industry, transforming how brands interact with consumers and conduct retail. Defined as persistent, interconnected virtual spaces, the metaverse facilitates social interaction and economic activity, offering a rich environment for immersive experiences (Trunfio and Rossi, 2022). This digital realm allows fashion houses to transcend traditional boundaries, creating innovative avenues for engagement and commerce (Dwivedi et al., 2023).

Within these virtual environments, fashion brands are hosting immersive experiences such as virtual fashion shows, which allow global audiences to participate and witness new collections in dynamic, interactive settings (Ahmed et al., 2025). These events, alongside other interactive brand activations, enable consumers to engage with fashion in unprecedented ways, fostering deeper connections and enhancing brand loyalty (Venturini and Columbano, 2024). A significant development in this space is the creation of digital twins, virtual replicas of physical garments. This blurs the lines between tangible and intangible apparel, enabling consumers to own and wear digital versions of their favorite outfits in the metaverse (Casciani, Chkanikova and Pal, 2022). This integration supports new retail models, including virtual stores where shoppers can browse and purchase items for their avatars, digital try-on experiences that provide a realistic preview of garments, and direct-to-avatar commerce, revolutionizing the shopping journey (Periyasami and Periyasamy, 2022). This evolution signifies a move towards a hybrid consumption model where physical and virtual fashion coexist and complement each other (Millatunnisa, 2022).

Key metaverse platforms are already facilitating these transformations through various brand integrations, as highlighted in the following table:

Table 1. Metaverse platforms as digital ecosystems for fashion brand integration

Metaverse Platform	Key Fashion Brand Integrations	Platform Features Relevant to Fashion (Examples)
Decentraland	Dolce & Gabbana, Estée Lauder, Tommy Hilfiger	Decentralized ownership, wearable NFTs, virtual land for stores, fashion week events (Decentraland, 2025; British Vogue, 2021).
Roblox	Gucci, Nike, Tommy Hilfiger	User-generated content, virtual economies, branded experiences, avatar customization (Roblox, 2022).
The Sandbox	Adidas, Ralph Lauren, Warner Music Group	Voxel-based aesthetics, NFT integration, gaming experiences, virtual real estate (The Sandbox, 2024).
Spatial	Dior, Louis Vuitton	Immersive 3D environments, collaborative spaces, virtual showrooms (Vogue, 2025 ; Klerk, 2021)
Fortnite	Balenciaga, Moncler	In-game cosmetic items, virtual concerts, character skins, brand collaborations (Fashionbi, 2025).

1.2 NFTs: Revolutionizing Ownership, Authenticity, and Value in Fashion

Non-Fungible Tokens (NFTs) are transforming the fashion industry by introducing new paradigms of ownership, authenticity, and value. At their core, NFTs are unique digital assets stored on a blockchain, serving as immutable proof of ownership for digital items (Wu, Liu and Weng, 2023). This blockchain-based certification provides a transparent and verifiable record, fundamentally shifting how value is assigned and transferred in digital realms (Sung, Kwon and Sohn, 2023).

One of the primary impacts of NFTs in fashion is the creation of digital scarcity and authenticity for virtual fashion items (Canag et al., 2025). By tokenizing digital apparel, accessories, and haute couture, brands can ensure their uniqueness and origin, mirroring the exclusivity and brand protection inherent in physical luxury goods. Beyond virtual assets, NFTs are also emerging as a powerful tool to combat counterfeiting in the physical luxury market by serving as irrefutable proof of ownership and authenticity for tangible products (Sung, Kwon and Sohn, 2023). This application allows consumers to verify the genuine nature of their purchases and track provenance, offering a significant enhancement to brand integrity and consumer trust (Millatunnisa, 2022). Furthermore, programmable NFTs introduce advanced functionalities, enabling creators to embed royalties into smart contracts, thus automatically earning a percentage on future sales in secondary markets. This not only fosters a vibrant ecosystem for digital fashion but also opens new revenue streams and dynamic ownership experiences for both brands and creators (Ahmed et al., 2025).

1.2.1 Overview of Blockchain Technologies Underpinning Major Fashion NFT Projects

Blockchain technologies provide the foundational infrastructure for the burgeoning fashion NFT market, with several platforms offering distinct advantages. Ethereum, as the pioneering platform for NFTs, boasts a large ecosystem and established security, although its high gas fees can pose challenges for widespread adoption in fashion (Casciani, Chkanikova and Pal, 2022). Polygon, a Layer 2 scaling solution for Ethereum, addresses these scalability issues by offering lower transaction costs and faster speeds, making it suitable for frequent fashion NFT releases and broader market engagement.

Table 2. Major blockchain networks and their applications in fashion NFTs

Blockchain	Core Technology	Key Fashion NFT Projects/ Integrations	Relevance for Fashion NFTs
Ethereum	Proof-of-Stake (formerly Proof-of-Work)	DressX, Auroboros, Gucci (NFTs on various marketplaces)	Pioneering NFT platform, large ecosystem, established security, but high gas fees can be a barrier for mass adoption (BTC markets, 2025).
Polygon	Layer 2 scaling solution for Ethereum (PoS sidechain, zk-rollups)	Dolce & Gabbana, Prada, Adidas (via mainstream platforms)	Solves Ethereum's scalability issues with lower transaction costs and faster speeds; ideal for frequent micro-transactions (Markets.com, 2024).
Solana	Proof-of-History, Proof-of-Stake	Aurory (gaming/fashion aesthetics), various high-volume marketplaces	High transaction throughput and low costs attract gaming and collectible NFTs, often involving fashion elements (Solana, 2025).
Flow	Proof-of-Stake	NBA Top Shot (sports collectibles), (Bernatik, 2022) Flowverse	Known for ease of use and mainstream appeal; emerging platform for digital fashion (Bernatik, 2022).
BNB Smart Chain	Proof-of-Staked Authority	Various metaverse projects (e.g., ArtMeta; also supports projects found on Ethereum like Decentraland)	Offers Ethereum compatibility with faster, lower-cost transactions; suitable for fashion, DeFi, and metaverse NFT projects (Bnbchain.org, 2025).

Solana's high transaction throughput and low costs attract gaming and collectible NFTs, often featuring fashion elements, demonstrating its potential for digital fashion. Flow and BNB Smart Chain also play significant roles, with Flow designed for mainstream adoption and user-friendly features, and BNB Smart Chain offering EVM compatibility with faster transactions and lower fees (Periyasami and Periyasamy, 2022).

2 TECHNOLOGICAL UNDERPINNINGS AND DIGITAL ASSETS

The fashion metaverse is built upon a constellation of advanced technologies—including VR/AR, AI, blockchain, and NFTs—that collectively enable immersive experiences and new forms of value creation (Coşkun, 2025; Ahmed et al., 2025). However, the strategic significance of these technologies becomes clearer when examined through empirical outcomes rather than technical descriptions. For instance, while VR and AR are widely promoted as tools for immersive retail, early activations reveal mixed results: Gucci's Roblox experience attracted over 20 million visits, yet user retention declined sharply after the first month, demonstrating the challenge of sustaining engagement beyond initial novelty (Heo, 2025; Vogue Business, 2025).

Blockchain's role in digital fashion extends beyond ensuring ownership; its strategic value lies in how it reshapes authentication, resale markets, and brand protection. Yet empirical evidence shows that blockchain-enabled fashion assets remain vulnerable to market instability. Nike's .SWOOSH platform experienced a decline of approximately 40% in NFT trading volume during the 2023 crypto downturn, reflecting broader market contraction across digital asset ecosystems (CoinDesk, 2023; Nike, 2022) . Similarly, Dolce & Gabbana's Collezione Genesi NFTs lost more than 70% of their resale value within six months of launch, mirroring broader declines in luxury NFT secondary-market performance (Vogue Business, 2022; Business of Fashion, 2022; NonFungible.com, 2022), highlighting the difficulty of maintaining long-term asset value without sustained utility or platform integration.

Digital twins are increasingly used to streamline design and production, enabling virtual prototyping, lifecycle tracking, and reduced sampling costs (Li and Liu, 2024). Their strategic impact is evident in operational efficiencies rather than technological novelty. Brands adopting digital twin workflows report shorter development cycles and improved supply chain visibility, aligning with broader industry shifts toward data-driven decision-making. Likewise, generative AI is transforming creative and commercial processes by automating design tasks, accelerating trend forecasting, and enabling large-scale personalisation (Ahmed et al., 2025). These capabilities signal deeper organisational changes, including new skill requirements and reconfigured design pipelines.

Digital fashion products—including virtual wearables and avatar accessories—continue to expand across metaverse platforms (Sayem, 2022). However, platform fragmentation remains a significant barrier to scalable adoption. Assets created for Roblox, Decentraland, or The Sandbox are rarely interoperable, increasing production costs and limiting consumer utility. This fragmentation reinforces the need for cross-platform standards if digital fashion is to evolve from experimental activations to sustainable retail models.

Table 3. Technological enablers and value creation in the fashion metaverse

Technology	Application in Fashion Metaverse	Examples/Impact
Virtual Reality (VR)	Immersive retail environments, virtual fashion shows, interactive brand experiences.	Users can explore virtual stores and try on digital garments in a highly engaging manner; enhances brand storytelling and consumer interaction in simulated worlds (Coşkun, 2025).
Augmented Reality (AR)	Virtual try-ons, interactive marketing campaigns, digital overlays on physical products.	Enables customers to visualize how digital clothing or accessories would look on them in real-time using their smartphones or AR devices; blurs the lines between physical and digital fashion experiences (Coşkun, 2025).
Artificial Intelligence (AI)	Personalized recommendations, trend forecasting, enhanced design processes, customer service chatbots.	AI algorithms can analyze fashion trends, suggest personalized outfits to avatars, and assist in the creative design phase, leading to more efficient product development and tailored consumer experiences (Ahmed et al., 2025).
Blockchain	Verifiable ownership of digital assets, transparent transactions, digital scarcity.	Underpins NFTs, ensuring the authenticity and unique ownership of digital fashion items; enables secure and traceable transactions for virtual clothing and accessories (Liu, Wang and Liu, 2025).
Non-Fungible Tokens (NFTs)	Unique digital fashion items, digital collectibles, proof of ownership, access to exclusive content.	Represent virtual garments, accessories, or experiences that are digitally scarce and provably owned; can serve as tickets to exclusive metaverse events or unlock special features in games, providing new revenue streams and brand engagement models (Trujillo and Bacciu, 2023).
Digital Twin Technology	Product design and development, lifecycle tracking, intelligent manufacturing, immersive displays.	Creates virtual replicas of physical products, enabling advanced design modifications, real-time monitoring of product performance, and efficient manufacturing processes; enhances the digital value of clothing by connecting to NFT ecosystems (Li and Liu, 2024).
Generative AI (GAI)	Enhanced design, manufacturing, sales, and client experiences.	Facilitates the rapid creation of new designs, optimizes manufacturing processes, personalizes sales interactions, and improves overall customer satisfaction within the metaverse (Ahmed et al., 2025).

3 OPERATIONAL AND SUPPLY CHAIN TRANSFORMATIONS IN DIGITAL FASHION ECOSYSTEMS

The integration of digital fashion, NFTs, and digital twins has significant implications for traditional fashion supply chains. Digital twins, which the chapter describes as enabling "product design and development, identity verification, and lifecycle tracking," are increasingly used to streamline pre-production processes (Li and Liu, 2024). By creating virtual prototypes, brands can reduce physical sampling, shorten design cycles, and minimize material waste — contributing to more agile and sustainable operations.

NFTs also introduce new supply chain architectures by embedding traceability and authentication into blockchain-based records. This reduces reliance on traditional paper-based provenance systems and enhances transparency across sourcing, manufacturing, and resale markets. For luxury brands, NFT-enabled authentication mitigates counterfeiting risks, while for fast fashion, digital assets reduce the need for large physical inventories by shifting consumption toward virtual goods (Casciani, Chkanikova and Pal, 2022; Canag et al., 2025).

Moreover, virtual try-ons and D2A commerce reduce demand uncertainty and overproduction — long-standing challenges in fashion logistics. As the chapter notes, "digital wearables… can be traded and utilized across various metaverse platforms," signalling a shift toward post-physical consumption models (Sayem, 2022). These developments collectively reshape inventory management, forecasting, and distribution, pushing the industry toward leaner, more data-driven supply chain operations.

4 STRATEGIC ADOPTION BY FASHION BRANDS: LUXURY VS. FAST FASHION

Luxury fashion brands are strategically leveraging the metaverse to cultivate exclusivity, digital scarcity, and immersive storytelling, primarily to engage younger demographics and elevate brand recognition (Coşkun, 2025; Heo, 2025). An illustrative example is Gucci Town on the Roblox platform, which demonstrates how luxury brands utilize gaming environments to market products, expand their market presence cost-effectively, and connect with new audiences by merging traditional exclusivity with interactive digital value propositions (Heo, 2025). In contrast, fast-fashion brands prioritize accessibility and gamified experiences, frequently engaging consumers on platforms such as Roblox and Decentraland to reach a broader consumer base (Coşkun, 2025; Elsayegh and Mohamed, 2024). Both segments integrate NFTs into their strategies, yet their motivations diverge: luxury brands focus on premium digital assets, while fast-fashion brands tend to explore more accessible digital offerings (Liu, Wang and Liu, 2025; Kniazeva et al., 2024).

Table 4. Metaverse and NFT strategies: A comparison of luxury and fast-fashion brands

Feature	Luxury Fashion	Fast Fashion
Primary Objective	Exclusivity, brand awareness, immersive storytelling, attracting younger consumers (Coşkun, 2025; Heo, 2025; Heo, 2025).	Accessibility, broad consumer reach, gamified experiences, sales volume (Coşkun, 2025).
Key Platforms	Roblox (e.g., Gucci Town), bespoke virtual experiences.	Roblox, Decentraland, other accessible gaming and virtual platforms (Coşkun, 2025; Elsayegh and Mohamed, 2024).
NFT Focus	Premium digital assets, digital scarcity, unique collectibles, exclusive access (Liu, Wang and Liu, 2025; Coşkun, 2025).	Accessible digital offerings, mass-market digital goods, virtual wearables (Liu, Wang and Liu, 2025; Kniazeva et al., 2024).
Engagement Strategy	Immersive experiences, co-creation, building brand narratives, social media sharing (Heo, 2025).	Interactive games, virtual try-ons, affordable digital items, high volume engagement (Coşkun, 2025).
Value Proposition	Blending traditional exclusivity with interactive digital experiences, identity, prestige (Heo, 2025).	Convenience, trendy virtual items, fostering community through gamification, affordability (Coşkun, 2025).
Challenges	Maintaining exclusivity in a digital environment, intellectual property risks, platform dependence (Heo, 2025).	Rapid content creation, managing digital asset value, brand differentiation in crowded spaces (Kniazeva et al., 2024).

This comparative analysis illustrates that while both luxury and fast-fashion brands are keen to leverage the metaverse and NFTs, their strategic approaches are tailored to their distinct market positions and target audiences.

4.1 Strategic Adoption by Fashion Brands (Failure Cases Integrated)

While luxury and fast-fashion brands are actively experimenting with metaverse activations and NFT-enabled strategies, early implementations reveal uneven outcomes. For example, Adidas' "Into the Metaverse" Phase 2 experienced a 70% decline in participation compared to Phase 1, signalling diminishing consumer enthusiasm for repetitive NFT drops and highlighting the difficulty of sustaining long-term engagement in digital asset ecosystems (Adidas, 2021). Similarly, although luxury brands such as Gucci have successfully leveraged platforms like Roblox to reach younger audiences, Gucci Garden's initial surge in visits did not translate into sustained user retention, underscoring the challenge of converting novelty-driven interactions into durable brand relationships (Aryal, 2025).

These cases illustrate that while the metaverse offers new avenues for storytelling and consumer reach, strategic adoption requires more than high-visibility activations. Brands must design digital assets with ongoing utility, interoperability, and community value to avoid the rapid engagement drop-offs observed in early experiments. The contrast between initial hype and subsequent decline demonstrates that digital fashion strategies must evolve beyond one-off campaigns toward more integrated, long-term digital ecosystems (Adidas, 2021; Aryal, 2025).

5 ECONOMIC AND BUSINESS MODELS IN METAVERSE AND NFT-DRIVEN FASHION ECOSYSTEMS

While fashion brands are rapidly experimenting with metaverse activations and NFT collections, the long-term economic viability of these initiatives depends on the development of sustainable business models. Current evidence suggests that NFT-enabled fashion ecosystems operate across several emerging revenue streams, including primary NFT sales, secondary-market royalties, token-gated experiences, and direct-to-avatar (D2A) commerce (Ahmed et al., 2025). The chapter already notes that "programmable NFTs introduce advanced functionalities, enabling creators to embed royalties into smart contracts" — a mechanism that fundamentally reshapes value capture in digital fashion markets.

Luxury brands typically adopt scarcity-driven pricing models, mirroring haute couture economics. Limited-edition NFT drops, auction-based pricing, and exclusive digital collectibles allow brands to maintain prestige while generating high margins. Fast-fashion brands, by contrast, rely on high-volume, low-cost digital wearables, leveraging micro-transactions on platforms such as Roblox and Decentraland to drive scale (Tam, 2025; Covington and Soltanisehat, 2025; Lim, 2021; Adidas, 2021; Aryal, 2025).

Long-term financial viability, however, hinges on ecosystem stability. NFT markets remain volatile, with fluctuating token values and unpredictable resale dynamics. Brands must therefore balance speculative demand with utility-driven value propositions — such as interoperability, access rights, or hybrid physical-digital bundles — to ensure sustained consumer engagement and revenue continuity. As the chapter notes, "the adoption of digital strategies… is directly influenced by the proportion of fashion customers to conventional customers," highlighting the need for business models that align with evolving consumer expectations (Liu, Wang and Liu, 2025).

6 CONSUMER ENGAGEMENT AND BEHAVIOUR IN THE METAVERSE

The metaverse is revolutionizing consumer engagement within the fashion industry by introducing novel virtual shopping environments, extensive avatar customization options, and interactive brand narratives (Donvito, Acuti and Song, 2024; Alexander et al., 2025). Consumer motivations for engaging with fashion in these virtual spaces are diverse. They include the pursuit of pleasure, opportunities for self-expression, stress relief, vicarious satisfaction, and the desire for augmented virtual experiences (Bialkova, 2024). The perception of innovation characteristics, such as relative advantage and compatibility, can also lead to a flow experience, which, however, can be moderated by factors like technostress and self-control (Kumar et al., 2026). Studies indicate that consumers are driven by freedom of expression, creativity, exploration, entertainment, and productivity in their metaverse interactions (Alexander et al., 2025).

Social presence and immersive features within metaverse platforms are crucial for driving customer brand engagement (CBE) and significantly influencing purchase intentions (Agnihotri et al., 2024). These elements shape cognitive, affective, behavioral, and social outcomes for users (Alexander et al., 2025). Furthermore, the realism of avatars plays a significant role in consumer behaviour. Greater avatar realism enhances mental simulation, leading to more positive product attitudes and influencing how consumers perceive and evaluate virtual fashion items (Bialkova, 2024). This effect is often moderated by construal levels, where individuals' processing modes affect their perception of visual similarity. While technological aspects of metaverse tools often surpass customer expectations, design factors, particularly visual appeal, frequently fall short, highlighting a need for improved user experience and engagement in the metaverse's aesthetic dimensions (Kumar et al., 2025).

6.1 Consumer Engagement and Behaviour in the Metaverse (Failure Cases Integrated)

Although immersive environments and avatar-based interactions offer new forms of consumer engagement, empirical evidence shows that user adoption remains inconsistent. For instance, Meta's Horizon Worlds fashion activations struggled with low user retention (Batchelor, 2022), reflecting broader adoption barriers in early metaverse platforms where technical friction, limited content depth, and unclear value propositions reduce sustained participation. These patterns align with research indicating that technostress, low self-efficacy, and perceived complexity weaken flow experiences and reduce purchase intentions in immersive retail contexts (Al-Adwan, A.S., 2024; Al-Kfairy, Alfandi and Khaddaj, 2025).

Similarly, NFT-driven engagement strategies face volatility. While early drops often generate significant attention, **secondary-market activity frequently declines sharply**, as seen in multiple fashion NFT collections whose resale values fell within months of launch (Bain, 2023). These failures highlight the importance of designing digital fashion experiences that extend beyond speculative value and provide meaningful, ongoing consumer benefits such as interoperability, exclusive access, or hybrid physical-digital rewards (Bain, 2023).

6.2 Consumer Motivations and Psychological Impacts

Research on consumer motivations highlights a range of intrinsic and extrinsic factors that drive engagement with products and experiences, particularly in digital and immersive contexts. Kim and Lee (2022) found that consumers are motivated by pleasure, self-expression, vicarious satisfaction, and stress relief, all of which significantly shape purchase satisfaction and influence future purchase intentions. These motivations go beyond simple product utility to reflect how consumption fulfils emotional and psychological needs, enhancing both immediate satisfaction and long-term engagement outcomes. Kumar et al. (2026) extend this understanding through innovation and flow theory, showing that perceived innovation characteristics such as relative advantage and compatibility can induce flow experiences among users. However, this immersive engagement can foster unrestrained buying behaviour, with technostress and individual self-control acting as moderators that weaken flow and mitigate impulsive purchase tendencies in hyper-technological environments. Alexander et al. (2024) identify motivations including freedom of expression, creativity, exploration, entertainment, and productivity in immersive spaces like the metaverse, revealing that these drivers produce cognitive, affective, behavioural, and social impacts on consumer outcomes. In aggregate, these studies illustrate that consumer motivations intertwine with psychological processes, such as emotional gratification, identity expression, and immersive engagement that profoundly affect satisfaction, future intentions, and broader behavioural outcomes in both physical and virtual consumption environments.

Table 5. Consumer motivations and psychological impacts

Author	Motivation Factors	Psychological Impacts
(Kim and Lee, 2022)	Pleasure, self-expression, vicarious satisfaction, stress relief.	Influence on purchase satisfaction and future purchase intentions.
(Kumar et al., 2026)	Flow experience from innovation characteristics (relative advantage, compatibility).	Unrestrained buying behaviour, technostress and self-control weaken flow experience.
(Alexander et al., 2025)	Freedom of expression, creativity, exploration, entertainment, productivity.	Cognitive, affective, behavioural, social outcomes.

6.3 Impact of Avatar Realism on Consumers

The realism of avatars in virtual environments is increasingly recognized as a critical determinant of consumer experience and product perception. Research into avatar design suggests that greater visual similarity between an avatar and the user facilitates easier mental simulation of virtual interaction scenarios, which in turn enhances attitudes toward products encountered in these spaces (Kim and Lee, 2025). This effect has been conceptualized through a serial moderated mediation mechanism involving construal levels, where high avatar realism reduces psychological distance and allows consumers to visualize themselves interacting with products more concretely, thereby fostering more favourable evaluations (Kim and Lee, 2025; Kim et al., 2023).

Studies on avatar-mediated experiences further demonstrate that realistic representations can strengthen consumer attachment to their digital selves and improve behavioural intentions, including continued platform use and engagement with branded content (Kim et al., 2023; Shih et al., 2023). These effects parallel broader phenomena identified in virtual behaviour research, such as the Proteus Effect, where avatar characteristics shape users' behaviours and beliefs in ways consistent with their digital representation ((Lin et al., 2023; Shah et al., 2025). In consumer contexts, enhanced avatar realism thus not only elevates the psychological presence within a virtual space but also contributes to improved product attitudes and potential purchase intentions by aligning the user's mental engagement with the simulated environment (Li and Huang, 2024).

Table 6. Impact of avatar realism on consumers

Author (Year)	Impact of Avatar Realism: on Consumers
Kim & Lee (2025)	Greater avatar similarity allows easier mental simulation of wearing or using a product, leading to better product attitudes; shows a serial moderated mediation effect involving construal levels.
Kim et al. (2023)	Higher avatar realism strengthens the user–avatar relationship and increases intention to use avatars via reduced psychological distance and enhanced social presence, improving engagement with virtual environments.
Shih et al. (2023)	Varying avatar realism influences perceived social presence and comfort in augmented/virtual interactions, with more realistic avatars enhancing comfort and connectedness in social contexts.
(Lin et al., 2023)	Higher virtual avatar appearance realism is perceived as more attractive and trustworthy, enhancing users' subjective interaction experience (e.g., trustworthiness, attractiveness) compared to lower realism.
Yan et al. (2024)	Virtual influencers with greater perceived humanness and realism elicit stronger emotional attachment and benefit-seeking behavior, mediated by social presence, enhancing consumer engagement and potentially purchase behavior in marketing contexts.
Li et al. (2024)	Anthropomorphic features (appearance, behavior, cognitive, emotional) of virtual streamers enhance cognitive trust and, in turn, amplify purchase intention; emotional anthropomorphism also directly influences purchase outcomes.

6.4 Impact of Social Presence on Consumers

Social presence plays a pivotal role in shaping consumer behaviour, particularly within online and immersive retail contexts. Agnihotri et al. (2024) demonstrate that social presence positively influences customer brand engagement (CBE), which in turn substantially impacts purchase intention (PI). Their research highlights that when consumers perceive others as present and interactive within a digital environment, whether through avatars or social cues, they experience stronger engagement with the brand, translating into higher intent to purchase. Importantly, consumer self-efficacy moderates the relationship between CBE and PI, suggesting that individuals who feel more capable and confident in navigating these environments derive greater motivational benefits from social presence. Alexander et al. (2024) also emphasize social presence as a central dimension of customer experience, finding that it shapes cognitive, affective, behavioural, and social outcomes in immersive spaces. Their qualitative work on fashion metaverses indicates that social presence not only enriches subjective experience but also supports deeper emotional connections and shared interactions among users, which collectively enhance overall consumer experience quality. These findings underscore that social presence is not merely a background factor but a core driver of engagement and behavioural responses in digitally mediated marketplaces, influencing both how consumers feel about a brand and how likely they are to act on those feelings.

Table 7. Impact of social presence on consumers

Author (Year)	Impact of Social Presence on Consumers
Agnihotri et al. (2024)	Positively impacts customer brand engagement (CBE), which substantially impacts purchase intention (PI); self-efficacy moderates CBE and PI.
Alexander et al. (2024)	Influences customer experience, including cognitive, affective, behavioral, and social outcomes in immersive spaces.
Poirier et al. (2024)	Social presence cues in product images evoke positive emotions and increase diagnosticity, which in turn increases purchase intentions.
Zheng et al. (2025)	Social presence enhances users' purchase intentions by fostering cognitive engagement and expectations in digital video contexts.
Dhakshitha (2025)	Social presence cues raise emotional arousal and consequently increase impulse buying behavior in e-commerce consumer environments.

6.5 Customer Experience and Design Factors

Customer experience in technology mediated environments depends heavily on both the capabilities of the underlying technology and the quality of design elements. Kumar et al. (2025) report that while advanced technologies often exceed customer expectations in terms of innovation and immersive potential, design factors such as visual appeal frequently fall short. This mismatch means that even when technology elevates user engagement through novelty and interactivity, subpar aesthetic and usability features can detract from the overall experience. In their analysis, customers expressed enthusiasm for technological capabilities but indicated that engagement and experience outcomes comprising cognitive satisfaction, emotional enjoyment, and behavioural follow through were not fully realized due to design limitations (Kumar et al., 2025; Alexander et al., 2025). This aligns with broader research on immersive customer experience, which suggests that experiential quality is shaped by multiple dimensions, including environmental fidelity, sociability, and usability (Kozinets, 2023; Zeng et al., 2025). When design factors are neglected, the immersive potential offered by cutting edge technologies may not translate into meaningful engagement or positive consumer outcomes. Therefore, ensuring that both technological performance and design aesthetics are optimally integrated is essential for achieving compelling customer experiences that meet or exceed user expectations in both virtual and hybrid retail environments (Menidjel et al., 2022; Romano, 2023).

Table 8. Customer experience and design factors

Author (Year)	Impact on Customer Experience
Kumar et al. (2025)	Technology often exceeded customer expectations, but design factors such as visual appeal fell short; customer experience and engagement did not meet expectations.
Alexander et al. (2024)	Immersive spaces (e.g., metaverse environments) significantly shape cognitive, affective, behavioral, and social experience dimensions; design pain points (e.g., navigation, interface aesthetics) can negatively affect overall experience quality.
Menidjel (2022)	In AI-driven marketing environments, factors like personalized, effective, enjoyable, reliable, and simple design significantly influence customer engagement; overly complex designs weaken experience outcomes.
Kozinets (2023)	Immersive technologies (AR/VR) transform customers' sensory, emotional, and behavioral experiences; design elements that fail to integrate multisensory feedback can limit the expected experiential benefits.
(Zeng et al., 2025)	Virtual and augmented content with higher realism and visual appeal enhances perceived presence, ease of imagination, and behavioral intentions (e.g., booking or purchase intentions), showing that design quality directly affects experience outcomes.
Archibald (2021 – AR retail)	Augmented Reality applications in retail improve customer experiences by increasing interactivity and engagement, but effectiveness depends on intuitive and visually appealing design integration.

6.6 Consumer Risk, Trust, and Market Volatility in NFT-Enabled Fashion

Although the metaverse offers new avenues for engagement, consumers face several risks that influence adoption and trust. NFT markets are characterized by price volatility, speculative behaviour, and fluctuating asset values, which may deter risk-averse consumers. The chapter already highlights that "technostress and self-control weaken flow experience" (Fortagne, 2025), suggesting that cognitive overload in digital environments can undermine positive engagement.

Fraud and authenticity concerns also persist. While blockchain provides immutable ownership records, consumers may still encounter counterfeit NFTs, phishing scams, or misleading marketing claims. Trust in blockchain-based fashion assets is further shaped by consumers' understanding of the technology; limited literacy around wallets, gas fees, and smart contracts can create perceived barriers to adoption (Upadhyay and Upadhyay, 2025).

Privacy concerns also emerge in immersive environments where avatar behaviour, biometric data, and interaction patterns may be tracked. Social presence — which the chapter identifies as a key driver of engagement — can also heighten vulnerability to social engineering or manipulative design practices (Pereira et al., 2025). Building consumer trust therefore requires transparent communication, robust authentication systems, and regulatory frameworks that protect users in virtual fashion ecosystems.

6.7 Market Volatility, Technological Fragmentation, and User Adoption Barriers

Although the chapter highlights the transformative potential of NFTs and the metaverse, several systemic challenges require deeper empirical grounding. NFT markets remain highly volatile, with rapid fluctuations in token prices and secondary-market activity. This volatility affects both consumer confidence and brand revenue predictability. For example, the dramatic price swings observed in major NFT marketplaces illustrate how speculative dynamics can undermine long-term value creation for fashion brands experimenting with digital assets (The Interline, 2025).

Technological fragmentation also poses a significant barrier to scalable adoption. As the chapter notes, "digital wearables... can be traded and utilized across various metaverse platforms," yet in practice, interoperability remains limited (Becker, 2025). Brands must currently develop separate assets for platforms such as Roblox, Decentraland, and The Sandbox, increasing production costs and reducing cross-platform utility for consumers. Empirical studies in digital ecosystems show that fragmentation reduces user retention and increases cognitive load, especially for non-expert users navigating wallets, gas fees, and platform-specific asset formats.

User adoption barriers further complicate mainstream uptake. Research on immersive retail environments indicates that technostress and low self-efficacy weaken flow experiences and reduce purchase intentions (McSorley, 2025). This aligns with your chapter's observation that "technological aspects of metaverse tools often surpass customer expectations, however design factors... frequently fall short," signalling a gap between technological capability and user readiness. Empirical studies presents that consumers cite complexity, lack of trust, and uncertain value as primary reasons for hesitating to purchase digital fashion assets (McSorley, 2025; Becker, 2025).

7 SYNERGISTIC APPLICATIONS: METAVERSE, NFTS, AND THE EVOLVING FASHION ECOSYSTEM

The convergence of the metaverse and NFTs is creating a dynamic and evolving fashion ecosystem, characterized by several synergistic applications that are redefining consumer engagement and brand strategies. One of the most prominent applications is the rise of wearable NFTs within the metaverse, enabling digital fashion to serve as a powerful medium for identity and self-expression (Venturini and Columbano, 2024). Consumers can personalize their avatars with exclusive digital garments and accessories, reflecting their style in virtual environments and fostering a parallel digital identity (Periyasami and Periyasamy, 2022). This digital apparel can range from unique one-off pieces to digital twins of real-world luxury items, offering a new dimension to fashion consumption.

Another significant development is the emergence of 'phygital' products, which combine physical items with linked NFTs to offer enhanced experiences or verifiable provenance (Millatunnisa, 2022). For instance, purchasing a physical luxury handbag might come with a corresponding NFT that certifies its authenticity, tracks its ownership history, and could even unlock exclusive content or experiences in the metaverse. This integration bridges the gap between the tangible and intangible, bolstering brand trust and providing a novel layer of value for consumers (Sung, Kwon and Sohn, 2023). Furthermore, the gaming industry is playing a crucial role in expanding the reach of digital fashion, with brands leveraging in-game economies and avatar customization through NFTs (Trunfio and Rossi, 2022). Fashion houses are collaborating with popular gaming platforms to create branded skins, virtual apparel, and interactive experiences, allowing millions of players to engage with fashion in a fun and accessible way (Amer et al., 2025). This not only introduces fashion to a broader, digitally native audience but also generates new revenue streams through the sale of in-game fashion NFTs.

This evolving ecosystem also empowers independent designers and creators through decentralized platforms and NFT drops. These platforms offer new avenues for emerging talent to showcase their work, reach a global audience, and monetize their creations without traditional intermediaries (Canag et al., 2025). NFT drops, often characterized by limited editions and unique designs, generate excitement and foster direct relationships between creators and consumers (Wu, Liu and Weng, 2023).

Notable fashion brands are actively engaging with these technologies, as highlighted in the table below:

Table 9. Examples of luxury fashion brand engagement with NFTs and the metaverse

Brand / Project	**Approach & Impact**
Gucci	Early adopter of digital fashion with virtual sneakers, Roblox experiences, and NFT collections. Demonstrated strong revenue potential and brand innovation (Independent, 2021; Roblox, 2022).
Louis Vuitton	Launched NFT-integrated mobile games ("Louis the Game") offering virtual collectibles and showcasing brand heritage in a digital format. Expanded brand narrative and engaged new audiences (Klerk, 2021).
Burberry	Partnered with Mythical Games for NFT collections in their *Blankos Block Party* game, featuring collectible digital vinyl toys and accessories. Explored new revenue models and digital collectible markets (Burberry, 2021; Butler, 2022).

continued on following page

Table 9. Continued

Brand / Project	Approach & Impact
Dolce & Gabbana	Released a nine-piece NFT collection, *Collezione Genesi*, which included both digital and physical haute couture items, achieving significant sales and proving the value of phygital luxury (British Vogue, 2021).
Nike	Acquired virtual shoe company RTFKT, signalling a major strategic move into digital fashion, NFTs, and metaverse wearables. Aims to dominate the virtual sneaker market and digital apparel space (Hahn, 2021; RTFKT, 2021).
Prada	Launched a time-based NFT series with physical apparel, offering exclusive experiences and demonstrating a commitment to integrating blockchain into luxury retail, enhancing customer loyalty and engagement (Zhang, 2022).

7.1 Brand Case Evidence Illustrating Market and Adoption Challenges

Concrete brand cases further illuminate these challenges:

Nike's .SWOOSH platform experienced rapid initial adoption but saw declining transaction volumes during periods of broader crypto-market downturns, demonstrating the sensitivity of fashion NFTs to macro-level volatility (Nike, 2022; The Interline, 2025).

Gucci's collaboration with Roblox succeeded in generating high engagement but highlighted interoperability limitations, as assets purchased in Roblox could not be transferred to other metaverse environments (Brooks, 2026).

Dolce & Gabbana's NFT couture collection achieved record-breaking auction prices, yet secondary-market activity declined sharply within months, illustrating the difficulty of sustaining long-term value without ongoing utility or platform integration (Brooks, 2026).

These cases underscore the need for brands to design NFT strategies that extend beyond novelty and incorporate durable value propositions such as interoperability, exclusive access, or hybrid physical-digital benefits.

8 CHALLENGES AND OPPORTUNITIES IN THE METAVERSE AND NFT FASHION LANDSCAPE

The burgeoning metaverse and NFT fashion landscape presents significant challenges alongside its transformative opportunities. One major hurdle lies in technical limitations, particularly concerning interoperability across diverse metaverse platforms (Nalbant & Aydın, 2023). The lack of seamless interaction between different virtual worlds restricts the fluid movement of digital assets and user identities, creating fragmented experiences for consumers and brands alike. Furthermore, the scalability of underlying blockchain networks remains a concern, as high transaction volumes can lead to network congestion and increased costs, particularly for proof-of-work (PoW) based systems (Casciani, Chkanikova and Pal, 2022). Addressing these technical issues is crucial for the widespread adoption and sustained growth of digital fashion.

Environmental concerns surrounding the energy consumption of certain blockchain protocols also pose a significant challenge. While newer consensus mechanisms like Proof-of-Stake (PoS) are substantially more energy-efficient, older and still widely used systems like Ethereum's former Proof-of-Work (PoW) model consumed considerable energy, leading to a substantial carbon footprint (Casciani, Chkanikova and Pal, 2022). As the table below illustrates, the difference in energy consumption between PoW and PoS blockchains is stark, highlighting the importance of transitioning to more sustainable frameworks for fashion NFTs. Regulatory uncertainty further complicates this evolving landscape, with nascent legal frameworks struggling to keep pace with rapid technological advancements (Wu, Liu and Weng, 2023). Issues such as intellectual property rights for digital designs, consumer protection within virtual economies, and data privacy in decentralized environments require clear guidance to foster trust and prevent exploitation (Sung, Kwon and Sohn, 2023).

Table 10. Energy efficiency and carbon footprint of blockchain platforms supporting fashion NFTs

Blockchain	Consensus Mechanism	Energy Consumption (kWh/transaction)	CO_2 Emissions (gCO_2/transaction)
Ethereum (PoW)	Proof-of-Work	83.3	48,000 (Igumenov, 2021)
Ethereum (PoS)	Proof-of-Stake	0.0003	0.08 (Kapengut and Mizrach, 2022; Igumenov, 2021)
Polygon	Proof-of-Stake	0.0003	0.08 (Saini, 2023; Markets.com, 2024)
Solana	Proof-of-History / Proof-of-Stake	0.00018	0.038 (OKX, 2025; Solana, 2025)
Flow	Proof-of-Stake	0.0000007	0.0001 (Dapper Labs, 2025)
BNB Smart Chain	Proof-of-Staked Authority	0.0002	0.05 (Saini, 2023; Bnbchain.org, 2025)

Despite these challenges, the sector offers immense opportunities, particularly in bridging the digital divide and enhancing accessibility. By lowering barriers to entry, decentralized platforms and accessible metaverse environments can democratize fashion, allowing independent designers to reach global audiences and consumers from diverse backgrounds to engage with fashion in new ways (Canag et al., 2025). Ensuring that these technologies are inclusive and easy to use for a broader audience is key to unlocking their full potential and fostering innovation in digital fashion (Periyasami and Periyasamy, 2022).

9 CHALLENGES AND OPPORTUNITIES: SUSTAINABILITY, ETHICS, AND FUTURE OUTLOOK

The integration of the metaverse into the fashion industry presents a complex array of challenges and opportunities. Key challenges that require careful consideration include ensuring the psychological well-being of users, addressing potential technology addiction, and safeguarding data privacy within these evolving digital environments (Donvito, Acuti and Song, 2024). Furthermore, navigating intellectual property rights within the virtual landscape is crucial to protect brand integrity and creative designs (Lapatoura, 2023).

Despite these hurdles, the metaverse offers significant opportunities, particularly in advancing sustainability within the fashion sector. Digital fashion can reduce the need for physical samples and minimize the environmental impact associated with traditional production and logistical processes (Yan, 2022). This innovative approach extends to business-to-business applications, where concepts like the "Mixed Tangible Catalog" combine physical and mixed reality elements to offer an efficient and sustainable alternative to conventional showrooms, thereby enhancing both economic and ecological aspects (Vangi et al., 2024). The overall strategy for utilizing metaverse technologies is anticipated to contribute to a sustainable trajectory for the apparel industry by fostering a circular economy and minimizing environmental damage (Periyasamy and Periyasami, 2023). However, there is a critical need to develop inclusive design solutions within digital fashion, especially for disabled consumers, to address existing accessibility gaps and the lack of adaptive clothing features in the metaverse (Shalini and Kavipriya, 2024).

The long-term success of metaverse integration in fashion hinges on a delicate balance between fostering innovation and implementing robust risk management strategies (Heo, 2025). Future research directions should focus on understanding consumer acceptance, the psychological impact of virtual engagement, and the evolving role of virtual influencers (Coşkun, 2025). Additionally, continued investigation into the intersection of luxury fashion with technology, sustainability practices, and talent retention within metaverse ventures is essential (Heo, 2025). Industry players need to address regulatory and ethical concerns, and work collaboratively to establish agreed-upon standards for space, interface, IT infrastructure, monetary systems, digital ID, and energy (Zalan and Barbesino, 2023). The frequency of identified challenges and opportunities in the metaverse fashion literature is visualized below, highlighting key areas of focus.

10 FUTURE TRAJECTORIES AND SOCIETAL IMPACT

The metaverse is set to become a foundational cultural space, significantly influencing fashion trends and subcultures. Within this digital realm, users can experiment with diverse aesthetics, fostering new forms of self-expression through digital apparel and accessories that may not be feasible or accessible in the physical world (Venturini and Columbano, 2024). This virtual experimentation can then translate into real-world fashion, creating a cyclical relationship where digital trends inspire physical ones and vice versa. This shift democratizes fashion by offering new avenues for diverse representation, allowing individuals to explore identities and subcultures without traditional gatekeepers or geographical limitations (Canag et al., 2025). Emerging designers and niche communities gain unprecedented platforms to showcase their work and connect with global audiences, fostering a more inclusive and varied fashion landscape (Wu, Liu and Weng, 2023).

However, this transformative potential comes with significant ethical considerations. Data privacy is paramount, as the collection and utilization of user data within immersive metaverse environments raise concerns about surveillance, targeted advertising, and the potential for misuse of personal information (Abilkaiyrkyzy et al., 2023). Digital identity and ownership, particularly with NFTs, also present complex challenges. While NFTs provide provenance, establishing clear legal frameworks for intellectual property rights, secondary market royalties, and the long-term value of digital assets is crucial (Sung, Kwon and Sohn, 2023). Responsible innovation is necessary to ensure that these technologies are developed and deployed in ways that benefit users and society, addressing issues like digital addiction and potential psychological impacts of prolonged engagement in virtual worlds (Orazi and Laer, 2023).

The long-term economic implications for traditional fashion retail and supply chains are profound. The rise of direct-to-avatar commerce and virtual storefronts could lead to a re-evaluation of physical retail spaces and shift consumer spending toward digital goods (Periyasami and Periyasamy, 2022). Supply chains may become more digitized, with an increased focus on 3D design software, virtual prototyping, and on-demand digital production, potentially reducing waste and speeding up market delivery (Casciani, Chkanikova and Pal, 2022). However, this also presents challenges for established manufacturing processes and necessitates significant investment in new infrastructure and skills. The seamless integration of physical and digital products, often referred to as 'phygital' experiences, will likely be a key strategy for traditional brands to maintain relevance and capture new revenue streams in this evolving ecosystem (Millatunnisa, 2022).

11 MANAGERIAL IMPLICATIONS AND PRACTITIONER INSIGHTS

Fashion brands navigating the metaverse and digital asset landscape must approach adoption strategically rather than reactively. Three key managerial implications emerge from this chapter's analysis:

11.1 Prioritise Value Beyond Novelty

Short-term engagement spikes from NFT drops or virtual pop-ups rarely translate into sustained value. Managers should design digital fashion initiatives that offer ongoing utility—such as access, interoperability, authentication, or community benefits—rather than relying on hype-driven campaigns (Schaub, 2025).

11.2 Align Digital Strategies With Brand Positioning

Luxury brands should leverage blockchain authentication, provenance tracking, and limited-edition digital assets to reinforce exclusivity and heritage. Fast fashion brands, by contrast, can focus on accessibility, gamification, and high-volume digital wearables that resonate with younger, platform-native consumers.

11.3 Build Organisational Capabilities for Long-Term Transformation

Successful digital fashion adoption requires investment in 3D design, immersive UX, AI-driven personalisation, and cross-functional collaboration. Managers must treat these technologies as enablers of new business models—such as digital twins for supply chain optimisation or generative AI for accelerated design—rather than isolated marketing tools.

Together, these insights provide a practical roadmap for brands seeking to translate emerging technologies into sustainable competitive advantage within evolving digital ecosystems.

12 SHORT-TERM TRENDS vs. LONG-TERM STRUCTURAL SHIFTS IN DIGITAL FASHION

To provide conceptual clarity, it is essential to distinguish between short-term experimental trends and long-term structural transformations in fashion retail.

12.1 Short-Term Trends (2021–2026)

These are characterised by experimentation, hype cycles, and platform-specific activations:

- Limited-edition NFT drops and speculative resale markets.
- Branded virtual pop-ups in Roblox, Fortnite, and Decentraland.
- Gamified experiences and avatar-based micro-transactions.
- High volatility in NFT pricing and consumer interest.
- Fragmented technological ecosystems requiring platform-specific asset creation.

These trends reflect early-stage innovation but lack stable infrastructure or widespread consumer adoption.

12.2 Long-Term Structural Shifts (2026–2035)

These represent deeper, more durable transformations supported by technological maturation:

- **Digital twins:** integrated into supply chains for lifecycle tracking and sustainable production.
- **Blockchain-based authentication:** becoming standard in luxury fashion.
- **Direct-to-avatar (D2A) commerce:** evolving into a mainstream retail channel.
- **Interoperable digital assets:** enabled by cross-platform standards.
- **Hybrid physical-digital ownership models:** (e.g., phygital products).
- **AI-driven design and personalisation:** embedded across retail operations.

13 RESEARCH PROPOSITIONS

This subsection builds directly on the chapter's interdisciplinary framework, which integrates technological capabilities, strategic decision-making, consumer psychology, and the specific dynamics of the fashion sector. By drawing together these perspectives, the analysis highlights how digital fashion adoption is shaped not only by technological affordances but also by brand positioning, consumer motivations, and organisational readiness. The propositions developed here extend this integrative logic, offering testable pathways that reflect the distinct strategic imperatives of luxury and fast-fashion brands while capturing the broader behavioural and technological mechanisms driving engagement in emerging digital ecosystems.

13.1 Proposed Research Propositions

Building on the integrative framework developed in this chapter, several testable research propositions emerge that can guide future empirical inquiry into digital fashion ecosystems:

P1: Avatar Realism & Consumer Psychology Higher levels of avatar realism will increase perceived social presence, which in turn enhances consumer engagement and purchase intention in virtual fashion environments.
P2: Digital Twins & Perceived Risk The integration of digital twin technology in virtual try-on experiences will reduce perceived product risk and increase consumer trust, particularly in categories with high fit uncertainty.
P3: Blockchain Authentication & Brand Type Blockchain-based authentication will have a stronger positive effect on perceived value and purchase intention for luxury brands than for fast fashion brands, due to higher consumer sensitivity to provenance and exclusivity.
P4: Interoperability & Long-Term Engagement The interoperability of digital fashion assets across multiple metaverse platforms will positively influence long-term consumer engagement and repeat purchase behaviour.
P5: Generative AI & Co-Creation Consumer participation in AI-assisted co-creation processes will increase brand attachment and willingness to pay for personalised digital fashion items.

These propositions extend the chapter's conceptual contributions into empirically testable pathways, strengthening its relevance for future academic research.

14 STRATEGIC IMPLEMENTATION FRAMEWORK FOR DIGITAL FASHION ADOPTION

To translate emerging technologies into meaningful organisational value, fashion brands can follow a three-stage strategic implementation pathway:

Step 1: Capability Assessment

Brands must evaluate their readiness across three core capability areas:

- **Creative capabilities:** (3D design, immersive UX, avatar modelling).
- **Data and technical infrastructure:** (blockchain integration, AI pipelines, asset interoperability).
- **Organisational alignment:** (cross-functional collaboration, digital literacy, governance structures).

This assessment identifies capability gaps and informs investment priorities.

Step 2: Value Proposition Design

Digital fashion initiatives must be anchored in a clear value proposition. Brands should determine whether the digital asset or experience delivers:

- **Utility value:** (access, authentication, interoperability).
- **Identity value:** (luxury signalling, exclusivity, self-expression).
- **Experiential value:** (virtual events, gamified interactions, immersive storytelling).

Aligning the value proposition with brand positioning is essential, especially when differentiating between luxury and fast fashion strategies.

Step 3: Ecosystem Integration

Brands must select platforms and partnerships that support long-term strategic goals. Key considerations include:

- **Audience fit**:(e.g., Roblox for Gen Z, Spatial for creative communities).
- **Interoperability potential:** (cross-platform asset use).
- **Sustainability and scalability:** (digital twin integration, AI-driven workflows).
- **Governance and compliance:** (IP protection, data ethics, regulatory alignment).

This step ensures that digital fashion initiatives evolve from isolated experiments into sustainable ecosystem strategies.

15 CONCLUSION: FASHION'S DIGITAL RENAISSANCE

The convergence of the metaverse and Non-Fungible Tokens (NFTs) marks a pivotal moment for the fashion industry, signalling not only emerging trends but deeper structural transformations (Casciani, Chkanikova and Pal, 2022). These technologies are reshaping how creativity is expressed, how fashion is consumed, and how ownership is conceptualised in digital environments (Millatunnisa, 2022). The metaverse provides an expansive, immersive canvas for brands to conceptualise and present fashion in ways unconstrained by physical limitations, enabling new forms of artistic and experiential expression (Venturini and Columbano, 2024). Simultaneously, NFTs are redefining value and authenticity by transforming digital assets into verifiable, ownable items that mirror the exclusivity of physical luxury goods (Sung, Kwon and Sohn, 2023). Together, these developments are reconfiguring the fashion value chain—from design and production to marketing and retail—establishing new paradigms for engagement and commercialisation (Periyasami and Periyasamy, 2022).

This digital renaissance empowers both established luxury houses and emerging designers to explore novel avenues for brand building and consumer interaction (Canag et al., 2025). Through virtual fashion shows, digital twins, and phygital products, brands can cultivate deeper connections with digitally native audiences, fostering community and loyalty in virtual spaces (Ahmed et al., 2025). Yet the trajectory of digital fashion is shaped by a mix of short-term experimentation and long-term structural change. Short-term trends—such as NFT drops, virtual pop-ups, and avatar skins—reflect early hype cycles that generate rapid attention but often lack durability, as evidenced by declining resale values and short-lived engagement on platforms like Roblox and Horizon Worlds. In contrast, long-term structural shifts—including blockchain-based authentication, digital twin-enabled supply chains, interoperable digital assets, and AI-driven design—represent enduring transformations that will reshape fashion retail over the next decade. These shifts alter organisational capabilities, operational models, and value creation processes, signalling a deeper evolution in how fashion brands operate within digital ecosystems.

Successfully navigating this transformative era will require continuous innovation, strategic integration of these technologies, and a nuanced understanding of evolving consumer behaviours and ethical considerations (Ahmed et al., 2025). Brands that embrace these changes, address technical and regulatory hurdles, and prioritise sustainable and inclusive practices will be best positioned to lead fashion's digital future.

REFERENCES

Abilkaiyrkyzy, A., Elhagry, A., Laamarti, F., & El Saddik, A. (2023). Metaverse key requirements and platforms survey. *IEEE Access : Practical Innovations, Open Solutions, 11*, 117765–117787. DOI: 10.1109/ACCESS.2023.3325844

Adidas. (2021). *Into the metaverse: How we got here and where we are headed.* Adidas News. https://news.adidas.com/originals/into-the-metaverse--how-we-got-here-and-where-we-are-headed/s/6ccb61cb-2135-453e-8626-ac3d56faab30

Agnihotri, R., De Keyser, A., Noble, S. M., Schillewaert, N., & Thakur, R. (2024). Purchase spillovers from the metaverse to the real world: The roles of social presence, trialability and customer experience. *Journal of Business Research, 170*, 114073. https://www.researchgate.net/publication/380600660_Purchase_spillovers_from_the_metaverse_to_the_real_world_The_roles_of_social_presence_trialability_and_customer_experience

Ahmed, R., Ahmed, E., Elbarbary, A., Darwish, A., & Hassanien, A. E. (2025). Fashion industry in the age of generative artificial intelligence and metaverse: A systematic review. *arXiv*. https://arxiv.org/abs/2505.17141

Al-Adwan, A. S. (2024). The meta-commerce paradox: Exploring consumer non-adoption intentions. *Online Information Review*, *48*(6), 1270–1289. DOI: 10.1108/OIR-01-2024-0017

Al-Kfairy, M., Alfandi, O., & Khaddaj, S. (2025). Psychological barriers to metaverse-based education: Examining the impact of technophobia and digital fatigue. *Journal of Information. Communication and Ethics in Society*, *23*(4), 550–566. DOI: 10.1108/JICES-04-2025-0080

Alexander, B., Blazquez, M., Chrimes, C., & Boardman, R. (2025). The role of immersive spaces on the customer experience: An exploration of fashion metaverses. *Psychology and Marketing*, *42*(2), 539–553. DOI: 10.1002/mar.22140

Aryal, A. (2025). *Gucci marketing strategy: Crisis-to-comeback (2025).* Blankboard.studio. https://www.blankboard.studio/originals/blog/gucci-marketing-strategy-crisis-to-comeback

Bain, M. (2023). *How fashion NFTs fizzled.* The Business of Fashion. https://www.businessoffashion.com/articles/technology/how-fashion-nfts-fizzled/

Batchelor, J. (2022). *Meta's flagship metaverse Horizon Worlds struggling to attract and retain users*. GamesIndustry.biz. https://www.gamesindustry.biz/metas-flagship-metaverse-horizon-worlds-struggling-to-attract-and-retain-users

Becker, K. (2025). *NFT fashion and virtual self re-creation* [Manuscript or unpublished work].

Bernatik, J. (2022). *Introduction to Flow blockchain*. Flow. https://flow.com/engineering-blogs/introduction-to-flow-blockchain

Bialkova, S. (2024). *The rise of AI user applications: Chatbots integration foundations and trends.*

British Vogue. (2021). *Dolce & Gabbana's NFT experiment is a million-dollar success story*. https://www.vogue.co.uk/news/article/fashion-nft-dolce-and-gabbana

Brooks, A. (2026). *Is the metaverse fashion trend dying or evolving into something new?* Alibaba.com. https://www.alibaba.com/product-insights/is-the-metaverse-fashion-trend-dying-or-evolving-into-something-new.html

Burberry. (2025). *Burberry X Blankos Block Party: New NFT collection and social space*. https://www.burberryplc.com/news/brand/2022/burberry-x-blankos-block-party--new-nft-collection-and-social-sp

Business of Fashion. (2022). *Why luxury NFT collections are struggling in the secondary market*. https://www.businessoffashion.com

Butler, C. (2022). *How to play and win: Blankos Block Party*. DappRadar. https://dappradar.com/blog/how-to-play-and-win-blankos-block-party

Canag, L., Loquinario, A., Pepano, R., & Peninoy, R. (2025). Democratizing fashion consumer engagement and redefining style in the digital age. *International Journal of Research*, *8*(XII), 935–949. DOI: 10.47772/IJRISS.2024.8120078

Casciani, D., Chkanikova, O., & Pal, R. (2022). Exploring the nature of digital transformation in the fashion industry: Opportunities for supply chains, business models, and sustainability-oriented innovations. *Sustainability : Science, Practice, & Policy*, *18*(1), 773–795. DOI: 10.1080/15487733.2022.2125640

Chain, B. N. B. (2025). *BNB Chain: Empowering the future of decentralized applications*. https://docs.bnbchain.org

CoinDesk. (2023). *NFT market faces steep decline as trading volumes drop across major platforms*. https://www.coindesk.com

Coşkun, M. (2025). Metaverse and the future of fashion: A comparative analysis of fast-fashion and luxury fashion brands. *Sosyal Mucit Academic Review*, *6*(3), 392–418. DOI: 10.54733/smar.1665701

Covington, S. M., & Soltanisehat, M. (2025). Enhancing transparency and traceability in fast fashion: The role of blockchain technology. *International Journal of Sustainable Fashion & Textiles*, *4*(2), 257–280. DOI: 10.1386/sft_00060_1

Decentraland. (2025). *Decentraland*. https://decentraland.org/

Dhakshitha, B. K. (2025). Exploring the impact of social presence cues on consumer impulse buying in E-commerce platforms. *Journal of Marketing & Social Research*, *2*, 119–125.

Donvito, R., Acuti, D., & Song, S. (2024). Fashion and the metaverse: Implications for consumers and firms. *Journal of Global Fashion Marketing*, *15*(1), 1–5. DOI: 10.1080/20932685.2023.2293290

Dwivedi, Y. K., Hughes, L., Wang, Y., Alalwan, A. A., Ahn, S. J., Balakrishnan, J., Barta, S., Belk, R., Buhalis, D., Dutot, V., Felix, R., Filieri, R., Flavián, C., Gustafsson, A., Hinsch, C., Hollensen, S., Jain, V., Kim, J., Krishen, A. S., & Wirtz, J. (2023). Metaverse marketing: How the metaverse will shape the future of consumer research and practice. *Psychology and Marketing*, *40*(4), 750–776. DOI: 10.1002/mar.21767

Elsayegh, O. A. E., & Mohamed, H. (2024). A proposed model for startup in the field of textile design and fashion through an analytical study in digital transformation of the global startups. *Journal of Design Sciences and Applied Arts*, *5*(2), 474–494. DOI: 10.21608/jdsaa.2023.232038.1370

Fashionbi. (2025). *Fashion brands penetrating into video games*. https://www.fashionbi.com/insights/fashion-brands-penetrating-into-video-games

Fortagne, M. A. (2025). *Human-app interaction for disruptive innovations: User-centered contributions to technology acceptance and effects* [Doctoral dissertation].

Hahn, J. (2021). *Nike gets "metaverse-ready" with acquisition of virtual sneaker company RTFKT*. Dezeen. https://www.dezeen.com/2021/12/15/rtfkt-nike-acquisition/

Heo, J. (2025). Marketing luxury fashion in the metaverse: The case of Gucci Town on Roblox. *Thunderbird International Business Review*, tie.70024. DOI: 10.1002/tie.70024

Igumenov, A., Filatovas, E., Medvedev, V., & Paulavičius, R. (2021). Estimating energy consumption of Ethereum network. In *DAMSS 2021: 12th conference on data analysis methods for software systems*.

Independent. (2021). *Gucci digitally outfits Gen-Z in metaverse foray with Roblox*. https://www.independent.co.uk/news/gucci-digitally-outfits-genz-in-metaverse-foray-with-roblox-gucci-nfts-italian-florence-alessandro-michele-b1862331.html

Kapengut, E., & Mizrach, B. (2022). *An event study of the Ethereum transition to proof-of-stake*. arXiv. https://www.researchgate.net/publication/369670342_An_Event_Study_of_the_Ethereum_Transition_to_Proof-of-Stake

Kim, D. Y., & Lee, H. K. (2025). "I am wearing what my avatar is wearing": The effect of avatar realism on mental simulation in the metaverse. *Journal of Fashion Marketing and Management*, *29*(5), 895–914. DOI: 10.1108/JFMM-04-2024-0116

Kim, D. Y., Lee, H. K., & Chung, K. (2023). Avatar-mediated experience in the metaverse: The impact of avatar realism on user-avatar relationship. *Journal of Retailing and Consumer Services*, *73*, 103382. DOI: 10.1016/j.jretconser.2023.103382

Kim, J., & Lee, H. (2022). How virtual try-on influences purchase intention: The role of self-expressiveness, vicarious satisfaction, and stress relief. *Frontiers in Psychology*, *13*, 938060. Advance online publication. DOI: 10.3389/fpsyg.2022.938060

Kim, N. E., & Lee, J. R. (2022). The purchasing status of the avatars and digital fashion items in metaverse and consumers' purchase satisfaction and the future purchase intentions according to usage motivation. *Journal of Korean Home Economics Education Association*, *34*(3), 133–148. DOI: 10.19031/jkheea.2022.9.34.3.133

Klerk, A. D. (2021). *Louis Vuitton has launched a game as part of its founder's birthday celebrations*. Harper's Bazaar. https://www.harpersbazaar.com/uk/fashion/a37218196/louis-vuitton-game/

Kniazeva, M., Aiello, G., Dasmi, C., Mazzoli, V., Nechaeva, O., & Syed, F. U. (2024). Why fashion brands enter the metaverse: Exploring the motivations of fast fashion and luxury fashion brands. *Journal of Global Fashion Marketing*, *15*(1), 62–89. DOI: 10.1080/20932685.2023.2269952

Kozinets, R. V. (2023). Immersive netnography: A novel method for service experience research in virtual reality, augmented reality and metaverse contexts. *Journal of Service Management*, *34*(1), 100–125. DOI: 10.1108/JOSM-12-2021-0481

Kumar, S., Malhotra, D., & Kathuria, G. (2026). Unrestrained buying behavior in the metaverse: Exploring the moderating role of technostress and self-control. *Spanish Journal of Marketing - ESIC*, *30*(2), 167–191. DOI: 10.1108/SJME-07-2024-0217

Lapatoura, I. (2023). NFTs, digital worlds and brand protection in fashion: A UK/European intellectual property perspective. *International Journal of Law in Changing World*, *2*(3), 141–152. DOI: 10.54934/ijlcw.v2i3.54

Li, C., & Huang, F. (2024). The impact of virtual streamer anthropomorphism on consumer purchase intention: Cognitive trust as a mediator. *Behavioral Sciences (Basel, Switzerland)*, *14*(12), 1228. DOI: 10.3390/bs14121228 PMID: 39767370

Li, J., & Liu, X. (2024). Empowering fashion design and intelligent manufacturing with digital twins in the metaverse era. *Asian Social Science*, *20*(4), 1–39. DOI: 10.5539/ass.v20n4p39

Lim, J. (2021). *In review 2021: The year that fashion truly embraced NFTs*. The Industry Fashion. https://www.theindustry.fashion/in-review-2021-the-year-that-fashion-truly-embraced-nfts/

Lin, J., Cronjé, J., Wienrich, C., Pauli, P., & Latoschik, M. E. (2023). Visual indicators representing avatars' authenticity in social virtual reality and their impacts on perceived trustworthiness. *IEEE Transactions on Visualization and Computer Graphics*, *29*(11), 4589–4599. DOI: 10.1109/TVCG.2023.3320234 PMID: 37788202

Liu, Y., Wang, W., & Liu, J. (2025). To NFT or not: A strategic analysis for fashion brands developing digital products in the metaverse. *Journal of Theoretical and Applied Electronic Commerce Research*, *20*(3), 155. DOI: 10.3390/jtaer20030155

Markets, B. T. C. (2025). *Ethereum's role in the NFT boom: A deep dive*. https://www.btcmarkets.net/blog/ethereums-role-in-the-nft-boom-a-deep-dive

Markets.com. (2024). *The role of Polygon (MATIC) in the NFT boom*. https://www.markets.com/education-centre/the-role-of-polygon-in-the-nft-boom

McSorley, J. (2025). *Fashioning the digital self: A critical examination of identity and sustainable user practices within digital fashion ecosystems* [Doctoral dissertation or unpublished manuscript].

Menidjel, C., Hollebeek, L. D., Leppiman, A., & Riivits-Arkonsuo, I. (2022). Role of AI in enhancing customer engagement, loyalty and loyalty programme performance. In *Handbook of research on customer loyalty* (pp. 316–331). Edward Elgar Publishing. DOI: 10.4337/9781800371637.00032

Millatunnisa, S. (2022). The 4.0 industry technology in fashion industries. In *4th International Conference on Innovation in Engineering and Vocational Education (ICIEVE 2021)* (pp. 336–340). Atlantis Press. DOI: 10.2991/assehr.k.220305.068

Nalbant, K. G., & Aydın, S. (2023). Development and transformation in digital marketing and branding with artificial intelligence and digital technologies dynamics in the Metaverse universe. *Journal of Metaverse*, *3*(1), 9–18. DOI: 10.57019/jmv.1148015

Naveen Kumar, R., Janani, M., Ramadevi, V., & Pavithra, R. (2025). *How the metaverse and virtual worlds are redefining fashion marketing and brand engagement*. Digital Transformation and Society. DOI: 10.1108/DTS-09-2024-0177

Nike. (2022). *Nike launches. SWOOSH, a new digital community and experience.* https://about.nike.com/en/newsroom/releases/nike-launches-swoosh-a-new-digital-community-and-experience

NonFungible.com. (2022). *Market trends report: Luxury NFT collections.* https://nonfungible.com

OKX. (2025). *XRP: The most energy efficient cryptocurrencies.* OKX Learn. https://www.okx.com/en-gb/learn/is-xrp-the-most-energy-efficient-cryptocurrency

Orazi, D. C., & Van Laer, T. (2023). There and back again: Bleed from extraordinary experiences. *The Journal of Consumer Research, 49*(5), 904–925. DOI: 10.1093/jcr/ucac022

Pereira, C., Marto, A., Ribeiro, R., Gonçalves, A., Rodrigues, N., Rabadão, C., Costa, R. L. D. C., & Santos, L. (2025). Security and privacy in physical–digital environments: Trends and opportunities. *Future Internet, 17*(2), 83. DOI: 10.3390/fi17020083

Periyasami, S., & Periyasamy, A. P. (2022). Metaverse as future promising platform business model: Case study on fashion value chain. *Businesses, 2*(4), 527–545. DOI: 10.3390/businesses2040033

Periyasamy, A. P., & Periyasami, S. (2023). Rise of digital fashion and metaverse: Influence on sustainability. *Digital Economy and Sustainable Development, 1*(1), Article 16.

Poirier, S. M., Cosby, S., Sénécal, S., Coursaris, C. K., Fredette, M., & Léger, P. M. (2024). The impact of social presence cues in social media product photos on consumers' purchase intentions. *Journal of Business Research, 185*, 114932. DOI: 10.1016/j.jbusres.2024.114932

Roblox. (2022). *Fashion metaverse trends.* https://corp.roblox.com/pdf/2022-metaverse-fashion-trends-report

Romano, B. (2023). *Retail technology and the customer experience: The use of augmented reality as a shopping tool* [Doctoral dissertation, Swinburne University].

RTFKT. (2025). *About RTFKT Studios.* https://grokipedia.com/page/RTFKT_Studios

Saini, R. (2023). *Energy consumption in different blockchains.* Medium. https://medium.com/@rahuleum/energy-consumption-in-different-blockchains-9edfe4458609

Sayem, A. S. M. (2022). Digital fashion innovations for the real world and metaverse. *International Journal of Fashion Design, Technology and Education, 15*(2), 139–141. DOI: 10.1080/17543266.2022.2071139

Schaub, M. (2025). *NFTs as the interface of cryptocurrency: Utility values and the act of wasteful spending.*

Shah, T. R., Purohit, S., Das, M., & Arulsivakumar, T. (2025). Do I look real? Impact of digital human avatar influencer realism on consumer engagement and attachment. *Journal of Consumer Marketing, 42*(4), 416–430. DOI: 10.1108/JCM-05-2024-6853

Shalini, S., & Kavipriya, T. (2024). Perceived risk of disabled's consumption and behavior on clothing under digital fashion-metaverse. In *2024 IEEE 4th International Conference on ICT in Business Industry & Government (ICTBIG)* (pp. 1–11). IEEE.

Shih, M. T., Lee, Y. C., Huang, C. M., & Chan, L. (2023). "A feeling of déjà vu": The effects of avatar appearance-similarity on persuasiveness in social virtual reality. *Proceedings of the ACM on Human-Computer Interaction, 7*(CSCW2), Article 1–31.

Solana. (2025). *Solana blockchain fundamentals.* Oboe. https://oboe.com/learn/solana-blockchain-fundamentals-1qeokyv

Sung, E., Kwon, O., & Sohn, K. (2023). NFT luxury brand marketing in the metaverse: Leveraging blockchain-certified NFTs to drive consumer behavior. *Psychology and Marketing, 40*(11), 2306–2325. DOI: 10.1002/mar.21854

Tam, F. Y. (2025). Blockchain technology adoption in the fashion supply chain: A review on the practices of fashion brands. *Journal of Global Operations and Strategic Sourcing.*

Tarek Amer, N., Rahman, S., Baker, J., & Ashill, N. J. (2025). From the internet dilemma to the metaverse dilemma: Exploring luxury brands' adoption of metaverse technologies. *Communications of the Association for Information Systems, 57*(1), 66. Advance online publication. DOI: 10.17705/1CAIS.05754

The Interline. (2025). *In the aftershock of the NFT collapse, fashion has a lot to discover about digital ownership.* https://www.theinterline.com/2025/04/29/in-the-aftershock-of-the-nft-collapse-fashion-has-a-lot-to-discover-about-digital-ownership/

The Sandbox. (2024). *Unbox: How top brands push fashion forward and thrive in the Sandbox.* https://www.sandbox.game/en/blog/unbox-how-top-brands-push-fashion-forward-and-thrive-in-the-sandbox/3418/

Trujillo, A., & Bacciu, C. (2023). Toward blockchain-based fashion wearables in the metaverse: The case of Decentraland. In *2023 IEEE International Conference on Metaverse Computing, Networking and Applications (MetaCom)* (pp. 653–657). IEEE. DOI: 10.1109/MetaCom57706.2023.00115

Trunfio, M., & Rossi, S. (2022). Advances in metaverse investigation: Streams of research and future agenda. *Virtual Worlds*, *1*(2), 103–129. DOI: 10.3390/virtualworlds1020007

Upadhyay, N., & Upadhyay, S. (2025). The dark side of non-fungible tokens: Understanding risks in the NFT marketplace from a fraud triangle perspective. *Financial Innovation*, *11*(1), 62. DOI: 10.1186/s40854-024-00684-6

Vangi, F., Ricci, M., Dastan, M., & Fiorentino, M. (2024). The mixed tangible catalog: Toward tangible and sustainable B2B metaverse fashion showrooms. *IEEE Computer Graphics and Applications*, *44*(4), 40–52. DOI: 10.1109/MCG.2024.3424247 PMID: 38968010

Venturini, A., & Columbano, M. (2024). 'Fashioning' the metaverse: A qualitative study on consumers' value and perceptions of digital fashion in virtual worlds. *Journal of Global Fashion Marketing*, *15*(1), 6–22. DOI: 10.1080/20932685.2023.2234918

Vogue. (2025). *What is fashion metaverse: A complete guide*. Vogue Fashion Institute. https://www.voguefashioninstitute.com/what-is-fashion-metaverse-a-complete-guide/

Vogue Business. (2022). *Luxury NFTs lose momentum as secondary market demand falls*. https://www.voguebusiness.com

Vogue Business. (2025). *Gucci's evolving metaverse strategy: From Roblox to digital ecosystems*. https://www.voguebusiness.com

Wu, C. H., Liu, C. Y., & Weng, T. S. (2023). Critical factors and trends in NFT technology innovations. *Sustainability (Basel)*, *15*(9), 7573. DOI: 10.3390/su15097573

Yan, R. (2022). A sustainable fashion industry business model revolution based on the metaverse: Practices and reciprocal processes. *Highlights in Business. Economics and Management*, *4*, 363–369.

Zalan, T., & Barbesino, P. (2023). Making the metaverse real. *Digital Business*, *3*(2), Article 100059.

Zeng, N., Ryding, D., Vignali, G., & Pantano, E. (2025). AR atmospherics and virtual social presence impacts on customer experience and customer engagement behaviours. *International Journal of Retail & Distribution Management*, *53*(1), 58–73. DOI: 10.1108/IJRDM-08-2023-0520

Zhang, R. (2022). *Prada launches new Timecapsule NFT collection*. Medium. https://medium.com/marketing-in-the-age-of-digital/prada-launches-new-timecapsule-nft-collection-60787fa268f

Zheng, C., Ling, S., Huang, Y., & Li, X. (2025). The impact of social presence on purchase intentions of knowledge products among knowledge-based short video users: A moderated mediation model. *Behavioral Sciences (Basel, Switzerland)*, *15*(8), 1140. DOI: 10.3390/bs15081140 PMID: 40867496

Chapter 2
Anthropomorphic Service Robotics in 2030:
Strategic Leadership and Organizational Readiness

Hayiath Qureshi
http://orcid.org/0009-0007-7030-6549
Global Banking School, Canterbury Christ Church University, UK

Kartinah Ayupp
Universiti Malaysia Sarawak, Malaysia

ABSTRACT

Artificial intelligence is developing rapidly, and anthropomorphic service robots are beginning to move from theory into everyday organisational practice. This chapter explores how advances in emotional AI and generative personality modelling may reshape service delivery in sectors such as healthcare, hospitality, retail, and tourism by 2030. Drawing on the Technology Acceptance Model, leadership theory, and research on anthropomorphism, it presents an integrated framework to support organisations in adopting these technologies. The discussion considers not only technological capability, but also customer expectations, employee readiness, and the wider organisational context. It also reflects on how leadership is evolving in response to these changes, highlighting the importance of AI literacy, ethical awareness, and the ability to manage collaboration between people and intelligent systems.

DOI: 10.4018/979-8-3373-6640-1.ch002

INTRODUCTION

The growing intersection of artificial intelligence and anthropomorphic robotics marks a significant shift in how services are designed and delivered across industries. What was once confined to speculative discussions is now becoming increasingly tangible, as organisations begin to explore how human-like robots can enhance customer experiences while also responding to ongoing labour shortages and pressures for greater operational efficiency.

At the heart of this shift is the integration of advanced machine learning, natural language processing, and human-centred design, which together enable robots to display behaviours and interactions that resemble those of humans in increasingly convincing ways (Belanche et al., 2020).

Importantly, this development is not simply about adopting new technology. It is reshaping the nature of service encounters themselves, requiring organisations to rethink how services are delivered, how employees and technologies work together, and how leadership approaches the challenges of implementation and organisational change.

Contemporary advances in artificial intelligence suggest that the technical foundations required for sophisticated anthropomorphic service robots are becoming increasingly commercially viable. Developments in affective computing, predictive modelling, and generative interaction systems have enhanced robots' capacity to recognise, interpret, and respond to human emotional cues with growing precision (Broadbent, 2017). When combined with improvements in embodied design, natural language processing, and contextual awareness architectures, these capabilities enable service robots to engage in more adaptive and socially responsive interactions.

The strategic implications of such developments are significant. Organisations must navigate the intersection of technological capability, customer trust and acceptance, workforce integration, and institutional readiness. The challenge is no longer confined to technical feasibility; rather, it concerns how anthropomorphic robotics can be embedded within service ecosystems in ways that align with organisational strategy, operational structures, and evolving expectations of human–machine interaction.

CHAPTER OBJECTIVES AND SCOPE

This chapter offers a forward-looking analysis of anthropomorphic service robotics, examining how emerging developments in artificial intelligence may reshape the human-like attributes of service robots over the coming decade. Its primary objective is to develop analytically grounded predictive frameworks and strategic guidance for organisations considering the implementation of emotionally intelligent robotic systems across diverse service contexts. Through a structured examination of technological trajectories, sector-specific applications, and organisational adoption dynamics, the chapter seeks to bridge the gap between current capabilities and plausible future developments. In doing so, it provides a theoretically informed basis for strategic planning and organisational decision-making in relation to anthropomorphic service technologies.

The scope of analysis encompasses four interrelated dimensions of anthropomorphic service robot evolution. First, the chapter examines the technological convergence underpinning enhanced human like capabilities, including developments in affective artificial intelligence, generative interaction models, and integration with complementary technologies such as augmented reality, virtual reality, and the Internet of Things. Second, it develops forward looking scenarios for 2030 across four major service sectors, namely healthcare, hospitality, retail, and tourism, exploring how human like robotic systems may reshape service processes and customer engagement within each context.

Third, the chapter proposes an integrated adoption, leadership and anthropomorphism framework that synthesises insights from the Technology Acceptance Model, contemporary leadership theory, and anthropomorphism research to inform organisational implementation strategies. Finally, it considers the evolution of leadership capabilities required to manage this technological transition, including artificial intelligence literacy, human and robot team coordination, ethical oversight, and strategic foresight competencies.

SIGNIFICANCE: WHY PREDICTIVE FRAMEWORKS ARE ESSENTIAL FOR ORGANISATIONAL SUCCESS

The need to develop robust predictive frameworks for anthropomorphic service robotics is closely tied to the rapid pace at which these technologies are evolving, as well as the complex ways in which they interact with human behaviour and organisational systems. Unlike earlier waves of technological change, the introduction of anthropomorphic robots brings a distinct set of challenges, particularly around emotional interaction between humans and machines, the development of trust, and how customers form expectations about what these systems can and cannot do (Huang and Rust, 2021).

In this context, organisations cannot afford to treat adoption as a purely technical decision. Without a deeper understanding of these dynamics, there is a real risk of unsuccessful implementation, unmet customer expectations, and ultimately a loss of competitive position as such technologies become more common across service settings.

The adoption of anthropomorphic service robotics presents a level of complexity that extends beyond the explanatory capacity of traditional technology adoption frameworks. Organisations must simultaneously address technological feasibility, psychological responses, and organisational readiness. The human propensity to attribute social and emotional qualities to human like robots creates both strategic opportunities and operational risks. Although anthropomorphic design can enhance engagement, trust, and perceived service quality, it may also generate expectations regarding emotional intelligence and social competence that exceed current technological limitations.

Consequently, predictive frameworks must account for these psychological dynamics alongside realistic assessments of technological development. Without such alignment, implementation strategies risk generating expectation gaps, diminished user acceptance, and organisational strain. Effective adoption therefore depends not only on technical capability, but also on managing perception, readiness, and the socio organisational implications of human robot interaction.

The strategic implications of anthropomorphic service robotics extend beyond individual organisations to shape broader industry structures and competitive dynamics. Early adopters of advanced human like robotic systems may secure competitive advantage through enhanced customer experience, improved operational efficiency, and differentiation within increasingly standardised service markets. However, successful implementation demands substantial investment in technological infrastructure, workforce development, and organisational change processes. These commitments require careful sequencing, capability alignment, and risk assessment.

Predictive frameworks therefore play a critical role in informing adoption timing, technology selection, and implementation design. By aligning technological possibilities with organisational capacity and market conditions, such frameworks support more strategically coherent and sustainable integration of anthropomorphic service systems.

CHAPTER STRUCTURE OVERVIEW

This chapter is structured to provide readers with a comprehensive journey from theoretical foundations through practical implementation guidance. Following this introduction, the literature review and theoretical foundation section establishes the conceptual groundwork by examining the evolution of anthropomorphism in service robotics, the application of the Technology Acceptance Model to robotic contexts, and the integration of leadership theory with technology adoption processes. This theoretical foundation provides the conceptual framework for understanding the complex dynamics surrounding anthropomorphic service robotics adoption.

The subsequent section on future anthropomorphic capabilities and technological convergence explores the cutting-edge developments that will define the next generation of human-like service robots. This includes detailed examination of emotional AI and advanced empathy systems, generative personality models that enable consistent and brand-aligned robot personas, and the technological convergence of augmented reality, virtual reality, Internet of Things, and metaverse technologies that will create hybrid physical-digital anthropomorphic experiences.

The predictive scenarios section provides concrete visualisations of how anthropomorphic service robotics will transform specific industry sectors by 2030. Through detailed analysis of healthcare, hospitality, retail, and tourism applications, readers gain practical understanding of sector-specific opportunities, challenges, and implementation considerations. These scenarios serve as strategic planning tools for organisations operating in these sectors whilst illustrating broader principles applicable across industries.

The framework development sections focus on practical implementation guidance, beginning with a future-oriented organisational adoption framework that extends traditional Technology Acceptance Model approaches to address the unique characteristics of emotionally intelligent robots. This is followed by examination of leadership evolution requirements, including expanded AI literacy needs, human-robot team management capabilities, and future leadership competency development priorities.

The chapter concludes with strategic futures and innovation trajectories that provide predictive frameworks, assessment tools, and future research priorities essential for long-term strategic planning. Throughout each section, the analysis maintains focus on evidence-based recommendations whilst acknowledging the inherent uncertainties associated with emerging technology forecasting. The comprehensive structure ensures that readers develop both theoretical understanding and practical capabilities for navigating the complex landscape of anthropomorphic service robotics implementation.

LITERATURE REVIEW AND THEORETICAL FOUNDATION

The theoretical foundation for understanding anthropomorphic service robotics requires examination of three interconnected domains: the evolution of anthropomorphism in service robotics, the application of technology acceptance frameworks to robotic contexts, and the integration of leadership theory with technology adoption processes. This multidisciplinary approach provides the conceptual groundwork necessary for developing comprehensive frameworks that address the complex dynamics surrounding human-robot interaction in service environments.

The Evolution of Anthropomorphism in Service Robotics

The concept of anthropomorphism in robotics has evolved significantly from early mechanical automata to contemporary emotionally intelligent service robots capable of sophisticated human-like interaction. Duffy (2003) provides foundational insight into this evolution, arguing that anthropomorphism in robotics represents more than mere physical resemblance to humans, encompassing behavioural, cognitive, and emotional dimensions that facilitate natural human-robot interaction. This multidimensional approach to anthropomorphism recognises that effective service robots must exhibit not only human-like appearance but also human-like social cognition, emotional responsiveness, and adaptive behaviour patterns that align with human expectations and preferences.

The theoretical foundations of anthropomorphism in technology extend beyond robotics to encompass broader human tendencies to attribute human-like qualities to non-human entities. Epley et al. (2007) propose a three-factor theory of anthropomorphism that identifies elicited agent knowledge, effectance motivation, and sociality motivation as the primary drivers of anthropomorphic attribution. Elicited agent knowledge refers to the accessibility and applicability of human-like concepts when interpreting non-human behaviour, suggesting that robots designed with human-like characteristics more readily trigger anthropomorphic responses. Effectance motivation involves the desire to understand and predict the behaviour of other agents, whilst sociality motivation reflects the fundamental human need for social connection and interaction. These psychological mechanisms provide crucial insights for designing anthropomorphic service robots that effectively leverage human cognitive tendencies whilst avoiding uncanny valley effects that can undermine acceptance.

The application of anthropomorphism principles to service robotics has demonstrated significant impact on customer perceptions and service experiences. Van Doorn et al. (2017) examine the emergence of automated social presence in organisational frontlines, highlighting how anthropomorphic design elements enable service robots to establish emotional connections with customers that enhance service quality perceptions and satisfaction outcomes. Their research demonstrates that anthropomorphic service robots can effectively supplement or replace human service employees in specific contexts, particularly where consistent service delivery, 24/7 availability, and multilingual capabilities provide operational advantages. However, they also identify important boundary conditions related to task complexity, emotional intensity, and customer preferences that influence the effectiveness of anthropomorphic service robots.

Contemporary developments in anthropomorphic service robotics reflect increasingly sophisticated integration of physical design, behavioural programming, and artificial intelligence capabilities. Modern service robots incorporate advanced facial recognition systems, natural language processing capabilities, and machine learning algorithms that enable real-time adaptation to customer preferences and contextual factors. These technological advances have expanded the range of service tasks that anthropomorphic robots can effectively perform whilst maintaining human-like interaction quality. The evolution from simple task-oriented robots to emotionally intelligent service companions represents a fundamental shift in the role of robotics within service organisations, requiring new theoretical frameworks that account for the complex dynamics of human-robot emotional interaction.

Technology Acceptance Model in Robotics Context

The Technology Acceptance Model (TAM), originally developed by Davis (1989), provides a foundational framework for understanding user adoption of information technologies based on perceived usefulness and perceived ease of use. In the context of anthropomorphic service robotics, TAM offers valuable insights into the factors that influence customer and employee acceptance of robot-mediated service delivery. However, the unique characteristics of anthropomorphic robots, particularly their human-like appearance and behaviour, introduce additional complexity that requires extension of traditional TAM constructs to adequately capture the nuanced dynamics of human-robot acceptance.

Heerink et al. (2010) developed the Almere model specifically for assessing acceptance of assistive social agent technology, extending TAM to include constructs particularly relevant to robotic contexts. Their model incorporates social influence, hedonic motivation, and trust as additional factors that significantly influence acceptance of social robots. Social influence reflects the impact of others' opinions and experiences on individual acceptance decisions, recognising that robot adoption often occurs within social contexts where peer attitudes and cultural norms significantly affect individual perceptions. Hedonic motivation captures the intrinsic enjoyment and pleasure derived from interacting with robots, which may be particularly relevant for anthropomorphic service robots designed to provide engaging and entertaining customer experiences. Trust emerges as a critical factor in robotic contexts, as customers must develop confidence in robots' capabilities, reliability, and safety before accepting them as legitimate service providers.

The application of TAM frameworks to service robotics contexts reveals important considerations regarding perceived usefulness and perceived ease of use in anthropomorphic robot interactions. Perceived usefulness in robotic service delivery encompasses both functional benefits such as efficiency, accuracy, and availability, and experiential benefits related to entertainment, personalisation, and emotional engagement. Anthropomorphic design elements can enhance perceived usefulness by making robot interactions feel more natural and intuitive, reducing cognitive effort required for human-robot communication. However, anthropomorphic characteristics may also raise customer expectations regarding robot capabilities, potentially leading to disappointment when robots fail to meet human-like performance standards in complex or emotionally charged service situations.

Perceived ease of use in anthropomorphic robotics contexts involves both technical usability and social comfort factors. Technical usability encompasses traditional interface design considerations such as navigation simplicity, response accuracy, and system reliability. Social comfort addresses customers' psychological ease when interacting with human-like robots, including factors such as conversation naturalness, appropriate social behaviour, and emotional appropriateness. Yoganathan et al. (2019) demonstrate through multi-site field experiments that customer service trade-offs between efficiency and personalisation significantly influence acceptance of automated service technologies. Their findings suggest that anthropomorphic service robots must carefully balance human-like characteristics with functional performance to optimise customer acceptance and satisfaction.

The evolution of TAM for anthropomorphic robotics also requires consideration of temporal factors that influence acceptance over time. Initial novelty effects may positively or negatively influence early adoption decisions, whilst longer-term acceptance depends on sustained positive experiences and demonstrated value delivery. The learning capabilities of modern AI-powered service robots introduce dynamic elements to the acceptance process, as robots that improve performance through experience may demonstrate increasing usefulness and ease of use over time. This temporal dimension requires longitudinal approaches to understanding robot acceptance that account for changing customer perceptions and evolving robot capabilities.

Leadership Theory and Technology Adoption

The integration of leadership theory with technology adoption processes provides essential insights for understanding how organisational leaders can effectively facilitate the implementation of anthropomorphic service robotics. Traditional leadership theories, including transformational and transactional leadership models developed by Bass and Riggio (2006), offer valuable frameworks for understanding how leaders can influence employee attitudes and behaviours during technology adoption processes. However, the unique characteristics of anthropomorphic service robotics—particularly the need to manage human-robot teams and address employee concerns about job displacement—require new leadership competencies and approaches.

Transformational leadership characteristics, including idealised influence, inspirational motivation, intellectual stimulation, and individualised consideration, are particularly relevant for anthropomorphic robotics implementation. Leaders who demonstrate idealised influence can serve as role models for positive human-robot collaboration, showing employees how to effectively work alongside anthropomorphic service robots. Inspirational motivation involves articulating compelling visions for human-robot service delivery that emphasise enhanced customer experiences and employee empowerment rather than job replacement. Intellectual stimulation encourages employees to think creatively about how anthropomorphic robots can augment human capabilities, whilst individualised consideration addresses specific employee concerns and adaptation needs during the transition process.

The emergence of AI literacy as a critical leadership competency reflects the increasing importance of technological understanding in contemporary organisational leadership. Leaders implementing anthropomorphic service robotics must develop sophisticated understanding of AI capabilities and limitations, ethical considerations in human-like robot deployment, and strategic implications of anthropomorphic technology investments. This AI literacy extends beyond technical knowledge to encompass strategic decision-making frameworks that account for the complex interdependencies between technological capabilities, customer expectations, and organisational culture factors.

Karakose and Tülübaş (2024) demonstrate the relationship between leadership styles and teacher attitudes toward AI technologies, highlighting how transformational leadership approaches can positively influence technology acceptance within educational contexts. Their findings suggest that leaders who demonstrate enthusiasm for AI technologies, provide adequate support for skill development, and maintain open communication about technology implementation challenges are more successful in fostering positive attitudes toward AI adoption. These insights are directly applicable to anthropomorphic service robotics contexts, where leadership support and communication strategies significantly influence employee acceptance and effective utilisation of robot capabilities.

The management of change processes during anthropomorphic robotics implementation requires sophisticated understanding of change management frameworks and their application to technology adoption contexts. Venkatesh and Davis (2000) provide theoretical extensions of TAM that incorporate social influence processes, cognitive instrumental processes, and temporal factors that influence technology acceptance over time. Their research highlights the importance of managerial interventions that address both rational and emotional factors influencing technology adoption, suggesting that leaders must attend to both functional benefits and psychological comfort factors when implementing anthropomorphic service robotics.

Contemporary leadership challenges in anthropomorphic robotics implementation include developing new competencies for managing mixed human-robot service teams, establishing communication strategies that effectively address employee concerns about job security and role changes, and building organisational cultures that embrace human-robot collaboration. These challenges require leaders to develop new skills in technology integration, change communication, and team management that extend beyond traditional leadership competencies. The successful implementation of anthropomorphic service robotics depends critically on leadership effectiveness in navigating these complex organisational dynamics whilst maintaining focus on customer experience enhancement and operational performance improvement.

FUTURE ANTHROPOMORPHIC CAPABILITIES AND TECHNOLOGICAL CONVERGENCE

The next generation of anthropomorphic service robotics will be characterised by unprecedented integration of advanced artificial intelligence capabilities, sophisticated emotional processing systems, and convergent technologies that create seamless hybrid physical-digital experiences. This technological evolution represents a fundamental shift from current task-oriented robots to emotionally intelligent companions capable of forming meaningful relationships with customers whilst delivering highly personalised service experiences. Understanding these emerging capabilities is essential for organisations preparing to implement next-generation anthropomorphic service robotics systems.

Emotional AI and Advanced Empathy Systems

The development of emotional artificial intelligence represents perhaps the most significant advancement in creating truly anthropomorphic service robots capable of sophisticated human-like interaction. Contemporary emotional AI systems integrate multiple modalities of human emotional expression, including facial expression analysis, vocal tone recognition, physiological monitoring, and contextual behaviour interpretation to create comprehensive understanding of customer emotional states. These systems extend far beyond simple emotion recognition to encompass predictive empathy capabilities that anticipate customer emotional needs based on historical interaction patterns, situational context, and real-time behavioural cues.

Advanced empathy systems utilise sophisticated machine learning algorithms trained on vast datasets of human emotional expression and response patterns to develop nuanced understanding of emotional context and appropriate responsive behaviour. Picard (2000) established the foundational principles of affective computing that inform contemporary emotional AI development, emphasising the importance of computers' ability to recognise, interpret, and respond to human emotions in ways that enhance rather than diminish human emotional experience. Modern applications of these principles in anthropomorphic service robotics involve real-time emotional state assessment, empathetic response generation, and adaptive interaction strategies that respond dynamically to changing customer emotional needs.

Ekman (2016) identifies fundamental agreements among emotion researchers regarding the universality of basic emotional expressions, providing crucial insights for developing emotional AI systems capable of cross-cultural operation. These insights inform the development of micro-expression analysis capabilities that enable anthropomorphic robots to detect subtle emotional cues that human service providers might miss, particularly in high-stress or emotionally charged service situations. The integration of micro-expression analysis with contextual awareness systems creates opportunities for proactive emotional support that anticipates customer distress before it becomes overtly expressed.

Predictive empathy algorithms represent a significant advancement beyond reactive emotional response systems, enabling anthropomorphic robots to anticipate customer emotional needs based on comprehensive analysis of individual history, situational factors, and population-level behavioural patterns. These algorithms utilise advanced machine learning techniques to identify emotional trajectory patterns that predict likely future emotional states, enabling proactive intervention and support strategies. For example, a healthcare service robot might recognise early indicators of patient anxiety and proactively implement calming interventions before anxiety escalates to problematic levels.

The implementation of emotional mirroring technologies enables anthropomorphic robots to reflect appropriate emotional responses that validate customer feelings whilst maintaining professional boundaries. Tapus et al. (2007) demonstrate how socially assistive robotics can provide emotional support through appropriate empathetic responses that acknowledge customer emotional states without overwhelming or inappropriate emotional displays. This capability requires sophisticated balance between emotional authenticity and professional appropriateness, ensuring that robot emotional responses enhance rather than complicate customer interactions.

Generative Personality Models

The development of generative personality models represents a revolutionary advancement in creating consistent, engaging, and brand-aligned anthropomorphic service robots capable of forming meaningful relationships with customers over extended periods. These systems utilise advanced artificial intelligence to generate coherent personality characteristics that remain consistent across interactions whilst adapting to individual customer preferences and cultural contexts. Unlike scripted responses or simple decision trees, generative personality models create dynamic personality expressions that evolve naturally through interaction experience whilst maintaining core personality traits aligned with organisational brand values.

Contemporary generative personality models draw upon established personality psychology frameworks, particularly the Five-Factor Model articulated by McCrae and Costa (2008), to create psychologically coherent robot personalities. These models incorporate dimensions of openness to experience, conscientiousness, extraversion, agreeableness, and neuroticism in ways that create recognisable and predictable personality patterns whilst allowing for situational adaptation. The integration of personality psychology principles with artificial intelligence enables anthropomorphic robots to exhibit personality-consistent behaviour across diverse service situations whilst maintaining the flexibility necessary for effective customer service delivery.

Gosling et al. (2003) provide empirical validation for brief personality assessment measures that inform the development of personality models for anthropomorphic robots. Their research demonstrates that coherent personality impressions can be formed through relatively brief interactions, suggesting that anthropomorphic robots can establish recognisable personality characteristics efficiently whilst building customer relationships over time. This insight is particularly valuable for service contexts where initial interactions must quickly establish trust and rapport whilst laying foundations for longer-term customer relationships.

Brand-aligned persona creation represents a critical application of generative personality models in commercial service robotics contexts. Organisations can utilise these systems to create anthropomorphic robots that embody brand values and communication styles whilst maintaining individual personality characteristics that enhance customer engagement. Araujo (2018) examines how anthropomorphic design cues and communicative agency framing influence perceptions of conversational agents, demonstrating the importance of alignment between robot personality characteristics and organisational brand identity. This alignment ensures that anthropomorphic robots enhance rather than dilute brand experiences whilst providing distinctive service delivery that differentiates organisations from competitors.

Adaptive personality traits based on customer preferences represent an advanced capability that enables anthropomorphic robots to modify personality expression in response to individual customer characteristics and preferences. These systems learn from customer interaction patterns to identify preferred communication styles, humour preferences, formality levels, and emotional expression patterns that optimise customer satisfaction. However, this adaptability must be carefully balanced with personality consistency to avoid creating confusing or inauthentic interaction experiences that undermine trust and relationship development.

The technical implementation of generative personality models requires sophisticated integration of natural language processing, behavioural generation algorithms, and memory systems that maintain personality consistency across interactions. These systems must manage the complex challenge of expressing consistent core personality traits whilst adapting surface-level expressions to match customer preferences and situational requirements. The successful implementation of these models creates anthropomorphic robots capable of forming genuine emotional connections with customers that enhance service quality and customer loyalty over extended periods.

Technological Convergence: AR/VR, IoT, and Metaverse Integration

The convergence of anthropomorphic robotics with augmented reality, virtual reality, Internet of Things, and metaverse technologies represents a fundamental expansion of service delivery possibilities that transcend traditional physical-digital boundaries. This technological convergence enables the creation of hybrid anthropomorphic experiences where physical robots seamlessly integrate with digital environments to provide comprehensive, immersive service experiences that adapt to customer location, preferences, and interaction modalities.

Augmented reality integration enables anthropomorphic robots to overlay digital information and interactive elements onto physical environments, enhancing their ability to provide contextual information and guidance. Barsom et al. (2016) demonstrate the effectiveness of augmented reality applications in medical training contexts, highlighting principles applicable to anthropomorphic service robotics in healthcare settings. AR-enabled anthropomorphic robots can provide visual overlays that highlight important information, display personalised content, and create interactive experiences that combine physical presence with digital enhancement capabilities.

The integration of anthropomorphic robotics with Internet of Things systems creates opportunities for contextual awareness and environmental responsiveness that significantly enhance service delivery capabilities. IoT integration enables robots to access real-time information about environmental conditions, customer preferences, and operational status across connected systems, facilitating proactive service delivery and personalised experiences. Xu et al. (2018) examine Industry 4.0 developments that demonstrate how connected systems create opportunities for intelligent automation and personalised service delivery that inform anthropomorphic robotics applications.

Metaverse integration represents perhaps the most revolutionary aspect of technological convergence, enabling anthropomorphic robots to maintain consistent personalities and relationships across physical and virtual environments. Mystakidis (2022) provides comprehensive analysis of metaverse capabilities that illuminate potential applications in anthropomorphic service robotics, including persistent digital identities, cross-platform presence, and immersive interaction capabilities. This integration enables customers to continue relationships with anthropomorphic service robots across multiple channels and environments, creating seamless omnichannel experiences that maintain relationship continuity regardless of interaction modality.

Cross-platform personality and memory consistency represents a critical technical challenge in implementing convergent anthropomorphic robotics systems. These systems must maintain coherent personality expression and relationship memory across physical robots, virtual avatars, mobile applications, and metaverse environments whilst adapting to the unique capabilities and constraints of each platform. The successful implementation of cross-platform consistency requires sophisticated data integration, personality model synchronisation, and interaction adaptation capabilities that ensure customers experience consistent service quality and relationship continuity regardless of interaction channel.

The development of hybrid physical-digital anthropomorphic experiences creates new possibilities for service delivery that combine the advantages of physical presence with the flexibility and scalability of digital systems. These hybrid experiences enable anthropomorphic robots to provide services that extend beyond their physical location through digital presence, whilst maintaining the emotional connection and trust associated with physical interaction. The integration of these technologies creates anthropomorphic service systems capable of providing personalised, contextually aware, and emotionally intelligent service experiences that adapt seamlessly to customer needs and preferences across multiple interaction modalities.

PREDICTIVE SCENARIOS FOR 2030 SERVICE LANDSCAPES

The transformation of service delivery through anthropomorphic robotics will manifest differently across various industry sectors, each leveraging unique capabilities to address sector-specific challenges and opportunities. By 2030, the convergence of emotional artificial intelligence, generative personality models, and technological integration will enable anthropomorphic robots to assume increasingly sophisticated roles that fundamentally reshape customer experiences and service delivery paradigms. The following scenarios provide concrete visualisations of how anthropomorphic service robotics will transform key service sectors whilst illustrating broader principles applicable across industries.

Healthcare: Emotionally Intelligent Care Companions

The healthcare sector will witness perhaps the most profound transformation through emotionally intelligent anthropomorphic care companions that integrate seamlessly with clinical care teams to provide comprehensive patient support extending far beyond traditional medical interventions. By 2030, these advanced care companions will utilise sophisticated emotional AI systems to monitor patient psychological wellbeing continuously, providing proactive mental health support that complements clinical treatment protocols. These systems will integrate physiological monitoring capabilities with emotional state assessment to create comprehensive patient wellness profiles that inform both immediate care decisions and long-term treatment planning.

Broadbent et al. (2009) identify critical factors influencing healthcare robot acceptance among older populations, including functionality, reliability, and social compatibility considerations that inform the development of emotionally intelligent care companions. Future care companions will address these acceptance factors through sophisticated personality models that adapt to individual patient preferences whilst maintaining professional therapeutic boundaries. These robots will form genuine therapeutic relationships with patients over extended periods, providing consistency and emotional support that supplements human care provider interactions.

Advanced patient monitoring capabilities will enable care companions to detect subtle changes in emotional state, cognitive function, and social engagement patterns that may indicate developing health concerns before they become clinically apparent. Robinson et al. (2013) demonstrate positive psychosocial effects of companion robots through randomised controlled trials, providing evidence for the therapeutic value of human-robot relationships in healthcare contexts. Future care companions will build upon these foundations to provide more sophisticated emotional support that includes grief counselling, anxiety management, and depression intervention capabilities delivered through empathetic, consistent, and culturally sensitive interactions.

Integration with healthcare teams and electronic health records will enable care companions to serve as comprehensive patient advocates who maintain detailed knowledge of individual medical histories, treatment preferences, and family dynamics whilst facilitating communication between patients and healthcare providers. Čaić et al. (2018) examine value co-creation and co-destruction processes in elderly care networks, highlighting the importance of balancing technological capabilities with human relationships to optimise care outcomes. Future care companions will navigate these complex dynamics by enhancing rather than replacing human connections whilst providing services that extend care capacity and improve patient outcomes.

The therapeutic relationship development capabilities of 2030 care companions will include sophisticated emotional intelligence that enables recognition of patient emotional needs, provision of appropriate comfort and reassurance, and facilitation of difficult conversations about health status and treatment options. These robots will maintain therapeutic relationships across extended periods, providing continuity of care that transcends individual healthcare provider schedules and availability constraints whilst ensuring consistent emotional support throughout patient care journeys.

Hospitality: Hyper-Personalised Guest Experiences

The hospitality sector will be revolutionised by anthropomorphic robots capable of delivering hyper-personalised guest experiences that surpass traditional service paradigms through comprehensive integration of guest preference data, predictive analytics, and emotionally intelligent interaction capabilities. By 2030, these sophisticated hospitality robots will maintain detailed guest profiles across multiple properties and brands, enabling seamless personalisation that recognises returning guests and adapts services to individual preferences regardless of location or previous interaction history.

Choi et al. (2020) examine service quality perceptions of human-robot interaction in hotel contexts, identifying factors that influence guest satisfaction and acceptance of robotic service delivery. Future hospitality robots will address these factors through sophisticated personality models that align with brand values whilst adapting to individual guest communication preferences, cultural backgrounds, and service expectations. These robots will provide consistent brand experiences whilst delivering personalised service that feels uniquely tailored to each guest's preferences and needs.

Multi-property memory systems will enable hospitality robots to recognise guests across different locations, remembering previous preferences, special occasions, dietary restrictions, and service feedback to create seamless experiences that demonstrate genuine care and attention. Kuo et al. (2017) investigate innovative service delivery through hospitality robots, highlighting opportunities for enhanced efficiency and personalisation that inform future service paradigms. Advanced hospitality robots will utilise these insights to provide proactive service delivery that anticipates guest needs before they are explicitly expressed, creating memorable experiences that exceed guest expectations.

Cultural adaptation and linguistic personalisation capabilities will enable hospitality robots to adjust communication styles, service approaches, and interaction patterns to match guest cultural backgrounds and preferences. Tussyadiah and Park (2018) examine consumer evaluation of hotel service robots, demonstrating the importance of appropriate cultural adaptation in international hospitality contexts. Future hospitality robots will incorporate sophisticated cultural intelligence that enables appropriate adaptation to diverse guest populations whilst maintaining consistent service quality standards across all interactions.

Integration with smart room technologies will create comprehensive guest experience ecosystems where anthropomorphic robots serve as personal concierges who coordinate all aspects of guest stays, from arrival and check-in through dining reservations, entertainment recommendations, and departure arrangements. These integrated systems will anticipate guest needs based on preference patterns, contextual factors, and real-time behaviour analysis whilst providing seamless service delivery that feels effortless and intuitive.

Retail: Emotional Connection Through Shopping Assistance

The retail sector will experience fundamental transformation through anthropomorphic shopping assistants that create emotional connections with customers whilst providing sophisticated product recommendations, personalised shopping experiences, and ongoing relationship management that extends far beyond individual transactions. By 2030, these advanced retail robots will utilise comprehensive customer data analysis, emotional intelligence, and predictive analytics to provide shopping assistance that feels genuinely helpful and personally relevant.

Grewal et al. (2020) examine the future of in-store technology, identifying opportunities for enhanced customer experiences through intelligent automation and personalisation capabilities that inform retail robot development. Future retail robots will leverage these technologies to provide advanced product recommendations that integrate emotional analysis with purchase history, lifestyle preferences, and contextual factors to suggest products that align with customer needs and preferences whilst introducing them to new possibilities they might not have considered independently.

Social shopping experiences will be enhanced through AI companions that facilitate group shopping decisions, provide entertainment and engagement during shopping processes, and create memorable experiences that encourage repeat visits and brand loyalty. Marinova et al. (2017) examine learning from technology-empowered frontline interactions, highlighting opportunities for retail robots to enhance rather than replace human service delivery whilst providing capabilities that extend beyond traditional human limitations.

Brand relationship building through consistent interactions will enable retail robots to form ongoing relationships with customers that extend beyond individual shopping trips to include post-purchase support, product education, and personalised recommendations for future needs. Roy et al. (2017) identify constituents and consequences of smart customer experience in retailing, demonstrating how technology-enabled personalisation creates value for both customers and retailers through enhanced satisfaction and loyalty outcomes.

Advanced retail robots will provide emotional support during shopping decisions, helping customers navigate complex choices whilst building confidence in purchase decisions through empathetic guidance and reassurance. These systems will recognise customer emotional states during shopping processes and adapt their assistance approaches to provide appropriate support that enhances rather than overwhelms the shopping experience.

Tourism: Culturally Adaptive Guide Systems

The tourism sector will be transformed through culturally adaptive anthropomorphic guide systems that provide personalised destination experiences whilst demonstrating sophisticated cultural intelligence and multilingual emotional expression capabilities. By 2030, these advanced tourism robots will serve as knowledgeable local ambassadors who adapt their guidance approaches to match visitor cultural backgrounds, interests, and interaction preferences whilst providing authentic insights into local culture and attractions.

Buhalis and Leung (2018) examine smart hospitality ecosystems that inform the development of integrated tourism guidance systems capable of coordinating multiple service providers to create seamless visitor experiences. Future tourism robots will serve as central coordination points for comprehensive destination experiences, managing reservations, transportation, dining recommendations, and cultural activities whilst maintaining awareness of visitor preferences and constraints.

Real-time cultural sensitivity adjustments will enable tourism robots to modify their communication styles, recommendation approaches, and interaction patterns to align with visitor cultural expectations whilst avoiding cultural misunderstandings or inappropriate suggestions. Gretzel et al. (2015) examine smart tourism foundations and developments, highlighting opportunities for technology-enabled personalisation that enhances rather than diminishes authentic cultural experiences.

Multilingual emotional expression capabilities will enable tourism robots to communicate effectively with diverse international visitors whilst maintaining emotional authenticity and cultural appropriateness across different languages and cultural contexts. Wang et al. (2013) examine smart tourism destination initiatives, demonstrating how technology integration can enhance service delivery whilst preserving authentic destination experiences that satisfy diverse visitor expectations.

The integration of continuous learning approaches into leadership development programmes ensures that leaders maintain current knowledge of technological developments, industry best practices, and emerging challenges in anthropomorphic service robotics management. These learning approaches should incorporate peer networks, research engagement, and industry collaboration that provide ongoing support for leadership development in this rapidly evolving field.

STRATEGIC FUTURES AND INNOVATION TRAJECTORIES

The strategic planning for anthropomorphic service robotics requires sophisticated forecasting frameworks and innovation assessment tools that enable organisations to navigate the complex landscape of emerging technologies whilst making informed investment decisions and preparing for future competitive challenges. As the pace of technological advancement accelerates and market dynamics continue to evolve, organisations must develop robust capabilities for assessing innovation trajectories, identifying strategic opportunities, and establishing research priorities that guide long-term organisational development. The strategic futures framework encompasses both predictive assessment tools for immediate implementation planning and forward-looking research agendas that address fundamental questions about the long-term implications of widespread anthropomorphic robotics adoption.

Predictive Frameworks and Assessment Tools

The development of comprehensive predictive frameworks for anthropomorphic service robotics requires integration of multiple analytical approaches that account for technological capabilities, market dynamics, organisational factors, and customer acceptance patterns. These frameworks must provide practical guidance for timing adoption decisions, selecting appropriate technologies, and assessing potential return on investment whilst acknowledging the inherent uncertainties associated with emerging technology forecasting. Brynjolfsson and McAfee (2017) examine the business implications of artificial intelligence adoption, highlighting the importance of strategic frameworks that balance technological potential with realistic assessment of implementation challenges and competitive dynamics.

Organisational readiness diagnostic instruments must be specifically designed to assess preparedness for anthropomorphic service robotics implementation across multiple dimensions including technological infrastructure, human resource capabilities, customer base characteristics, and cultural compatibility factors. These diagnostic tools should provide comprehensive evaluation of organisational strengths and weaknesses whilst identifying specific development priorities that enhance implementation success probability. Fountaine et al. (2019) examine approaches to building AI-powered organisations, emphasising the importance of systematic assessment and development processes that align technological capabilities with organisational strategic objectives.

Technology adoption timeline planning tools must account for the complex interdependencies between technological development, organisational capability building, and market readiness factors that influence optimal implementation timing. These planning tools should incorporate scenario analysis capabilities that enable organisations to evaluate multiple adoption pathways whilst considering various contingencies and external factors that may influence implementation success. The development of realistic timelines requires sophisticated understanding of technology maturation cycles, organisational learning requirements, and market acceptance patterns specific to anthropomorphic service robotics.

Return on investment assessment models for anthropomorphic robots must capture both quantifiable benefits such as operational efficiency improvements and cost savings, and qualitative benefits including customer satisfaction enhancement, brand differentiation, and employee engagement improvements. Traditional ROI models inadequately capture the full value proposition of anthropomorphic service robotics, particularly the emotional and relationship-building capabilities that create long-term customer loyalty and competitive advantage. Ransbotham et al. (2017) examine gaps between AI ambition and action, highlighting the importance of realistic value assessment that accounts for both implementation costs and long-term strategic benefits.

Predictive frameworks must also incorporate risk assessment capabilities that evaluate potential challenges including technology failure, customer acceptance issues, employee resistance, and competitive responses that may influence implementation outcomes. These risk assessment tools should provide guidance for mitigation strategies whilst enabling organisations to make informed decisions about acceptable risk levels and contingency planning requirements. The complexity of anthropomorphic robotics implementation requires comprehensive risk frameworks that address technical, operational, financial, and reputational risks associated with human-like robot deployment.

Assessment tools must incorporate performance benchmarking capabilities that enable organisations to evaluate their anthropomorphic robotics implementation against industry standards and best practices whilst identifying opportunities for continuous improvement and competitive advantage enhancement. These benchmarking tools should provide comparative analysis capabilities that account for sector-specific factors, organisational characteristics, and implementation contexts that influence performance outcomes.

FUTURE RESEARCH PRIORITIES AND AGENDAS

The continued development of anthropomorphic service robotics requires comprehensive research agendas that address fundamental questions about human-robot interaction dynamics, long-term societal implications, and optimal design approaches for maximising benefits whilst minimising potential negative consequences. These research priorities encompass psychological and social implications of human-like service robots, cross-cultural acceptance and adaptation studies, and longitudinal analyses of employment and social impacts that inform policy development and organisational strategy.

Psychological and social implications of human-like service robots represent critical research priorities that require interdisciplinary collaboration between robotics engineers, psychologists, sociologists, and organisational behaviour researchers. Savela et al. (2018) provide systematic literature review of social acceptance of robots across occupational fields, identifying research gaps that require attention to advance understanding of human-robot interaction dynamics. Future research must address questions regarding the psychological impact of long-term relationships with anthropomorphic robots, the development of emotional attachment to artificial agents, and the implications of these relationships for human social development and wellbeing.

Research priorities must also examine the potential for anthropomorphic robots to influence human behaviour and attitudes in ways that may have unintended consequences for individual development and social functioning. The capacity of emotionally intelligent robots to provide consistent emotional support and validation may create dependency relationships that interfere with human resilience development and social skill acquisition. Understanding these potential impacts requires longitudinal research designs that track individual and community outcomes over extended periods whilst controlling for various contextual factors that influence human development.

Cross-cultural acceptance and adaptation studies represent essential research priorities for organisations planning global deployment of anthropomorphic service robotics. Cultural differences in robot acceptance, communication preferences, emotional expression norms, and service expectations significantly influence implementation success across diverse international markets. Future research must examine how anthropomorphic robot design and behaviour should be adapted to align with cultural values whilst maintaining operational consistency and brand identity across global service operations.

The development of culturally intelligent anthropomorphic robots requires comprehensive research into cultural variation in emotional expression, social interaction norms, and service relationship expectations that inform robot personality development and behaviour programming. This research must address both surface-level cultural differences such as language and communication styles, and deeper cultural values regarding technology, human relationships, and appropriate boundaries between human and artificial agents.

Long-term impact on service industry employment represents perhaps the most significant research priority given the potential for anthropomorphic service robotics to fundamentally alter employment patterns across service sectors. Frey and Osborne (2017) examine job susceptibility to computerisation, providing frameworks for assessing employment impact that require extension to address the specific characteristics of anthropomorphic service robotics. Unlike previous automation technologies that primarily affected routine manual tasks, anthropomorphic robots have potential to impact emotional labour and interpersonal service roles that were previously considered automation-resistant.

Manyika et al. (2017) examine implications of technological change for jobs, skills, and wages, identifying research priorities for understanding how technological advancement will reshape employment whilst creating new opportunities for human workers. Future research must examine how anthropomorphic service robotics will transform rather than simply eliminate service employment, identifying new roles and skill requirements that emerge as human workers collaborate with emotionally intelligent robots.

Research agendas must also address the broader societal implications of widespread anthropomorphic robotics adoption, including effects on human social development, community relationships, and cultural values regarding human-artificial agent interaction. These studies require longitudinal research designs that track societal changes over extended periods whilst examining causal relationships between anthropomorphic robotics adoption and social outcomes. The complexity of these research questions necessitates interdisciplinary collaboration and innovative research methodologies that capture both quantitative outcomes and qualitative experiences of human-robot coexistence in service environments.

Future research priorities must incorporate evaluation of policy and regulatory frameworks necessary for governing anthropomorphic service robotics deployment whilst protecting consumer interests and ensuring ethical implementation practices. This research should inform the development of industry standards, certification processes, and regulatory guidelines that promote innovation whilst addressing potential risks associated with emotionally intelligent human-like robots in service contexts.

CONCLUSION AND IMPLICATIONS

This chapter has provided a comprehensive examination of future trajectories in anthropomorphic service robotics, establishing theoretical foundations, exploring technological capabilities, and developing practical frameworks for organisational implementation. The analysis demonstrates that the convergence of emotional artificial intelligence, generative personality models, and advanced technological integration will fundamentally transform service delivery across multiple sectors by 2030, creating both unprecedented opportunities for enhanced customer experiences and complex challenges for organisational adaptation and leadership development.

Summary of Key Insights and Contributions

The primary contribution of this chapter lies in the development of an integrated adoption-leadership-anthropomorphism framework that synthesises insights from technology acceptance theory, leadership research, and anthropomorphism studies to provide comprehensive guidance for organisational implementation of emotionally intelligent service robots. This framework recognises that successful anthropomorphic robotics adoption requires simultaneous attention to technological capabilities, human psychological factors, and organisational change management processes that traditional technology adoption models inadequately address.

The analysis reveals that anthropomorphic service robotics represents a qualitatively different category of technology adoption that requires expanded theoretical frameworks and implementation strategies. Unlike previous automation technologies that primarily addressed functional efficiency concerns, anthropomorphic robots engage emotional and social dimensions of human experience that create new categories of value creation and risk management. The emotional intelligence capabilities of future service robots will enable relationship formation and emotional support provision that extends far beyond traditional service delivery paradigms, requiring organisations to develop new competencies in managing human-robot emotional interactions.

The sectoral analysis demonstrates that whilst anthropomorphic robotics will transform all service industries, the specific applications and implementation strategies must be carefully tailored to address sector-specific requirements, customer expectations, and operational contexts. Healthcare applications will focus on therapeutic relationship development and emotional support capabilities, whilst hospitality implementations will emphasise personalisation and cultural adaptation. Retail applications will prioritise emotional connection and brand relationship building, whilst tourism deployments will emphasise cultural intelligence and comprehensive destination guidance capabilities.

The leadership evolution analysis reveals that successful anthropomorphic robotics implementation requires fundamental expansion of leadership competencies beyond traditional technology management skills. Leaders must develop sophisticated AI literacy, human-robot team management capabilities, and strategic foresight skills that enable navigation of the complex organisational dynamics associated with emotionally intelligent robot deployment. Jarrahi (2018) emphasises the importance of human-AI symbiosis in organisational decision-making, highlighting principles that inform leadership approaches to anthropomorphic robotics implementation.

Practical Implications for Organisations and Leaders

The practical implications of this analysis provide concrete guidance for organisations preparing to implement anthropomorphic service robotics whilst highlighting critical success factors that influence implementation outcomes. Organisations must begin immediate preparation for anthropomorphic robotics adoption through comprehensive readiness assessment, leadership development programmes, and strategic planning processes that address the unique characteristics of emotionally intelligent human-like robots.

Leadership development emerges as perhaps the most critical practical priority, requiring organisations to invest in comprehensive training programmes that develop AI literacy, human-robot team management skills, and change communication capabilities necessary for successful implementation. These development programmes must address both technical knowledge requirements and interpersonal skills necessary for managing the complex psychological and social dynamics associated with anthropomorphic robotics adoption.

Organisational culture development represents another essential practical priority, requiring systematic efforts to build acceptance of human-robot collaboration whilst addressing employee concerns about job security and role changes. Rust and Huang (2014) examine service revolution implications for marketing science, providing insights into organisational transformation requirements that apply broadly to anthropomorphic robotics adoption across service sectors.

Strategic planning processes must incorporate scenario analysis and contingency planning that account for various technological development trajectories, competitive responses, and market acceptance patterns that may influence implementation success. Organisations should develop flexible implementation strategies that enable adaptation to changing circumstances whilst maintaining progress toward anthropomorphic robotics capabilities.

Infrastructure development requirements necessitate substantial investment in data management systems, integration platforms, and privacy protection capabilities necessary for supporting emotionally intelligent robots. These infrastructure investments must be carefully planned and implemented in phases that align with organisational learning and adoption readiness to optimise resource utilisation and implementation success.

Future Research Directions and Limitations

This chapter identifies several critical research priorities that require attention to advance understanding of anthropomorphic service robotics implications and optimise implementation strategies. Longitudinal studies of human-robot relationship development represent essential research priorities that will inform understanding of how emotional connections with anthropomorphic robots evolve over time and their implications for customer loyalty, emotional wellbeing, and service satisfaction outcomes.

Cross-cultural research examining cultural variation in anthropomorphic robot acceptance, communication preferences, and service relationship expectations remains critically needed to guide global implementation strategies. Wirtz et al. (2018) examine service robots in frontline contexts, identifying research gaps that require attention to advance understanding of optimal human-robot service delivery models across diverse cultural contexts.

The limitations of this analysis include reliance on extrapolation from current technological trends and research findings to predict future capabilities and outcomes. The rapid pace of AI development creates inherent uncertainty regarding specific technological capabilities that will be available by 2030, requiring ongoing assessment and strategy adaptation as new developments emerge. Additionally, the analysis focuses primarily on positive implementation scenarios whilst acknowledging that negative outcomes including customer rejection, employee resistance, and technical failures represent significant risks that require careful management.

Call to Action for Strategic Preparation

The analysis presented in this chapter demonstrates that anthropomorphic service robotics represents both inevitable technological evolution and strategic imperative for service organisations seeking competitive advantage in increasingly automated markets. Organisations that delay preparation for anthropomorphic robotics adoption risk significant competitive disadvantage as early adopters establish market leadership through enhanced customer experiences and operational capabilities.

Immediate action priorities include comprehensive organisational readiness assessment, leadership development programme initiation, and strategic planning processes that prepare organisations for anthropomorphic robotics implementation. These preparation activities should begin immediately regardless of specific implementation timelines, as the complexity of organisational change required for successful adoption necessitates substantial lead time and systematic development efforts.

The strategic preparation for anthropomorphic service robotics requires commitment to ongoing learning, adaptation, and investment in human capital development that enables organisations to harness the full potential of emotionally intelligent human-like robots whilst maintaining focus on human employee empowerment and customer relationship enhancement. The future of service delivery will be characterised by sophisticated human-robot collaboration that enhances rather than replaces human capabilities, creating opportunities for service excellence that exceed current paradigms whilst addressing fundamental human needs for emotional connection and personalised attention.

Leadership Evolution for Anthropomorphic AI Futures

The emergence of anthropomorphic service robotics necessitates fundamental evolution in leadership competencies, organisational management approaches, and strategic decision-making frameworks that extend far beyond traditional technology adoption requirements. Leaders in the anthropomorphic AI era must develop sophisticated understanding of human-robot interaction dynamics, ethical implications of emotionally intelligent systems, and the complex organisational changes required for successful integration of human-like robots into service delivery teams. This leadership evolution encompasses expanded AI literacy requirements, new approaches to managing hybrid human-robot teams, and comprehensive competency development programmes that prepare leaders for the strategic and operational complexities of anthropomorphic service environments.

Expanding AI Literacy Requirements

Contemporary leadership in anthropomorphic service robotics contexts demands sophisticated AI literacy that encompasses technical understanding, strategic applications knowledge, and ethical considerations specific to emotionally intelligent human-like systems. This expanded literacy requirement extends beyond basic familiarity with artificial intelligence concepts to include deep comprehension of emotional AI capabilities, limitations, and appropriate applications within service delivery contexts. Leaders must develop nuanced understanding of how emotional artificial intelligence systems function, their accuracy limitations, and the potential consequences of inappropriate emotional responses in customer service situations.

Davenport and Ronanki (2018) identify critical factors for successful AI implementation in organisational contexts, emphasising the importance of leadership understanding regarding AI capabilities and realistic assessment of implementation challenges. Future leaders of anthropomorphic service robotics must develop expertise in evaluating emotional AI claims, assessing vendor capabilities, and making informed decisions about technology investments that balance potential benefits with implementation risks and costs. This evaluation capability requires understanding of current AI limitations, emerging technology trends, and the complex technical requirements for supporting sophisticated anthropomorphic robots in operational environments.

Strategic decision-making for anthropomorphic technology investments requires leaders to assess not only technical capabilities and costs but also broader organisational implications including employee acceptance, customer readiness, and competitive positioning considerations. Long and Magerko (2020) examine AI literacy competencies and design considerations that inform leadership development for AI-enhanced environments, highlighting the importance of critical thinking skills for evaluating AI applications and understanding their broader implications for organisational strategy and operations.

Ethical considerations in human-like robot deployment represent a critical component of AI literacy that requires leaders to navigate complex questions regarding emotional manipulation, privacy protection, and appropriate boundaries in human-robot relationships. Leaders must develop frameworks for ensuring that anthropomorphic robots enhance rather than exploit customer emotional vulnerabilities whilst maintaining transparency regarding robot capabilities and limitations. Wilson and Daugherty (2018) examine collaborative intelligence principles that guide ethical human-AI collaboration, providing insights applicable to anthropomorphic robotics management that emphasise complementary capabilities rather than replacement dynamics.

The rapid pace of AI development requires leaders to maintain current knowledge of emerging capabilities, industry best practices, and regulatory developments that influence anthropomorphic robotics implementation. This ongoing learning requirement necessitates commitment to continuous professional development, engagement with technology vendors and research communities, and participation in industry forums focused on responsible AI deployment in service contexts.

Leadership Styles for Human-Robot Teams

The management of mixed human-robot service teams requires leadership approaches that recognise the unique characteristics of anthropomorphic robots whilst maintaining focus on human employee development, motivation, and job satisfaction. Traditional leadership styles must be adapted to address the complex dynamics of human-robot collaboration, including employee concerns about job security, role clarity in human-robot teams, and the psychological adjustments required for working alongside human-like artificial colleagues.

Avolio et al. (2009) provide comprehensive analysis of leadership theories and research directions that inform approaches to managing technology-enhanced teams, emphasising the importance of adaptive leadership styles that respond to changing organisational contexts and team compositions. Leaders managing human-robot service teams must develop capabilities for facilitating effective collaboration between human employees and anthropomorphic robots whilst ensuring that human workers feel valued and supported throughout the transition process.

Communication strategies for anthropomorphic technology adoption must address both rational and emotional concerns that employees experience when working with human-like robots. Yukl (2012) examines effective leadership behaviours that include supportive communication, participative decision-making, and empowering approaches that enhance employee engagement during organisational change processes. These leadership behaviours are particularly important in anthropomorphic robotics contexts where employees may experience anxiety about job displacement, uncertainty about role changes, and psychological discomfort regarding human-like artificial colleagues.

Building trust and acceptance among employees requires leaders to demonstrate authentic commitment to human-robot collaboration whilst providing concrete evidence of benefits for both employees and customers. Zhang and Bartol (2010) examine the influence of empowering leadership on employee creativity and engagement, highlighting approaches that encourage innovative thinking about human-robot collaboration possibilities whilst addressing concerns about technology displacement. Leaders must create environments where employees feel empowered to contribute to human-robot team effectiveness whilst developing new skills that complement robot capabilities.

The development of role clarity and performance expectations for mixed human-robot teams requires leaders to establish clear guidelines regarding robot capabilities, appropriate task allocation between humans and robots, and performance standards that account for the unique characteristics of human-robot collaboration. These guidelines must balance efficiency optimisation with human employee satisfaction whilst ensuring that customers receive consistent, high-quality service regardless of whether they interact with human employees or anthropomorphic robots.

Leaders must also develop skills in managing the emotional dynamics of human-robot teams, including addressing employee feelings about working with human-like artificial colleagues, facilitating positive human-robot relationships, and ensuring that anthropomorphic robots enhance rather than undermine team cohesion and morale. This emotional management capability requires sophisticated understanding of human psychology, group dynamics, and the specific challenges associated with integrating artificial agents into human social systems.

Future Leadership Competency Development

The complexity of anthropomorphic service robotics implementation necessitates comprehensive leadership competency development programmes that prepare leaders for the multifaceted challenges of managing human-robot service environments. These development programmes must address technical knowledge requirements, strategic thinking capabilities, and interpersonal skills necessary for successful anthropomorphic robotics leadership whilst incorporating cross-cultural considerations relevant for global deployment contexts.

Training programmes for AI-enhanced service environments must provide leaders with practical experience in managing anthropomorphic robots, understanding their capabilities and limitations, and developing strategies for optimising human-robot collaboration. Day (2000) examines leadership development approaches that emphasise experiential learning, mentoring, and skill-building activities that inform comprehensive development programmes for anthropomorphic robotics leadership. These programmes should include hands-on experience with robot management systems, customer interaction monitoring, and performance optimisation techniques that enable leaders to develop practical expertise in anthropomorphic robotics operations.

Strategic foresight capabilities for technology evolution represent essential competencies for leaders navigating the rapidly changing landscape of anthropomorphic service robotics. Leaders must develop skills in technology trend analysis, scenario planning, and strategic adaptation that enable proactive responses to emerging capabilities and market developments. Hoch and Kozlowski (2014) examine leadership in virtual teams, providing insights into distributed leadership approaches that inform management of human-robot teams operating across multiple locations and service contexts.

Cross-cultural leadership capabilities for global anthropomorphic robot deployment require understanding of cultural differences in robot acceptance, communication preferences, and service expectations that influence implementation strategies in diverse international markets. Leaders must develop cultural intelligence that enables appropriate adaptation of robot personalities, interaction styles, and service approaches to align with local cultural norms whilst maintaining brand consistency and service quality standards.

Mumford et al. (2017) identify nine critical cognitive skills for leadership performance that provide frameworks for developing comprehensive competency programmes for anthropomorphic robotics leadership. These skills include problem-solving capabilities, strategic thinking, innovation management, and systems thinking that enable leaders to navigate the complex interdependencies between technology capabilities, human factors, and organisational outcomes in anthropomorphic service environments.

Future leadership competency development must also address the psychological and emotional demands of managing human-robot service teams, including skills in conflict resolution, motivation management, and change communication that account for the unique challenges associated with anthropomorphic technology adoption. Leaders must develop empathy and emotional intelligence capabilities that enable effective support for employees experiencing anxiety or uncertainty about human-robot collaboration whilst maintaining focus on customer experience enhancement and operational performance improvement.

The integration of continuous learning approaches into leadership development programmes ensures that leaders maintain current knowledge of technological developments, industry best practices, and emerging challenges in anthropomorphic service robotics management. These learning approaches should incorporate peer networks, research engagement, and industry collaboration that provide ongoing support for leadership development in this rapidly evolving field.

Future-Oriented Organisational Adoption Framework

The successful implementation of anthropomorphic service robotics requires sophisticated organisational frameworks that extend beyond traditional technology adoption models to address the unique psychological, social, and operational complexities associated with emotionally intelligent human-like robots. Contemporary organisational adoption frameworks must evolve to accommodate the distinctive characteristics of anthropomorphic robotics, including emotional interaction dynamics, trust development processes, and the complex interplay between human employees and robot colleagues. This evolution necessitates comprehensive integration of technology acceptance principles, organisational readiness assessment methodologies, and strategic implementation pathways specifically designed for anthropomorphic service robotics contexts.

Evolution of TAM for Emotionally Intelligent Robots

The traditional Technology Acceptance Model requires substantial evolution to adequately capture the complex dynamics surrounding emotionally intelligent robot adoption in service organisations. Whilst perceived usefulness and perceived ease of use remain relevant constructs, the emotional capabilities of anthropomorphic robots introduce additional dimensions that significantly influence acceptance decisions. Enhanced perceived usefulness in emotionally intelligent robotics contexts encompasses both functional benefits and emotional value creation that traditional TAM frameworks inadequately address.

Ghazizadeh et al. (2012) provide critical insights into extending TAM for automation contexts, demonstrating how trust, perceived reliability, and system transparency influence acceptance of automated technologies. These insights are particularly relevant for emotionally intelligent robots, where customers must develop confidence not only in functional capabilities but also in emotional appropriateness and social competence. The evolution of TAM for anthropomorphic robotics must incorporate trust as a central construct that mediates the relationship between perceived capabilities and acceptance decisions, recognising that emotional intelligence claims create higher expectations regarding robot behaviour and performance.

Enhanced perceived usefulness through emotional capabilities requires recognition that anthropomorphic robots provide value through both task completion and emotional experience enhancement. Traditional utility assessments focus primarily on efficiency, accuracy, and convenience factors that capture only partial value propositions of emotionally intelligent robots. Future TAM models must incorporate hedonic value, emotional satisfaction, and relationship development as core usefulness dimensions that acknowledge the emotional labour capabilities of anthropomorphic robots. These expanded usefulness perceptions recognise that customers may accept emotionally intelligent robots not merely for functional benefits but for the emotional experiences and relationships they facilitate.

Changing ease of use perceptions with intuitive emotional interfaces present both opportunities and challenges for anthropomorphic robot acceptance. Whilst human-like interaction modalities may reduce learning requirements for basic robot interaction, the complexity of emotional communication and social appropriateness expectations may create new usability challenges. Hoff and Bashir (2015) examine trust in automation, identifying factors that influence user confidence in automated systems that directly apply to emotionally intelligent robotics contexts. Their research highlights the importance of predictable behaviour, appropriate feedback mechanisms, and transparent communication regarding system capabilities and limitations.

Trust and emotional safety emerge as new TAM constructs specifically relevant to anthropomorphic service robotics, reflecting the heightened psychological vulnerability associated with emotional interaction with artificial agents. Lee and See (2004) provide comprehensive analysis of trust in automation design principles that inform the development of trustworthy anthropomorphic robots. Future TAM models must incorporate emotional safety perceptions that address customer concerns about emotional manipulation, privacy protection, and appropriate boundaries in human-robot emotional relationships. These constructs recognise that emotional intelligence capabilities create new categories of risk and benefit that traditional technology acceptance frameworks do not adequately capture.

The temporal dynamics of trust development in emotionally intelligent robot interactions require TAM extensions that account for relationship evolution over time. Unlike traditional technologies where acceptance decisions may be relatively stable, anthropomorphic robots engage in ongoing relationship development that continuously influences acceptance and utilisation patterns. Future TAM models must incorporate relationship quality, emotional bond strength, and satisfaction trajectory as dynamic constructs that evolve through continued interaction experience.

Organisational Readiness Assessment Models

Organisational readiness for anthropomorphic service robotics implementation requires comprehensive assessment of cultural, technological, and human resource factors that influence successful adoption and integration outcomes. Traditional technology readiness models must be substantially expanded to address the unique requirements of emotionally intelligent robots, including employee attitudes toward human-robot collaboration, organisational culture compatibility with anthropomorphic technology, and infrastructure capabilities necessary for supporting advanced AI systems.

Cultural readiness for emotionally intelligent robots encompasses organisational values, employee attitudes, and customer base characteristics that influence acceptance and successful integration of anthropomorphic service delivery. Holt et al. (2007) provide systematic development of organisational change readiness scales that inform the assessment of cultural preparedness for technology adoption. Future readiness assessment models must incorporate specific measures of employee comfort with emotional AI, customer openness to robot relationships, and organisational commitment to human-robot collaboration that go beyond general technology adoption attitudes.

Infrastructure requirements for advanced AI systems extend beyond traditional IT capabilities to encompass data management systems, privacy protection protocols, and integration platforms necessary for supporting emotionally intelligent robots. These requirements include sophisticated data analytics capabilities for processing emotional intelligence inputs, secure storage systems for sensitive customer interaction data, and integration platforms that enable seamless coordination between robots and existing service delivery systems. Weiner (2009) provides theoretical frameworks for organisational readiness that inform comprehensive assessment of infrastructure capabilities necessary for successful anthropomorphic robotics implementation.

Change management frameworks for anthropomorphic technology must address the unique psychological and social challenges associated with introducing human-like robots into service environments. These frameworks must account for employee concerns about job displacement, customer adjustment to robot service delivery, and the complex dynamics of human-robot team collaboration. Kotter (1996) provides foundational change management principles that require adaptation for anthropomorphic robotics contexts, including vision development, coalition building, and short-term wins that demonstrate the value of human-robot collaboration whilst addressing stakeholder concerns about technology displacement.

Assessment models must also incorporate evaluation of leadership capabilities necessary for guiding anthropomorphic robotics implementation, including AI literacy levels, change communication skills, and strategic vision development capabilities. These assessments should identify leadership development needs whilst evaluating organisational capacity for managing the complex transition to human-robot service delivery paradigms.

Strategic Implementation Pathways

Strategic implementation of anthropomorphic service robotics requires carefully planned phased adoption strategies that balance technological capabilities with organisational learning needs whilst minimising disruption to existing service delivery systems. Rogers (2003) provides foundational insights into innovation diffusion processes that inform the development of strategic implementation pathways for anthropomorphic robotics, emphasising the importance of relative advantage demonstration, compatibility with existing systems, and trialability in encouraging adoption decisions.

Phased adoption strategies for anthropomorphic robots should begin with pilot implementations in controlled environments where organisational learning can occur whilst minimising risks associated with customer dissatisfaction or operational disruption. These pilot phases enable organisations to develop experience with robot management, customer interaction protocols, and integration procedures before expanding to broader implementation across service operations. Tornatzky and Fleischer (1990) examine technological innovation processes that highlight the importance of organisational learning and adaptation during technology implementation phases.

Integration with existing service delivery systems requires sophisticated coordination between robot capabilities and human service provider roles to create complementary rather than competitive service delivery models. Successful integration strategies recognise that anthropomorphic robots are most effective when they augment human capabilities rather than simply replacing human workers, creating opportunities for enhanced service delivery that leverages the unique strengths of both human and robot service providers.

Performance measurement and optimisation frameworks must account for both traditional service metrics and new indicators specific to anthropomorphic robot effectiveness, including emotional satisfaction measures, relationship quality assessments, and long-term customer loyalty impacts. Zhu et al. (2006) examine innovation assimilation processes that demonstrate the importance of performance feedback systems in successful technology implementation. Future performance frameworks must incorporate measures of emotional intelligence effectiveness, personality consistency, and relationship development success that capture the unique value propositions of anthropomorphic service robotics.

Strategic implementation pathways must also address the complex technical requirements for supporting emotionally intelligent robots, including data integration systems, privacy protection protocols, and ongoing maintenance and update procedures necessary for maintaining optimal robot performance. These technical considerations require substantial organisational investment and long-term commitment that must be carefully planned and managed throughout implementation phases to ensure successful outcomes and sustained competitive advantage through anthropomorphic service robotics capabilities.

REFERENCES

Araujo, T. (2018). Living up to the chatbot hype: The influence of anthropomorphic design cues and communicative agency framing on conversational agent and company perceptions. *Computers in Human Behavior, 85*, 183–189. DOI: 10.1016/j.chb.2018.03.051

Avolio, B. J., Walumbwa, F. O., & Weber, T. J. (2009). Leadership: Current theories, research, and future directions. *Annual Review of Psychology, 60*(1), 421–449. DOI: 10.1146/annurev.psych.60.110707.163621 PMID: 18651820

Barsom, E. Z., Graafland, M., & Schijven, M. P. (2016). Systematic review on the effectiveness of augmented reality applications in medical training. *Surgical Endoscopy, 30*(10), 4174–4183. DOI: 10.1007/s00464-016-4800-6 PMID: 26905573

Bass, B. M., & Riggio, R. E. (2006). *Transformational leadership* (2nd ed.). Lawrence Erlbaum Associates. DOI: 10.4324/9781410617095

Belanche, D., Casaló, L. V., Flavián, C., & Schepers, J. (2020). Service robot implementation: A theoretical framework and research agenda. *Service Industries Journal, 40*(3–4), 203–225. DOI: 10.1080/02642069.2019.1672666

Broadbent, E. (2017). Interactions with robots: The truths we reveal about ourselves. *Annual Review of Psychology, 68*(1), 627–652. DOI: 10.1146/annurev-psych-010416-043958 PMID: 27648986

Broadbent, E., Stafford, R., & MacDonald, B. (2009). Acceptance of healthcare robots for the older population: Review and future directions. *International Journal of Social Robotics, 1*(4), 319–330. DOI: 10.1007/s12369-009-0030-6

Buhalis, D., & Leung, R. (2018). Smart hospitality: Interconnectivity and interoperability towards an ecosystem. *International Journal of Hospitality Management, 71*, 41–50. DOI: 10.1016/j.ijhm.2017.11.011

Čaić, M., Odekerken-Schröder, G., & Mahr, D. (2018). Service robots: Value co-creation and co-destruction in elderly care networks. *Journal of Service Management, 29*(3), 519–544.

Choi, Y., Choi, M., Oh, M., & Kim, S. (2020). Service robots in hotels: Understanding the service quality perceptions of human-robot interaction. *Journal of Hospitality Marketing & Management, 29*(6), 613–635. DOI: 10.1080/19368623.2020.1703871

Davenport, T. H., & Ronanki, R. (2018, January–February). Artificial intelligence for the real world. *Harvard Business Review, 96*(1), 108–116.

Davis, F. D. (1989). Perceived usefulness, perceived ease of use, and user acceptance of information technology. *Management Information Systems Quarterly, 13*(3), 319–340. DOI: 10.2307/249008

Day, D. V. (2000). Leadership development: A review in context. *The Leadership Quarterly, 11*(4), 581–613. DOI: 10.1016/S1048-9843(00)00061-8

Duffy, B. R. (2003). Anthropomorphism and the social robot. *Robotics and Autonomous Systems, 42*(3–4), 177–190. DOI: 10.1016/S0921-8890(02)00374-3

Ekman, P. (2016). What scientists who study emotion agree about. *Perspectives on Psychological Science : A Journal of the Association for Psychological Science, 11*(1), 31–34. DOI: 10.1177/1745691615596992 PMID: 26817724

Epley, N., Waytz, A., & Cacioppo, J. T. (2007). On seeing the human: A three-factor theory of anthropomorphism. *Psychological Review, 114*(4), 864–886. DOI: 10.1037/0033-295X.114.4.864 PMID: 17907867

Fountaine, T., McCarthy, B., & Saleh, T. (2019, July–August). Building the AI-powered organization. *Harvard Business Review, 97*(4), 62–73.

Frey, C. B., & Osborne, M. A. (2017). The future of employment: How susceptible are jobs to computerisation? *Technological Forecasting and Social Change, 114*, 254–280. DOI: 10.1016/j.techfore.2016.08.019

Ghazizadeh, M., Lee, J. D., & Boyle, L. N. (2012). Extending the Technology Acceptance Model to assess automation. *Cognition Technology and Work, 14*(1), 39–49. DOI: 10.1007/s10111-011-0194-3

Gosling, S. D., Rentfrow, P. J., & Swann, W. B.Jr. (2003). A very brief measure of the Big-Five personality domains. *Journal of Research in Personality, 37*(6), 504–528. DOI: 10.1016/S0092-6566(03)00046-1

Gretzel, U., Sigala, M., Xiang, Z., & Koo, C. (2015). Smart tourism: Foundations and developments. *Electronic Markets, 25*(3), 179–188. DOI: 10.1007/s12525-015-0196-8

Grewal, D., Noble, S. M., Roggeveen, A. L., & Nordfält, J. (2020). The future of in-store technology. *Journal of the Academy of Marketing Science, 48*(1), 96–113. DOI: 10.1007/s11747-019-00697-z

Heerink, M., Kröse, B., Evers, V., & Wielinga, B. (2010). Assessing acceptance of assistive social agent technology by older adults: The Almere model. *International Journal of Social Robotics, 2*(4), 361–375. DOI: 10.1007/s12369-010-0068-5

Hoch, J. E., & Kozlowski, S. W. J. (2014). Leading virtual teams: Hierarchical leadership, structural supports, and shared team leadership. *The Journal of Applied Psychology, 99*(3), 390–403. DOI: 10.1037/a0030264 PMID: 23205494

Hoff, K. A., & Bashir, M. (2015). Trust in automation: Integrating empirical evidence on factors that influence trust. *Human Factors, 57*(3), 407–434. DOI: 10.1177/0018720814547570 PMID: 25875432

Holt, D. T., Armenakis, A. A., Feild, H. S., & Harris, S. G. (2007). Readiness for organizational change: The systematic development of a scale. *The Journal of Applied Behavioral Science, 43*(2), 232–255. DOI: 10.1177/0021886306295295

Huang, M.-H., & Rust, R. T. (2021). Engaged to a robot? The role of AI in service. *Journal of Service Research, 24*(1), 30–41. DOI: 10.1177/1094670520902266

Jarrahi, M. H. (2018). Artificial intelligence and the future of work: Human-AI symbiosis in organizational decision making. *Business Horizons, 61*(4), 577–586. DOI: 10.1016/j.bushor.2018.03.007

Kotter, J. P. (1996). *Leading change.* Harvard Business School Press.

Kuo, C.-M., Chen, L.-C., & Tseng, C.-Y. (2017). Investigating an innovative service with hospitality robots. *International Journal of Contemporary Hospitality Management, 29*(5), 1305–1321. DOI: 10.1108/IJCHM-08-2015-0414

Lee, J. D., & See, K. A. (2004). Trust in automation: Designing for appropriate reliance. *Human Factors, 46*(1), 50–80. DOI: 10.1518/hfes.46.1.50.30392 PMID: 15151155

Long, D., & Magerko, B. (2020). What is AI literacy? Competencies and design considerations. In *Proceedings of the 2020 CHI Conference on Human Factors in Computing Systems* (CHI '20), Article 596, 1–16. Association for Computing Machinery. DOI: 10.1145/3313831.3376727

Manyika, J., Chui, M., Miremadi, M., Bughin, J., George, K., Willmott, P., & Dewhurst, M. (2017). *A future that works: Automation, employment, and productivity.* McKinsey Global Institute.

Marinova, D., de Ruyter, K., Huang, M.-H., Meuter, M. L., & Challagalla, G. (2017). Getting smart: Learning from technology-empowered frontline interactions. *Journal of Service Research, 20*(1), 29–42. DOI: 10.1177/1094670516679273

McCrae, R. R., & Costa, P. T.Jr. (2008). The five-factor theory of personality. In John, O. P., Robins, R. W., & Pervin, L. A. (Eds.), *Handbook of personality: Theory and research* (3rd ed., pp. 159–181). Guilford Press.

Mumford, M. D., Todd, E. M., Higgs, C., & McIntosh, T. (2017). Cognitive skills and leadership performance: The nine critical skills. *The Leadership Quarterly*, *28*(1), 24–39. DOI: 10.1016/j.leaqua.2016.10.012

Mystakidis, S. (2022). Metaverse. *Encyclopedia*, *2*(1), 486–497. DOI: 10.3390/encyclopedia2010031

Picard, R. W. (2000). *Affective computing*. MIT Press. DOI: 10.7551/mitpress/1140.003.0008

Ransbotham, S., Kiron, D., Gerbert, P., & Reeves, M. (2017). *Reshaping business with artificial intelligence: Closing the gap between ambition and action*. MIT Sloan Management Review and The Boston Consulting Group.

Robinson, H., MacDonald, B., Kerse, N., & Broadbent, E. (2013). The psychosocial effects of a companion robot: A randomized controlled trial. *Journal of the American Medical Directors Association*, *14*(9), 661–667. DOI: 10.1016/j.jamda.2013.02.007 PMID: 23545466

Rogers, E. M. (2003). *Diffusion of innovations* (5th ed.). Free Press.

Roy, S. K., Balaji, M. S., Sadeque, S., Nguyen, B., & Melewar, T. C. (2017). Constituents and consequences of smart customer experience in retailing. *Technological Forecasting and Social Change*, *124*, 257–270. DOI: 10.1016/j.techfore.2016.09.022

Rust, R. T., & Huang, M.-H. (2014). The service revolution and the transformation of marketing science. *Marketing Science*, *33*(2), 206–221. DOI: 10.1287/mksc.2013.0836

Savela, N., Turja, T., & Oksanen, A. (2018). Social acceptance of robots in different occupational fields: A systematic literature review. *International Journal of Social Robotics*, *10*(4), 493–502. DOI: 10.1007/s12369-017-0452-5

Tapus, A., Mataric, M. J., & Scassellati, B. (2007). Socially assistive robotics [grand challenges of robotics]. *IEEE Robotics & Automation Magazine*, *14*(1), 35–42. DOI: 10.1109/MRA.2007.339605

Tornatzky, L. G., & Fleischer, M. (1990). *The processes of technological innovation*. Lexington Books.

Tussyadiah, I., & Park, S. (2018). Consumer evaluation of hotel service robots. In Stangl, B., & Pesonen, J. (Eds.), *Information and communication technologies in tourism 2018* (pp. 308–320). Springer.

Van Doorn, J., Mende, M., Noble, S. M., Hulland, J., Ostrom, A. L., Grewal, D., & Petersen, J. A. (2017). Domo arigato Mr. Roboto: Emergence of automated social presence in organizational frontlines and customers' service experiences. *Journal of Service Research*, *20*(1), 43–58. DOI: 10.1177/1094670516679272

Venkatesh, V., & Davis, F. D. (2000). A theoretical extension of the Technology Acceptance Model: Four longitudinal field studies. *Management Science*, *46*(2), 186–204. DOI: 10.1287/mnsc.46.2.186.11926

Weiner, B. J. (2009). A theory of organizational readiness for change. *Implementation Science : IS*, *4*(1), 67. DOI: 10.1186/1748-5908-4-67 PMID: 19840381

Wilson, H. J., & Daugherty, P. R. (2018, July–August). Collaborative intelligence: Humans and AI are joining forces. *Harvard Business Review*, *96*(4), 114–123.

Wirtz, J., Patterson, P. G., Kunz, W. H., Gruber, T., Lu, V. N., Paluch, S., & Martins, A. (2018). Brave new world: Service robots in the frontline. *Journal of Service Management*, *29*(5), 907–931. DOI: 10.1108/JOSM-04-2018-0119

Yukl, G. (2012). Effective leadership behavior: What we know and what questions need more attention. *The Academy of Management Perspectives*, *26*(4), 66–85. DOI: 10.5465/amp.2012.0088

Zhang, X., & Bartol, K. M. (2010). Linking empowering leadership and employee creativity: The influence of psychological empowerment, intrinsic motivation, and creative process engagement. *Academy of Management Journal*, *53*(1), 107–128. DOI: 10.5465/amj.2010.48037118

Zhu, K., Kraemer, K. L., & Xu, S. (2006). The process of innovation assimilation by firms in different countries: A technology diffusion perspective on e-business. *Management Science*, *52*(10), 1557–1576. DOI: 10.1287/mnsc.1050.0487

Chapter 3
Trust Me, I'm a Robot: AI-Powered Personalisation, Consumer Trust, and Purchase Intention in Anthropomorphic Retail Service

Abneer E. Samuel
http://orcid.org/0000-0001-8422-9125
Global Banking School Ltd., UK

Philip Nukpe
Global Banking School Ltd., UK

Shaffan Zahid
http://orcid.org/0009-0002-4856-1053
Global Banking School Ltd., UK

Chitra Sankaran
http://orcid.org/0009-0005-8669-1125
Global Banking School Ltd., UK

ABSTRACT

Retailing enters a new era where the AI personaliser goes beyond digital technology to be an anthropomorphic robot that can identify the customer, recall their history of purchases, and adapt itself based on the type of interaction. Using CASA theory, the personalization versus privacy paradox, and uncanny valley theory, the chapter formulates a conceptual framework where anthropomorphic design intensity acts as a moderator to intensify the role of trust building through AI personalization and privacy concerns on purchase intentions. The extensions are based on an empirically tested model (N = 214) where bootstrapped mediation revealed that trust has

DOI: 10.4018/979-8-3373-6640-1.ch003

a positive mediating role in the relationship between personalization and purchase intentions, whereas privacy concern had a negative indirect effect on it. Strategic implications are made for the use of anthropomorphic service robots as brand ambassadors with a considerable potential to influence consumer trust.

1. INTRODUCTION

Our retail service landscape is changing more than ever. The reconfiguration of humans and machines is not the only sign of this transformation, but also the ways the consumers come in contact with the brands, the advice they get and the personal attention they receive in commercial spaces. From luxury fashion and consumer electronics to food retail, humanoid service robots have transitioned from demonstrations to the front lines. Pepper, the softbank robotics' POCD, has been presented to shoppers at Carrefour hypermarkets and in NatWest bank branches. Simbe Robotics has trialled autonomous shelf scanning robots in Walmart superstore. In the East Asian markets, the facial recognition and natural language dialogue and emotional response technologies utilised by robotic service agents are a reality of high-end retail settings. These systems are transactional based and their value is not just technological, they represent a blend of a social interaction, a service worker and a data-driven decision medium. These are powered by AI software and can identify even repeat customers, recall purchase and preference trends, provide real-time product suggestions and modify their communication style according to the detected emotional state of the consumer. It's a personalisation of the retail experience that's closer, more personal to actual users than traditional ecommerce personalisation, since the recommendation is reaching the end-user through voice, gaze, gesture and interpersonal proximity, rather than a simple Web page or an app notification. This shift from demonstration to deployment also alters the research question: when robots move into normal retail environments, consumers do not need to imagine themselves in a futuristic world, but think about how they will respond when confronted with a physical robot which is a real service agent and might impact their actual buying behavior.

This seems analytically quite significant because it combines two phenomena studied separately in both the marketing and human–robot interaction (HRI) literatures: technology and the perceived quality of the service. This is analytical challenge because of the known benefit/risk of personalisation is changed by the introduction of a body, a face and a set of human-like behavioural cues in the personalising system. This is the first: AI-driven personalisation as the ability of intelligent systems to shape commercial content, recommendations and experience for individual consumers, according to models of consumer preference, derived from these data. In retail practice, this ability extends from a more simple capability of matching with the recommended system up to more advanced real-time adaptations in a process of behavioural traces, inferences from loyalty data and demographies, purchase histories or interactions leaving evidences on the encounters. In digital environments personalisation has been well researched, and while it has been proven to lead to better relevance and conversion, has also been found to spur privacy concerns that may reduce trust and negate purchase intentions (Awad & Krishnan, 2006; Bleier & Eisenbeiss, 2015; Dinev & Hart, 2006). The second is anthropomorphism with respect to service robotics, which embodies the design principle whereby a robot is given anthropomorphic features such as facial expressions, gestural fluency, natural language prosody and apparent emotional responsiveness to help ease social interaction between the robot and the consumer (Bartneck et al, 2009; Epley et al, 2007; van Doorn et al, 2017). The advantages of these attributes are their commercial utility: if robotic service is perceived as more accessible, consumer byproducts of robotics are more beneficial. It is also an important psychological factor; consumers make overt evaluations of salt robots from their sets of social attributes such as helpfulness, sincerity, politeness, competence and discretion.

Implementing these two phenomena in physical retail is not sufficiently systematically explored. Existing research has studied technological acceptance of robotic services and the efficacy of a digitally personalized service encounter, but not much related to the two issues together in a physically embodied, socially expressive, data-rich service encounter. So far, research in the field of HRI has mainly been about technical functionality and behavioural acceptability, with personalisation being considered as a part of the procedural capabilities of the system rather than as a psychologically loaded process that gives rise to a sense of trust, privacy calculation and brand assessment. The marketing literature has, on the other hand, focused solely digital applications of AI personalisation in the context of ecommerce websites, recommendation agents and chatbots, neglecting to consider the qualitatively distinct psychological dynamics that occur when the personalising agent has a body, a face and a proven ability to engage in social and emotional reciprocity. This is relevant because the recommendation appearing suitable on a website may be read differently by consumers when it is spoken by a robot who is aware of the consumer, who is standing near the consumer, and feels like it's speaking to the consumer, rather than the common profiles associated with a particular website.

This chapter aims to fill in that gap. It does so by highlighting the embodied personalisation as a different theoretical problem, not an an easy extension of an online recommendation system. It created a conceptual framework that brings together the design of anthropomorphic-looking robots, the incorporation of artificial intelligence for personalisation, consumer trust, concerns over privacy and purchase intention within a retail service context. The framework has been based on an empirical validation of the authors' original structural model (N = 214), which tested the framework in an online retail setting. This empirical baseline is used with care: the chapter suggests that online and robot-mediated retail transactions are not the exact same, but the established relationships can be extended and elaborated upon by theorising how embodiment, anthropomorphism, and social presence can modulate and shape the consumer's responses. This model suggests that personalisation positively predicts trust and purchase intention, and that personalisation–purchase intention relationship was partially mediated by trust while purchase intention was negatively predicted by privacy concern. The chapter validated this baseline and applies this in a physical anthropomorphic robot context, adding anthropomorphic design intensity as a theoretically motivated moderating variable, and using the theory from CASA, the uncanny valley hypothesis, and social presence theory to provide explanations for the effect of embodied personalisation for each of the three structural pathways. The contribution is both empirical and conceptual, empirical from an experiment-based trust mediated model, conceptual from the development of an extended model specifically applicable to the robot.

The chapter is structured and organized as follows: The organisations follows a logic that goes from theory to empirical evidence and finally to the extension of robot-context, rather than a speculative one. The relevant literatures relevant to service robots, AI personalisation, consumer trust and privacy concern are reviewed in section 2. The conceptual model and formal hypotheses are given in Section 3. An empirical methodology is described in Section 4. The quantitative results are presented in section 5. Section 6 extends the theoretical development from the context of anthropomorphic robot. Ethical issues are explored in Section 7. Section 8 provides a managerial insight and some strategic recommendations. Finally, Section 9 presents a research agenda for future. In these sections the main theme that the chapter focuses on is that the very same aspects that enable personalisation of AI systems make the use of data personal feel intrusive and manipulative.

2. LITERATURE REVIEW

2.1 Anthropomorphic Service Robots in Retail

Service robots are increasingly being perceived less as machines to automate tasks and more as societal actors working in service settings, roles that traditionally were taken by human workers (Wirtz et al., 2018). The conceptualisation is significant because it transcends indicators of efficiency and cost reduction to seek to understand what customers perceive robotic behaviour means in a service relationship, especially if the robot is a visible, interactive representation of the retailer. Van Doorn et al.

(2017) coined the term of automated social presence, which refers to the process of how the AI embodied or virtual service agents establish a sense of social presence that impacts customer experience and service evaluation. The instantiation of this phenomenon that is particularly interesting, and dynamic, is anthropomorphic service robots: they activate social cognitive processes which would not be set off by a conventional automated system (Nass & Moon, 2000). Processes like making intentions, showing politeness, respect, even moral responsibility, in reference to the agent are being used despite his or her concurrent realization that the robot is still a programmed system. In the retail industry, the concept of automated social presence has gained immense significance as customers tend to form an opinion of the service based on minor nonverbal cues such as expressed acknowledgment, promptness and helpfulness, which are all potentially replicable and adjustable with a robot's design.

An anthropomorphic design literature reveals that certain aspects of anthropomorphism are important for the response of consumers to a service robot. Anthropomorphism is, therefore, not a simple design decision, but a set of cues that can work on different levels, which might be warm voice in the cutie mark that makes people feel socially at ease, highly realistic face in the cutie mark that makes them feel uncomfortable, and smoother gestural timing that make them feel competent on that mark. Bartneck et al. (2009) created the Godspeed questionnaire series to assess anthropomorphism, animacy, likeability, perceived intelligence, and perceived safety as separate, but inter-related, facets of the human response to robots. According to studies (Qiu & Benbasat, 2009; Tung & Au, 2018; Yoganathan et al., 2021), there is evidence linking each of the three factors facial expressiveness, voice naturalness and gestural fluency to perceived warmth and social competence which are key factors in service satisfaction and behavioural intentions. That does not necessarily translate to a positive response by consumers, though. This non-linearity forms the core of this chapter as it suggests that retailers can not blindly adopt the belief that the more human-like, the more accepted and/or trusted and the more likely the purchase behaviour will be. The uncanny valley hypothesis was popularly proposed by Mori (1970): There were times when Robots which were very similar to humans would generate a greater sense of affinity, but then, at some point in the range of near-human realisation, the response would suddenly turn to aversion or unease. Ho and MacDorman (2010) showed and quantified this trend, reaching the conclusion that the uncanny valley effect was strongest for hybrid forms a human face and non-human movement. In terms of retail robot design, this has relevance: the key is to avoid reaching the uncanny zone, so anthropomorphic features need to be tuned not to veer off-road into the uncanny valley. Also, there is sensitivity to service context since the degree of "humanlikeness" that works in a simple way-finding case might not be suitable in cases where individuals are buying at high values, having personal information that is relevant to a purchase, or emotionally sensitive. The significance of these design elements is because service in retail is typically quick and emotional: customers could judge on helpfulness or discomfort before it even hits their consciousness whether the robot recommended a technically good service.

Some consistent findings in the literature from a sector perspective are documented with regard to the retail and hospitality sector and service robots. The facts revealed that there is more than a technologic performance aspect to which users assess robots; they also assess the appropriateness, respect, usefulness of the robot's behaviour and its consistency with the expectations of the service situation. When involving robots in service functions, the information delivery and wayfinding kind of tasks are the most effective tasks to be performed, whereas customers' satisfaction levels have complex connections with perceived warmth and emotional responsiveness in complex service encounters, relative to the technical competence aspects (Xiao & Kumar, 2021; Choi et al., 2021). De Keyser et al. (2019) postulated a typology of service technology infusion, with service augmentation services and service substitution services, suggesting that the reaction of the consumers depends on whether they perceive the human–robot ratio as being suitable for the service context. This difference is especially significant when it comes to retail personalisation, where a robot can optionally collaborate with a human to give data-informed support, but also remove all previous human personalisation and raise different concerns about trust, authenticity, and control for the customer. The more personal and advisory the service task is the more crucial it is to not only assess whether the robot can carry out a functional task, but to comprehend the concepts of trust, privacy, and social presence.

2.2 AI-Powered Personalisation

In the commercial world, personalisation is the manipulation of the product, communication and experience according with the representations derived from data of the preferences, behaviours, needs of individual customers. The business case is that relevance minimises decision fatigue, enhances the sense of convenience and provides a feeling of knowing what the business wants for the consumer, and not offering a generic list of items. Personalisation technologies are the key function of electronic commerce value creation in delivering better search and decision quality, alleviating information overload, says Adomavicius and Tuzhilin (2005). The retail application of AI personalisation includes these or some other examples – recommendation systems powered by collaborative filtering and machine learning to anticipate customer preferences; dynamic pricing based on customer browsing and purchase patterns; and conversational agents that finely tailor their communication to each customer. Personalisation is also an interaction when inserted within a service robot, this may be complemented by explanation, persuasion, helping the consumer to state in an accepted declaration, and/or by reassuring, which would change both the meaning of the recommendation and the use of the data. This minimization of information overload holds value for businesses in high-choice retail settings where customers might not want to buy if they do not understand the differences in products or picture out of the many to choose from, how many they are considering.

The most commonly reported results from the personalisation literature suggest that relevance is a key factor in value: when consumers experience something personalised as being relevant and helpful, they find it valuable (Shumanov & Johnson, 2021). Relevancy, then, has a dual role as a cue of competence (appearance of knowing what he or she wants) and a cue of benevolence (appearance of reducing the burden of choice). However, there is also a privacy calculus, as consumers have to also consider "the cost of sharing that data with the company they get personalized service from" (Awad & Krishnan, 2006; Dinev & Hart, 2006, p. 8). Without consent, personalisation can cause discomfort and distrust leading to avoidance behaviour (Bleier & Eisenbeiss, 2015). Intrusiveness tends to arise because of a lack of understanding on behalf of the consumer about why a recommendation has been given, a disclosure of surprising knowledge from the recommendation about the consumer, or because the personalisation steps over the limits of what is considered 'service' and into the realm of surveillance. The personalisation – privacy paradox refers to the conflict between users' claims to a privacy desire and their willingness to disclose information for personalized experiences (Awad & Krishnan, 2006). It is these contextual factors that are precisely what is different when the personalisation is provided from an anthropomorphic robot, as opposed to a digital interface. When there is an embodied encounter, the recommendation cannot be separated from the agent who is providing the recommendation; the encounter is socially staged, physically close, and potentially more memorable than such a digital encounter. This is particularly the case when thinking about retail robotics; whilst the advantages of personalisation may manifest instantaneously in the experience, the disadvantages of data collection, profiling or secondary-use can be less apparent but equally potent.

2.3 Students' Understanding of the Service through AI

Trust is among the most universal factors deemed to have significant impact on consumers' behavior in technologically mediated exchanges. In the case of services where AI is used, users are invariably required to accept what the AI recommends, suggests, or decides.In particular, AI-driven services often involve users accepting decisions, recommendations, and interactions that stem from hidden algorithms. Gefen et al. (2003) made it clear that there are three interrelated components of trust placed in an online retailer: Ability (the extent to which the retailer is believed to be able to perform as he or she claims), Benevolence (the extent to which the retailer is believed to be kind in his or her treatment of the consumer), Integrity (the extent to which the retailer is believed to behave based on acceptable principles). Pavlou (2003) incorporated trust into the TAM and showed that perceived risk is decreased and transaction readiness in electronic commerce is increased when a transaction has trust. Much like in the current scenario trust this time is acting to mitigate risks with a more complex object of trust, such as consumers trusting the robot and/or the retailer and/or the algorithm and/or the data infrastructure and/or the overall brand ecosystem. They are also applicable to robot-mediated retail, albeit formulated differently, ability as recommendation accuracy, benevolence as the apparent helpfulness of the robot, and integrity in terms of transparency regarding how the data is being used and limitations.

Therefore, AI systems can alleviate the level of mistrust by using anthropomorphic design elements to trigger social cues that kick off cognitive and affective processes of trust normally engaged with human agents. While these cues do not remove any doubt about the algorithm, they can help create a more socially familiar experience, particularly for consumers that may feel that AI systems are cold, opaque, or impersonal. The theoretical account is what comes from CASA theory (Nass & Moon, 2000): Humans involuntary share social rules and social expectations when interacting with computers and robots who have human like qualities. However, Qiu and Benbasat (2009) applied CASA theory to product recommendation agents, finding that anthropomorphic features in the agents led to greater trust in recommendations and lower purchase uncertainty as well as to higher purchase intend. The personalisation achieved by the use of an AI can be reinforced by the use of an anthropomorphic robot, as the social presence of the robot contributes due to its communicative behavior, which gives the impression of attentiveness and care, builds the sense of trust and service. This amplification can be fragile, though, as the consumer can be misled by the robot's "warmness' and consider it as manipulated or false if the later communication fails to do so. This automaticity is important because people need not explicitly think that the robot is a human to make social reactions; when people think that a robot is human, they will think that robo (he, it) feels the same way as they do.

2.4 Privacy Concern and the Embodied Personalisation Context

In the context of personalised AI, privacy concern is the issue when the consumer knows that a system captures, stores and uses data from the user to produce individualised outputs. It thus displays not only a concern for misuse of data but a general sensitivity to the 'line' of 'rightly and rightly known knowledge' in a commercial enterprise. Dinev and Hart (2006) have found that privacy concern had direct and negative influence on the intention to use online service, mediated by trust. In the anthropomorphic robot arena, the robot has a physical body and faces the human agent; having it ask "how would you like to be called?" or "what would you like me to recommend for you?" in the face of the robot leaves a human agent with the cognitive impression that it is actively listening to him/her, recalling name and wishes, and projecting its own recommendations socially rather than algorithmically generated into a text. This creates a more immediate impression of privacy, as usage of the data wasn't behind a screen but was realised with verbal interaction, posture and eye-to-eye contact as well as conversational moments. The study of social presence and privacy has suggested that there is a stronger sense of intimacy when the entity collecting them is assumed to be social; in human–computer interaction, social norms of confidentiality and discretion are activated differently.The study of social presence and privacy has indicated that the perceived intimacy of how the data are gathered is greater when the data is supposed to be collected by a social entity as opposed to a human–computer interaction (Rosenbaum & Massiah, 2011). This indicates that a sense of 'privacy concern' may be more strongly engaged in anthropomorphic robot experiences compared to similar personalisation experiences in the digital domain. The message is that in order to be accurate and convenient, robot mediated personalisation should also be boundary managing, consent related and self-discretionary.

3. CONCEPTUAL MODEL AND HYPOTHESES

In this chapter, a conceptual model is tested, which is shown in Figure 1. The model is intentionally parsimonious, while omitting components that are only marginally pertinent to the personalisation's relationship to purchase intention; it does so because it assumes that the core psychological mechanism through which personalisation is supposed to affect purchase intention is the only one surveyed and included in the model, acknowledging privacy concern as a counterbalancing force that could reduce trust. The model expands upon the validated empirical framework used by the authors in the context of online retail, and adds anthropomorphic design intensity, which captures the extent to which the physical characteristics, speech and gesture, as well as emotional responses of a service robot resemble human form and behaviour. This moderating construct is particularly relevant as embodiment does not simply impact consumers in a consistent way, across all consumers and all retail contexts, but affects both the level of useful personalisation and the data intrusion in a way that is expected to amplify the effect of the latter. It starts with the belief that personalisation as a service stimulus is the first prerequisite; this is when the consumer perceives the service as tailored to her needs, relevant to her, and responsive in a personal manner, that sets the positive Trust pathway in motion and initiates the potential privacy Concern pathway.

Figure 1. Conceptual model: AI-powered personalisation in anthropomorphic retail service Encounters Note. H2 shows the direct effect of AI-Powered Personalisation on Purchase Intention, bypassing the trust pathway. H5 shows trust as the mediator of the Personalisation – Purchase Intention relationship (indirect effect = 0.166; 95% CI [0.104, 0.244]). H6 (Anthropomorphic Design Intensity moderator) is a theoretical extension to the robot context and is not empirically tested in the present study.

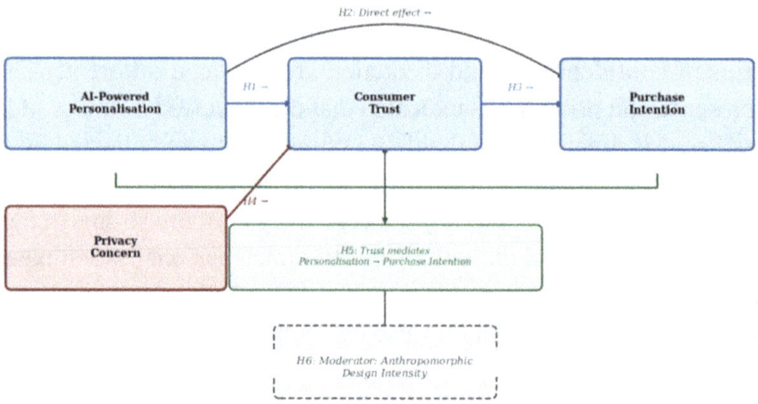

3.1 H1: AI-Powered Personalisation and Consumer Trust

> **H1:** *AI-powered personalisation delivered by an anthropomorphic service robot positively influences consumer trust.*

The items measured under privacy concern involve concern about collection, storage and use of data. Purchase intention items evaluated the likelihood of making a purchase with the platform's recommendations. All items were adapted from existing scales found in the personalisation and trust literatures (Awad & Krishnan, 2006; Gefen et al., 2003; Pavlou, 2003). As existing measures were adapted to incorporate personalised elements, the content validity was enhanced, yet the importance of the 'Personalisation aspect of AI' called for information that was contemporary with the retail environment and had data-driven aspects rather than content that solely addressed online shopping. People should trust the system generally, as well as trust it technically for instance, but also trust that it will be fair and in their interest with regards to data usage.

3.4 Analytical Approach

The five-step analysis was conducted as follows: Step 1: reliability analysis by revealing Cronbach's alpha for the data; Step 2: descriptive analysis; Step 3: Pearson bivariate correlation; Step 4: five hypotheses testing of H1–H4 using hierarchical multiple regression analysis; and Step 5: five hypotheses testing of H5 through bootstrapped mediation testing using PROCESS Model 4 (Hayes, 2018), indexed with 5,000 replications. This analytical approach was staged to get the study from measurement adequacy to association and prediction, and then to mediation, which minimized the possibility of doing structural pathways before establishing basic scale reliability and construct relationships. IBM SPSS Statistics (version 17) was used for all analyses. Using bootstrapping was useful, since indirect effects tend to be non-normal, and bootstrapping adds to a good test of mediation beyond what might be thought to be provided by traditional stepwise procedures.

4. RESULTS

This lesson focuses on reliability and descriptive statistics.This lesson deals with reliability & descriptive statistics.

All 4 constructs are presented in terms of reliability and descriptive statistics in Table 1 and in Figure 2. The results of the preliminary statistics serve as a guide to the interpretation of the following regression and mediation results: if the internal consistency of the measurement is weak or the distribution is skewed, then the structural results are not that reliable. The internal consistency of the Cronbach's alpha values was between .689 and .762, showing acceptable to good internal consistency for an exploratory research context (Nunnally, 1978). The lowest alpha ($\alpha = .689$) was obtained with the AI-powered personalisation construct, being just below the criterion of .70, which is indicated as a limitation for further fine-tuning of the instrument. Construct means were fairly evenly distributed around the scale's midpoint ($M = 3.38$-3.40) with no ceiling or floor effects. The pattern is beneficial as it means that instead of all participants having attitudes that were entirely positive or entirely negative toward the use of AI personalisation, trust, privacy concern, or purchase intention, there were enough differences within the sample to enable the meaningful testing of relationships

Table 1. Reliability and descriptive statistics (N = 214)

Construct	Items	Mean	SD	Cronbach's α
AI-Powered Personalisation	4	3.39	0.56	0.689
Consumer Trust	4	3.40	0.61	0.746
Privacy Concern	4	3.38	0.59	0.728
Purchase Intention	4	3.39	0.61	0.762

Figure 2. Internal consistency reliability by construct

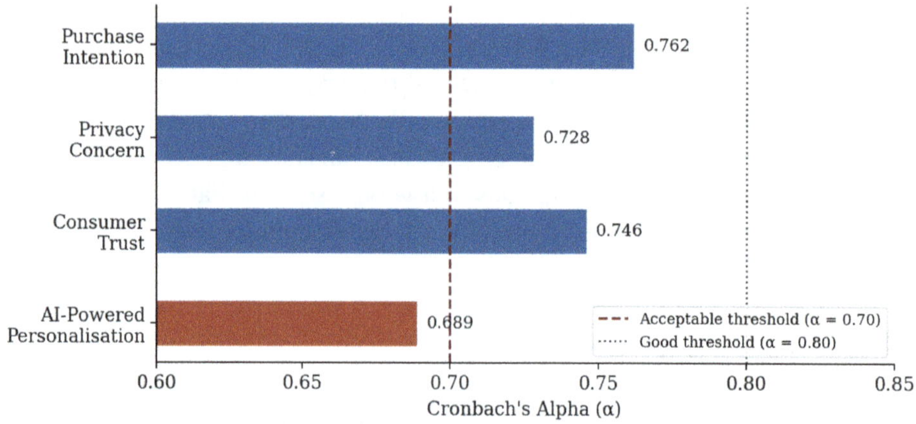

Note. Dashed line indicates the conventional acceptable threshold (α = .70).

4.2 Correlation Analysis

The Pearson correlation matrixs got provided in Table 2 and Figure 3. The correlation results do not represent causality or a causal relationship between the theoretical variables but do offer some initial indication that the variables are related in the perceived direction up to the point of the consideration of the more controlled regression models. This initial support for H1 and H2 through concrete results is linked with the positive and significant correlation between personalisation and both trust ($r = .382$, $p < .01$) and the purchase intention ($r = .336$, $p < .01$). As expected, H3 was supported by strong and positive association between the variables of trust and purchase intention ($r = .520$, $p < .01$). As hypothesized in H4, there was a negative correlation between privacy concern and trust ($r = -.274$, $p < .01$) as well as purchase intention ($r = -.243$, $p < .01$). Although expected, the correlation between the two constructs—personalisation and privacy concern—was not statistically significant (r = −.126, p > .05), which validates the empirical distinctiveness of personalisation and privacy concern. Theoretically, this indicates that personalisation, while clearly associated with high privacy concern, does not imply that this concern is necessarily reported by the consumer and that privacy concern is not an evaluative response but is rather an inverse of perceived personalisation. This was the largest bivariate correlation that could be found in the matrix, providing further evidence to support the idea that trust is not merely a peripheral mindset but a central explanatory construct in consumer willingness to engage in positive actions based on AI-mediated recommendations.

*Table 2. Pearson correlation matrix note. ** p < .01.*

Variable	**1**	**2**	**3**	**4**
1. Personalisation	—			
2. Trust	0.382**	—		
3. Privacy Concern	−0.126	−0.274**	—	
4. Purchase Intention	0.336**	0.520**	−0.243**	—

Figure 3. Pearson correlation matrix of lower triangle heatmap

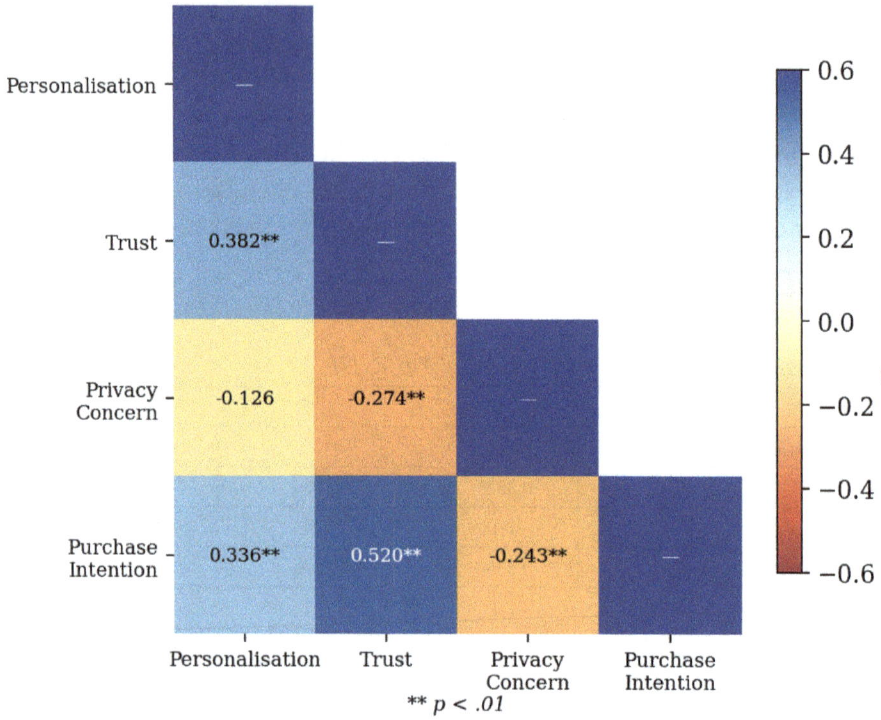

*Note. ** p < .01. Upper triangle suppressed. Diagonal = 1.000 (not shown).*

4.3 Regression Results

The summary of the models and the regression coefficients are provided in Tables 3 and 4, and in Figure 4. The regression analysis contributes to clarifying which relationship is retained when the variables are considered in relation to each other, providing a stronger meaning than the two variable level analysis provides. Model 1 regressed trust on personalisation and privacy concern, $F(2, 211) = 26.01$, $p < .001$, $R^2 = .198$. The explained variance gives an idea about the amount of trust formation that is explained by the model, and shows that there is still room for other factors in trust formation, including brand reputation, previous experience, perceived security, and readiness of the individual using the technology. Personalisation positively predicted trust ($B = 0.386$, $\beta = .353$, $p < .001$), supporting H1. Privacy concern negatively predicted trust ($B = -0.236$, $\beta = -.230$, $p < .001$), supporting H4. The results showed that personalisation directly positively influenced purchase intention ($B = 0.369$, $\beta = .336$, $p < .001$), thus, supporting H2. Model 3 explained

30.3% of variance in purchase intention, $F(3, 210) = 30.44$, $p < .001$. Trust was the strongest predictor ($B = 0.433$, $\beta = .431$, $p < .001$), supporting H3. It validates the pivotal nature of the trust relationship as the mediating pathway between consumers' positive attitudes toward AI-powered service and their intention to interact with the AI service. Partially mediated by the direct effect of personalisation ($B = 0.174$ and $p = .012$). With the entire model, privacy concern was not a direct and important predictor of purchase intent ($p = .083$), indicating its impact primarily via the pathway of trust. In the practical sense, this would imply that the effect of privacy concern on purchase intention could be lowered indirectly rather than by outright dissuasion from purchase, but by decreasing consumer willingness to accept and follow personalized recommendations. This reduction is really significant because the relationship between the effect of personalisation and trust is so strong that we cannot say that personalisation does not have a personalisation effect via trust, but there is still an effect of personalisation through by the service usefulness of personalized recommendations.

Table 3. Model summary

Model	Predictors	R	R²	Adj. R²	SE
Model 1: Trust	Personalisation, Privacy Concern	.451	0.198	0.190	0.547
Model 2: Purchase Intention	Personalisation	.336	0.113	0.109	0.576
Model 3: Purchase Intention	Personalisation, Trust, Privacy Concern	.571	0.303	0.293	0.513

Table 4. Regression coefficients

Model	Predictor	B	SE	β	t	Sig.
1	Constant	2.891	0.337	—	8.573	<.001
1	Personalisation	0.386	0.068	0.353	5.676	<.001
1	Privacy Concern	−0.236	0.064	−0.230	−3.698	<.001
2	Constant	2.138	0.244	—	8.773	<.001
2	Personalisation	0.369	0.071	0.336	5.199	<.001
3	Constant	1.692	0.367	—	4.606	<.001
3	Personalisation	0.174	0.068	0.158	2.540	0.012
3	Trust	0.433	0.065	0.431	6.707	<.001
3	Privacy Concern	−0.108	0.062	−0.104	−1.740	0.083

Figure 4. Standardised regression coefficients (β) across models

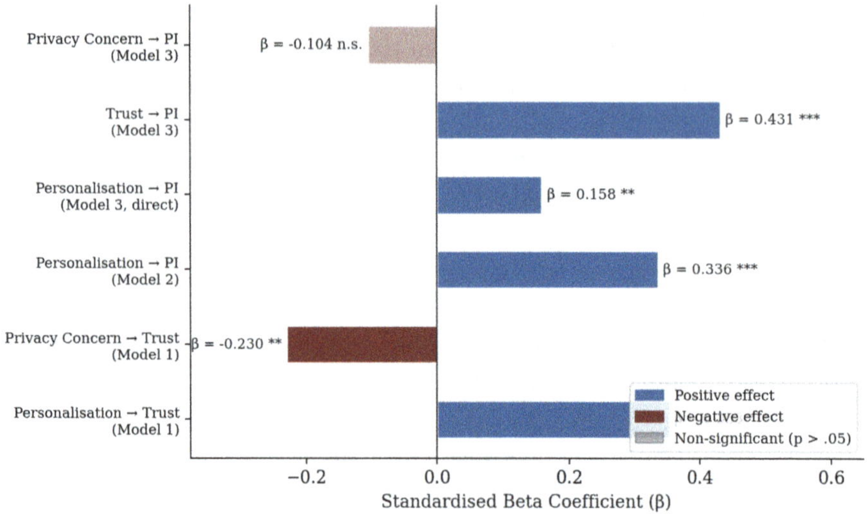

*Note. *** p < .001; * p < .05; n.s. = non-significant. Faded bars indicate p > .05.*

Figure 5. R^2 and adjusted R^2 by regression model

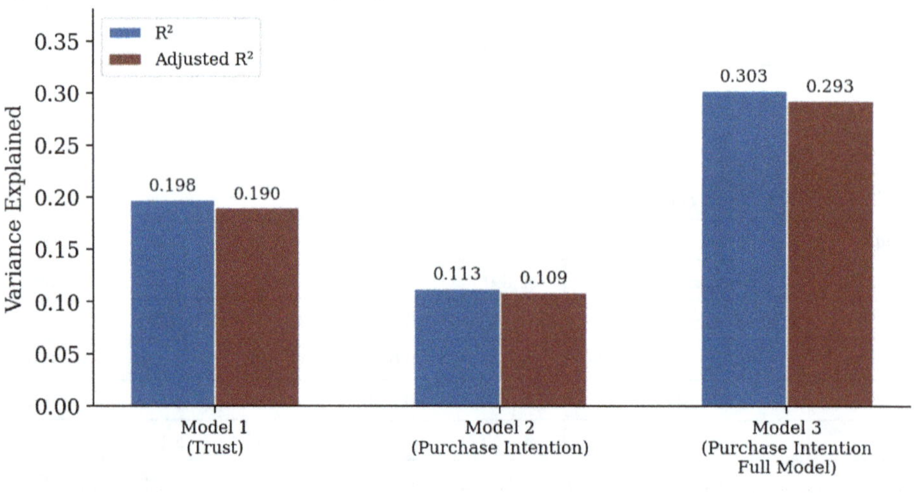

4.4 Mediation Analysis

The bootstrapped mediation analyses are provided in Table 5 and in Figure 5. The mediation analysis is the key element of the chapter, as it allows determining if trust works as the mediating mechanism between personalisation and purchase intention or rather as a correlated variable found in the relationship between the two. The indirect path personalisation had on purchase intention via trust was 0.166 (95% CI [0.104, 0.244]). However, the confidence interval does not contain 0, an indication of partial mediation, and thus H5 is supported. The partial mediation is very informative, as it indicates that there is both a functional mediating effect in terms of prediction of recommendation relevance and a relational mediating effect in terms of trusting personalisation. However, the personalisation effect remained significant ($B = 0.174$, $p = .012$), suggesting a partial rather than total mediation effect—the mediation effect of personalisation was found, in addition to the direct relevance pathway, to be through the trust building pathway. The dual pathway becomes significant for the anthropomorphic robot extension as the level of trust may be fed back into the robot, and the meaningfulness of the recommendation remains strong in terms of the fit of the product to the consumer.

Table 5. Bootstrap mediation results (5,000 bootstrap samples)

Effect	Estimate	LLCI	ULCI	Decision
Direct: Personalisation → Purchase Intention	0.174	—	—	Retained (p = .012)
Indirect: Personalisation → Trust → Purchase Intention	0.166	0.104	0.244	Significant mediation
Total Effect	0.369	—	—	Overall positive

Figure 6. Mediation model with standardised coefficients

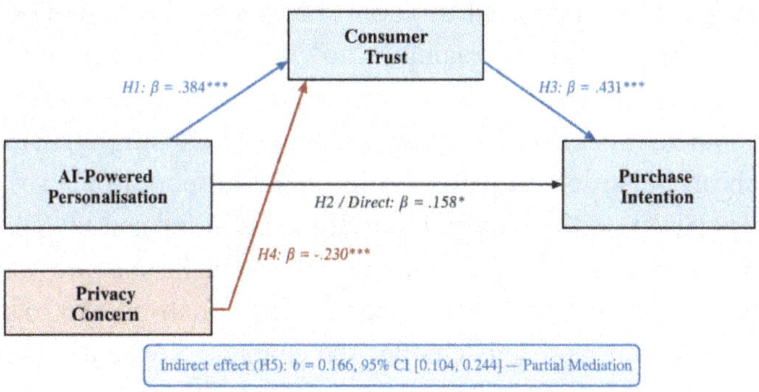

Note. Partial mediation confirmed: direct effect retained ($\beta = .158$, $p = .012$). Indirect effect = 0.166, 95% CI [0.104, 0.244].

Table 6. Hypothesis testing summary

H	Statement	Result	Evidence
H1	Personalisation → Trust (+)	Supported	B = 0.386, $p < .001$
H2	Personalisation → Purchase Intention (+)	Supported	B = 0.369, $p < .001$
H3	Trust → Purchase Intention (+)	Supported	B = 0.433, $p < .001$
H4	Privacy Concern → Trust (−)	Supported	B = −0.236, $p < .001$
H5	Trust mediates Personalisation → Purchase Intention	Supported	Indirect = 0.166; 95% CI [0.104, 0.244]
H6	Anthropomorphic design intensity moderates the personalisation–trust and privacy concern–trust pathways	*Proposed but not empirically tested*	Theoretical extension to robot context; empirical testing identified as a future research priority (see Section 9)

Note. H6 is a theoretical proposition developed for the anthropomorphic robot context and was not tested in the present empirical model, which is based on online retail data. Empirical testing of H6 is identified as a priority for future research (see Section 9).

5. THEORETICAL EXTENSION: FROM ONLINE RETAIL TO ANTHROPOMORPHIC ROBOT SERVICE ENCOUNTERS

5.1 How Embodiment Changes the Personalisation–Trust Pathway

The empirical model presented in section 5 proves that AI personalisation has a positive influence on trust and purchase intention in online retail. It is not to be suggested in passing that the same relationships would be entirely replicated in a robot-based retail environment, but to adopt the given relationships as a serious Starting Point for formulating theorising of how the embodied form would shape their strength and significance. This section builds on this work to make claims about how the embodied, human-like nature of the personalising agent qualitatively influences each of the structural pathways in the anthropomorphic robot context. The extension is based on the empirical baseline and informed by CASA theory, social presence research and HRI literature in order to make specific predictions as to how the dynamics of the model should change if personalisation is to be provided by an anthropomorphic robot. This enables the chapter to offer a conceptual contribution without being entirely speculative as to future technologies, but to relate it to actually measured consumer behaviour. The purpose is thus not necessarily to modify the original empirical model but to elucidate its explanatory range by uncovering the psychology mechanisms that are more relevant when the agent is physically present and socially expressive.

The most impactful shift relates to the way in which users link between personalisation and trust (H1). For a robot mediated encounter, the recommendation may be provided to the consumer in a socially expressive mode, thus where the interpersonal presentation is evaluated together with the evaluation of the recommendation accuracy. In online environment, when the outputs of personalisation are saliently relevant and accurate, the consumer can infer competence and benevolence in the personalisation process. In anthropomorphic robot setting, trust is also cultivated with social cues being expressed by the robot: expressions of attentiveness, voice of warmth, and gestures of affirmation that express care and responsiveness. CASA theory can predict that such social cues trigger processes of building trust that are apparently more immediate and affectively engaging than the cognitive inference process in digital interactions. This means that the pathway for personalization – trust should be stronger and the path to benevolence-based trust stronger compared to competence-based trust for the anthropomorphic robot context. Note that it does not mean that competence is not important, but it does mean that the feeling of being served or taken care of might be a more important factor in establishing trust than the screen in an entirely screen-based approach.

Qiu and Benbasat (2009) offer the first direct empirical validation of this (amplification) effect. Their work is relevant, because they cannot only demonstrate that anthropomorphic recommendation agents can modify the social interpretation of the recommendation system itself, but rather, that {social interpretation} is different than visual presentation. The results from the experiments in this paper that focused on contrast the text-based and anthropomorphic recommendation approach was in favor of anthropomorphic approach; the anthropomorphic approach generated greater consumer trust, lower purchase uncertainty and higher purchase intention, especially when applied to the audiences who are uncertain about the item purchases. The result has a direct application for retail personalisation – in a surfing demonstration using an anthropomorphic robot, a consumer who saw the robot showing some knowledge of his/her likes and dislikes was not only being given a recommendation, but a social communication, which has a stronger effect on increasing the consumers' trust in the robot. It may thus be seen as personalised care, promoting the retail experience as more consultative and relational than one based on a button click.

5.2 How Embodiment shifts the Privacy Concern – Trust Pathway

While predicting a high-to-high score in the anthropomorphic robot context, the pathway of privacy concern – trust is predicted to be stronger than in the online retail context. The answer is because this Embodiment creates a sensationally and socially different way to operate with data, one that becomes more visible, audible and personal. In digital situations, privacy concerns tends to be raised on the basis of inferential realisation that data is being collected, i.e., the activation takes predominantly cognitive form. Experientially, data collection with the anthropomorphic robot context means that the robot faces the consumer and recognizes him by name, remembers the previous visit to the shop and offers suggestions that are clearly calculated based on the information contained in the data. This is because the consumer has no notion that the data is being collected, and sees only what it has to offer as part of a social encounter with the collector that is action sharing with an emotional connotation of being 'in front of face'. When the robot reveals information like their name, previous purchase, loyalty status, prior product inquiry, and so on, without obtaining permission first, the information can lose the sense of privacy and become more of a personal identifier and influence. Confidentiality and discretion rules are triggered by the social norms. A review of social norm activation in the field of HCI suggests that humanlike agents initiated norms (cues) that call for stronger negative responses upon violation, while non-humanlike agents called for less negative responses (Nass & Moon, 2000; Rosenbaum & Massiah, 2011). What this means is that the invasion of privacy in a retail setting when done with a robot may not just be a technical nuisance, but a social faux pas. Whereas robot mediated personalisation can make the data relationship feel socially present, as the consumer experiences the impact of the data usage with the voice and body directed at them.

5.3 The Moderating Role of Anthropomorphic Design Intensity

Following the moderation hypothesis (H6), anthropomorphic design intensity across the personalisation–trust pathway was expected to increase the effect of personalisation on trust, as was the effect of privacy concerns on trust. This is the core theoretical conflict that can be found in the chapter, and that is that higher levels of humanlikeness can lead to higher levels of perceived warmth and responsiveness of personalisation, but humanlikeness can also lead to higher levels of perceived intimacy and intrusiveness of data use. Dual amplification structure will have significant implications for retail robot design. A higher anthropomorphic design intensity will lead to more trust of a personalised encounter; it will also evoke more privacy-driven mistrust of a consumer who is already taking privacy concerns with him. The net impact on purchase intention will thus rely on the initial level of privacy concerns among consumers. Therefore anthropomorphic intensity as an design attribute cannot not be seen as a universal design attribute value; is better regarded as a multiplier effect for both positive and negative aspects of the anthropomorphic psychological effect which is leadingly to amplify them both.

Regarding the trauma of low privacy concern, high anthropomorphic design is expected to be beneficial: the increased trust its personalisation accelerated should demonstrate to the increased purchase intension. In these instances, the robot can come across as an efficient and interesting assistant, helping the shopping process and lessening decision uncertainty. In certain segments (elevated privacy concern) elevated anthropomorphic design intensities can even be counterproductive: As the impact of the increase in personalisation to build trust for consumers can be overturned by the amplifying impact in the increase in privacy concern to decrease trust. This implies that retailers should not think in the same way when considering how to implement robot use, and should target robots interacting with customers in ways that respect their privacy such as opt-in personalisation, reduction of data recalled and making robot roles less socially connected. One direct practical implication of this prediction is that the anthropomorphic design intensity of a service robot that is deployed in an environment with a high privacy concern should be considered: perhaps a lower intensity would result in greater gains in the commercial aspects of service. The uncanny valley hypothesis also introduces another moderating factor; the high end of the anthropomorphic design intensity spectrum may lead to a negative revival of human-likeness effects due to a negative reaction to high levels of human realism (Mori, 1970; Ho & MacDorman, 2010).

6. ETHICAL DIMENSIONS OF ANTHROPOMORPHIC RETAIL ROBOTS

6.1 Algorithmic Transparency and Disclosure

The introduction of any anthropomorphic service robot (ASR) that can be driven by AI technology opens up sensitive questions of transparency, not limited to the field of digital personalisation ethics. Disclosure is more complicated when there is not just a digital interaction going on, but a live one in which it has to be understood, socially timely and appropriate. In digital environments, transparency tools like privacy policies, cookie reminders communication, etc. promote a nominal distance between the data collection system and the social interaction/social process of the service encounter by the consumer. In the anthropomorphic robotic context, this separation is merged into the social encounter and there is no formal disclosure moment as such: the personalisation is offered to the robot during the social interaction itself. When a purchase history can influence a recommendation, a consumer may not be aware that the robot is accessing a loyalty programme database, a behavioural tracking system or real time facial recognition feed. Such a lack of clarity can have a negative impact on trust even if access to/reuse of data is legal, as users are likely to hold an opinion about the experience based on their notions of respect, consent, and control of their personal boundaries, rather than the legal obligations alone.

Ethics of disclosure in this context is not only about the need for privacy policies to be present but ensuring that the personalisation ability of the robot is captured in the social register of the service encounter. In other words, disclosure needs to be designed to be included as part of the interaction with the person, not as part of the paper. Some commercial deployments have in fact resorted to letting the robot reveal its personalisation mechanism first, within the first paragraph, or the first few words, of the interaction, in a natural conversational way: a statement that no consumer could possibly read, but might be considered to be ethically sound – and potentially more business-charging – because it is in the first paragraph of the interaction, or in the first few words. It can also help to convey the impression of control if the consumer is left with an option of continuing without personalisation, explanation or limitation of information given to the robot.

6.2 Emotional Manipulation and Affective Computing

The ability of anthropomorphic robots to recognise and respond to emotion via affective computing (through facial expression recognition, voice pitch detection and physiological monitoring to deduce emotional state and adjust accordingly) raises questions around ethics that simply cannot be mirrored in digital personalisation. These are amplified because emotions are personal, unverifiable, and potentially game-changers at times when consumers are feeling unconfident, fatigued, or frightened or when they are under social pressure. A robot's detection of hesitation coupled with a higher level response (such as accentuating the positive, reassuring or suggesting an enhancement) is more direct and personally intimate than the equivalent algorithmic adjustment in the digital world. The moral dilemma is whether it is unfair to manipulate or simply the robotic version of the attentive human salesperson who reads into their emotional state, and adapts and responds accordingly. It is a response that is partially dependent on the level of disclosure, partially on the commercial interest, and partially on the vulnerability of consumers. It also relies on the consumers' understanding/consumer confidence, or on the intensification of persuasion, exploiting hesitation or undermining consumers' actual freedom of choice. Especially when the financial commitment, personal vulnerability, health, identity, or social status are at stake, the adaptation the robot takes may influence choices when people are plotting to buy, pushing them to make a decision at a vulnerable moment.

6.3 Brand Authenticity and the Automation of Care

Consumers who feel that a brand is consistent, authentic, and true to its values are more likely to become loyal and advocate for the brand (Huang & Rust, 2018). Anthropomorphising robots push the boundaries of authenticity by giving the impression of a personalised care, but functioning by using automated rules and rulesets, databases and algorithmic inferences. Brands known for their values of human connection, craft and personalised care may end up faced with an 'authenticity paradox', as the anthropomorphic robot can offer personalisation which is more consistently accurate and delivered more efficiently than the human customer service representative, but the warmth of the social behaviour of the robot may be seen as at odds with the brand's promise of human connection, craft and personalised care. A robot who seems a knowledgeable assistant, is transparent about his artificial intelligence and truly assists consumers in navigating their complex product assortment is more likely to be seen as authentic than one that appears to act warm to people based on techniques they can guess. Instead, the big question for any retailer is whether the shopping bot should look and act like a human... at any price.

7. MANAGERIAL IMPLICATIONS AND STRATEGIC RECOMMENDATIONS

7.1 Calibrating Anthropomorphic Design for Trust Optimisation

This core theoretical implication, drawn from the chapter, is the non-linear, context-dependent linkages between anthropomorphic design intensity and consumer outcomes. Therefore, managers are encouraged to make designing a robot a strategic decision related to the redesign of a service, and not just a novelty investment or a straightforward automation initiative. Intelligent retailers and robots designers should not fall for their intuitive sense that more human-like robots are better ones. The more relevant question is, what human-like cue works better for the service task, the product category, and the consumer likely's expectations of privacy. Retail bodies will require different levels of anthropomorphic design intensity in a particular deployment based on three factors: target customer population's privacy concern profile, data sensitivity of a product category, and brand's positioning on the human connection–efficiency continuum. In product categories that do not require consumers to spend much to evaluate their quality, a moderately anthropomorphic robot with strong personalisation capability and explicit disclosure process would probably generate the highest level of trust-building effect and most positive purchase intentions included in this chapter's empirical trust model. In high-suspense sectors, the premium retailers would find a lower degree of anthropomorphic design intensiveness with a greater focus on competence signalling stimuli rather than warmth signalling stimuli to yield better results. For instance, robots that are reliable, communicate parts of the product features and how the data will be utilized might have a better credibility than those trying to imitate the emotional dynamics of people. A friendly, moderately humanlike robot may be effective on a retailer into the sale of every day low threat products, whereas a company that sells a sensitive product or high involvement products could profit from a extra subdued and symbolic look that highlights the professionalism and control of the person.

The penetration of transparency into the commercial strategy.8.2 The commercial movement of transparency.

Furthermore, the ethical rationale for disclosure of robot personalisation capabilities is strengthened by a commercial rationale; the retailer's capacity to boost consumer confidence is demonstrated in showing transparency with regard to personalisation mechanisms, by giving consumers confidence in the retailer's respect of the consumers' right to know how their data is being used. Transparency should therefore not be looked at as a compliance obligation, but rather as part of the customer journey which will help build trust. Perceived control over data, or the consumer's sense of his or her ability to control, correct or opt out of their personalisation is always reported to have a stronger influence on trust than the fact that the data is not collected (Dinev & Hart, 2006). For the anthropomorphic robot scenario, this becomes a design guideline: robots must let the consumer know about their own personalisation capabilities, provide an ability for the consumer to have interactions without personalisation, and give the consumer a way to review and to manage the data they consume. The urgency of encountering a robot could otherwise result in consumers feeling that their use of the data was necessary or out of their control without the chance to input their consent.

Anthropomorphic service robots serve as brand agents, and their attitude, interaction and communicative behaviors represent and convey the brand values during all interactions with the consumer. They ought to be treated as seriously as their advertising campaigns, customer messages, staff training, and the user experience of their digital offerings. Retailers must attempt to design a robot as a brand governing process, not only as a technical procurement selection. The personae of the robot should be designed in the same manner as all of the other messaging assets of the brand, both with regard to its name and to its visual image, voice, gestural register and style of conversation. It involves checking for credibility of the persona, if the persona is familiar enough, and whether it uses personal data in a manner understandable to the target customers. Transitions from multiple persona and the brand's stated values are authenticity threats to brand equity, while consistency provides opportunities to connect with brand values at scale across consumers. Therefore, the cross-functional governance responsibilities across the marketing functions, data protection, customer experience, operations and technology functions should be considered.

This chapter has conceptualized and empirically derived an integrated model for analyzing the effects of anthropomorphic personalised service robot use on consumer trust, privacy concern, and purchase intentions in a retail service interaction. The framework indicates a personalisation effect that is not always a positive one; it is shaped by the consumers' understanding of data-based adaptation process: a helpful recognition versus an intrusive monitoring. The main services it brings are: First, it fills the gap between the literature on anthropomorphism and the literature on marketing personalisation by analyzing both the HC and marketing personalisation literatures in a similar framework; second, it provides an empirically validated structural model that shows the magnitude and direction of the key relationships and offers a baseline quantitative point of comparison to theoretise and compare robot-context effects; and third, it introduces anthropomorphic design intensity as a theoretically motivated design moderator variable that can generate novel and practically actionable predictions regarding conditions under which anthropomorphic personalisation will help versus hinder the commercial efficacy of AI personalisation. Those contributions have been made with the idea of aiding the design and development of academic research, and informing decision-making by managers doing pragmatic business, in a field where the business matures faster than the theory, measurement and governance.

The chapter also adds to ongoing debates about the ethics of anthropomorphic service robots, including this work by identifying three specific ethical challenges, namely algorithmic transparency, emotional manipulation and authenticity, where design and governance must take specific form and actions, not exclusively through generic frameworks of AI ethics. The analysis indicates that what counts as the ethical design and the design that is effective commercially are not necessarily at odds with one another; in the case of robot-mediated personalisation, it is possible that being created as an ethical design is one process in which trust and purchase intention may be maintained.

This chapter has stressed the theoretical progression and development, but it also has identified some priority areas for empirical studies. The moderating effects proposed in this manuscript should be evaluated in the context of real or simulated robot-life experiences: Experimental manipulations of anthropomorphic design intensity along with explicit measures of perceived social presence, privacy concern, and trust should be used in such studies. Trust in anthropomorphic service robots must be studied over repeated service encounters, which may result in trust being created through affective response to social cues in the first instance, but being more stable and cognitively elaborated in repeated service encounters, where differentially successful personalisation level and weighings on privacy concerns may create trust. The interaction between anthropomorphism and privacy concern varies with other cultures in which anthropomorphic robots have a longer commercial history and are more culturally familiar than in European contexts where norms of data privacy have raised privacy awareness and GDPR norms are more adhered to. Experimental designs across cultures are necessary in order to test how the expectations of the model apply to the different situations. Lastly, there is a need for validated, anthropomorphic personalisation-context-specific measurement instruments; the constructs of anthropomorphic design intensity, social presence–mediated trust and embodied privacy concern need psychometric development beyond the adaptation of scales of digital commerce. These would be able to differentiate between trust in the robot's ability to perform elevated social functions and trust in the data processing processes of the retailer and trust in the predictive algorithm's recommendations. Comparative work would be useful to determine the nature of the trust-construction proven to be available from anthropomorphism, whether it is culturally universal or culturally contingent, depending on local histories of technology acceptance, retail services norms and privacy regulation.

The anthropomorphic service robot is not just a new sensation of technology. It is a new, person-orientated AI application where automation, personalisation, embodiment, and brand communication are (as it were) flowing together in one experience. Through the activation of social cognitive architecture of consumer trust, it is a new class of brand agent that cannot be replicated by digital systems but involves risks of privacy aversion and authenticity scepticism to be controlled by design and governance. What is crucial for the countries' retailers is not only the academic relevance of this approach, but also the commercial urgency to implement these systems as the number of retailers in the countries making use of such systems grows. Retailers who comprehend such complexity will better be equipped in designing robotic service encounters that will not just be efficient and persuasive, but also trusted, transparent and suitable to consumers' expectations.

REFERENCES

Adomavicius, G., & Tuzhilin, A. (2005). Personalization technologies: A process-oriented perspective. *Communications of the ACM*, *48*(10), 83–90. DOI: 10.1145/1089107.1089109

Awad, N. F., & Krishnan, M. S. (2006). The personalization privacy paradox: An empirical evaluation of information transparency and the willingness to be profiled online for personalization. *Management Information Systems Quarterly*, *30*(1), 13–28. DOI: 10.2307/25148715

Bartneck, C., Kulić, D., Croft, E., & Zoghbi, S. (2009). Measurement instruments for the anthropomorphism, animacy, likeability, perceived intelligence, and perceived safety of robots. *International Journal of Social Robotics*, *1*(1), 71–81. DOI: 10.1007/s12369-008-0001-3

Bleier, A., & Eisenbeiss, M. (2015). The importance of trust for personalized online advertising. *Journal of Retailing*, *91*(3), 390–409. DOI: 10.1016/j.jretai.2015.04.001

Choi, S., Mattila, A. S., & Bolton, L. E. (2021). To err is human(-oid): How do consumers react to robot service failure and recovery? *Journal of Service Research*, *24*(3), 354–371. DOI: 10.1177/1094670520978798

De Keyser, A., Köcher, S., Alkire, L., Verbeeck, C., & Kandampully, J. (2019). Frontline service technology infusion: Conceptual archetypes and future research directions. *Journal of Service Management*, *30*(1), 156–183. DOI: 10.1108/JOSM-03-2018-0082

Dinev, T., & Hart, P. (2006). An extended privacy calculus model for e-commerce transactions. *Information Systems Research*, *17*(1), 61–80. DOI: 10.1287/isre.1060.0080

Epley, N., Waytz, A., & Cacioppo, J. T. (2007). On seeing the human: A three-factor theory of anthropomorphism. *Psychological Review*, *114*(4), 864–886. DOI: 10.1037/0033-295X.114.4.864 PMID: 17907867

Gefen, D., Karahanna, E., & Straub, D. W. (2003). Trust and TAM in online shopping: An integrated model. *Management Information Systems Quarterly*, *27*(1), 51–90. DOI: 10.2307/30036519

Hayes, A. F. (2018). *Introduction to mediation, moderation, and conditional process analysis: A regression-based approach* (2nd ed.). Guilford Press.

Ho, C. C., & MacDorman, K. F. (2010). Revisiting the uncanny valley theory: Developing and validating an alternative to the Godspeed indices. *Computers in Human Behavior, 26*(6), 1508–1518. DOI: 10.1016/j.chb.2010.05.015

Huang, M. H., & Rust, R. T. (2018). Artificial intelligence in service. *Journal of Service Research, 21*(2), 155–172. DOI: 10.1177/1094670517752459

Mori, M. (1970). The uncanny valley. Energy, 7(4), 33–35. (Trans. MacDorman, K. F., & Minato, T., 2005)

Nass, C., & Moon, Y. (2000). Machines and mindlessness: Social responses to computers. *The Journal of Social Issues, 56*(1), 81–103. DOI: 10.1111/0022-4537.00153

Nunnally, J. C. (1978). *Psychometric theory* (2nd ed.). McGraw-Hill.

Pavlou, P. A. (2003). Consumer acceptance of electronic commerce: Integrating trust and risk with the technology acceptance model. *International Journal of Electronic Commerce, 7*(3), 101–134. DOI: 10.1080/10864415.2003.11044275

Qiu, L., & Benbasat, I. (2009). Evaluating anthropomorphic product recommendation agents: A social relationship perspective to designing information systems. *Journal of Management Information Systems, 25*(4), 145–182. DOI: 10.2753/MIS0742-1222250405

Rosenbaum, M. S., & Massiah, C. (2011). An expanded servicescape perspective. *Journal of Service Management, 22*(4), 471–490. DOI: 10.1108/09564231111155088

Shumanov, M., & Johnson, L. (2021). Making conversations with chatbots more personalized. *Computers in Human Behavior, 117*, 106627. Advance online publication. DOI: 10.1016/j.chb.2020.106627

Tung, V. W. S., & Au, N. (2018). Exploring customer experiences with robotics in hospitality. *International Journal of Contemporary Hospitality Management, 30*(7), 2680–2697. DOI: 10.1108/IJCHM-06-2017-0322

van Doorn, J., Mende, M., Noble, S. M., Hulland, J., Ostrom, A. L., Grewal, D., & Petersen, J. A. (2017). Domo arigato Mr. Roboto: Emergence of automated social presence in organizational frontlines and customers' service experiences. *Journal of Service Research, 20*(1), 43–58. DOI: 10.1177/1094670516679272

Wirtz, J., Patterson, P. G., Kunz, W. H., Gruber, T., Lu, V. N., Paluch, S., & Martins, A. (2018). Brave new world: Service robots in the frontline. *Journal of Service Management, 29*(5), 907–931. DOI: 10.1108/JOSM-04-2018-0119

Xiao, L., & Kumar, V. (2021). Robotics for customer service: A useful complement or an unwelcome threat? *Journal of Service Research*, *24*(1), 138–155. DOI: 10.1177/1094670519878881

Yoganathan, V., Osburg, V. S., Kunz, W. H., & Toporowski, W. (2021). Check-in at the Robo-desk: Effects of automated social presence on social cognition and service implications. *Tourism Management*, *85*, 104309. Advance online publication. DOI: 10.1016/j.tourman.2021.104309

Chapter 4
Robotics and Retail:
A Study on Job Redesign and Role Redundancy in Smart Stores

Ruchika Rastogi
http://orcid.org/0000-0001-7144-4387
Pranveer Singh Institute of Technology, Kanpur, India

Swati Gupta
Canterbury Institute of Management, Australia

Sashwat Kumar Dixit
Pranveer Singh Institute of Technology, Kanpur, India

ABSTRACT

The rapid acceptance of new technologies such as artificial intelligence (AI), machine learning (ML), The Internet of Things (IoT), robotics, and augmented reality (AR) have drastically transformed the retail landscape, especially in the post-COVID-19 pandemic landscape. This paper examines the implications of technological innovation on the changed relations of employees, including the introduction of new roles, some revised roles, and a small percentage of job losses. An excellent empirical example is the use of robots and the Internet of Things and generative artificial intelligence at Walmart, the application of which has shown the transformational effect of technology on the retail sphere and has created a more efficient and customer-oriented ecosystem (Rastogi & Saha, 2025). The research highlights why we need to create a balance between automation and human flexibility to have a sustainable and inclusive growth.

DOI: 10.4018/979-8-3373-6640-1.ch004

1. INTRODUCTION

The retail market has always been dynamic and has acted in terms of the changing consumer demands and expectations, technological advances and competition in the market. In their evolution through the decades, retailers made several innovations, such as the barcoding systems, point-of-sale terminals, loyalty programmes, and online shopping websites. Later and more specifically in recent times especially in the time around the COVID -19 pandemic, the industry has witnessed an unseen pace in the integration of next-generation technologies. The retailers across the globe have embraced Artificial Intelligence (AI), Machine Learning (ML), the Internet of Things (IoT), robotics, and advanced data analytics at a rate that has never been witnessed before (Agrawal et al., 2024). This action was more than a strategic move but it was a necessity to stay afloat; lockdowns, disruptions in supply-chains, and physical-distancing guidelines forced retailers to turn to automation, digital processes, and contact-less technologies to keep operations afloat. The AI and ML help retailers predict consumer behaviour, tailor suggestions, streamline inventory, and automate procurement. Inventory transparency has been revolutionized and shrinkage has been reduced using IoT technologies such as smart shelves, RFID tags, connected sensors, and automated checkout systems. Robotics has become a strong tool of the warehousing sector, shelf scanning, order fulfilment, and the last-mile delivery. All these innovations in digital form have decreased manual labour, minimised errors, and increased operational efficiency significantly. Other than process optimisation, the effects of such technologies on the workforce division, job description and required skills are also likely. The formerly manual jobs, i.e. cashiers, stock clerks, and warehouse pickers, get more and more automated with self-checkout kiosks, robotic pick-and-place systems, and autonomous inventory drones. Thus, workers are moving to more analytic, supervisory and customer-related roles. The transformation of manual to digital work requires additional competences of data interpretation, human-machine interaction, and digital problem-solving, which were not historically linked to the job in the retail sphere. This change anticipates two important phenomena of job redesign and job redundancy. Job redesign refers to restructuring of the current jobs whereby the workers are reskilled to work with intelligent machines. In other words, a previously working cashier can now operate self-checkout machines or help customers with troubleshooting electronic interfaces. Conversely, job redundancy occurs when jobs are made redundant because of total automation or redesign (Foster et al., 2019). Despite the fact that redundancy can only impact on a small percentage of the working force, it has immense consequences on employee morale, organisational culture and labour-market stability. Therefore, the accelerated technological transformation in the retail sector is not simply a story of innovation, but it is a radical socio-economic transformation that is reshaping the work organization, provision and worth in the present retail ecosystem.

1.1 Scale of technology adoption: pre and post COVID-19

Prior to covid-19 the adoption of emerging technologies was relatively less and restricted to large retailers like: Walmart, Amazon etc (Taylor, 2022)- these folks were experimenting with automation tools like robotic warehouse pickers, AI driven recommendation engines and self-service kiosks. However, the majority of the industry specially the medium and small sized enterprises remained hesitant –reason being high initial cost of investment, lack of digital infrastructure and concern over workforce displacement (Kahn, 2021).

COVID-19 played the role of catalyst that drastically accelerated the pace and scale of adoption of these new technologies (Stewart et al., 2020). With the announcement of lockdown and strict regulation of social distancing put in place, traditional in-store retail operations were severely disrupted. Retailers were forced to pivot rapidly toward contactless and automated solutions to ensure business continuity and meet consumer expectation (According to a 2021 Mckinsey report, the pandemic accelerated digital adoption by upto five years in several sectors, including retail).

In the immediate aftermath of COVID-19, the use of robotics for inventory management, autonomous cleaning robots, contact less digital payment system and AI- powered customer services tools increased substantially (Banur et al., 2024). for example, Walmart expanded its use of shelf scanning. Small retailers too began adopting more accessible version of these technologies through cloud-based platforms and third-party solutions (Liew et al., 2023).

Today the scale of adoption continues to grow. Invest in emerging technologies has become a strategic priority. Retailers now view automation as the new normal to make sure they stay relevant and maintain competitive advantage in the market. As a result, robotics and AI- are becoming embedded in the operational core of modern smart stores (Soori et al., 2023).

2. LITERATURE REVIEW

Before automation, many retail jobs were task oriented and repetitive, such as stocking shelves, operating cash registers and manually tracking inventory. With the installation of smart system and robotics, these tasks are handled by machines

(Christl,2023). This shift allows human workers to move into roles that require more cognitive flexibility, emotional intelligence and technical literacy (Fuchs et al., 2023).

For example, rather than manually restocking shelves, retail employees now monitor smart inventory systems and responds to queries and alerts raised by IoT devices. cashiers once central to customers interactions, are re-assigned to roles such as customer engagement specialist or in-store tech support personnel. These redesigned roles are more aligned with evolving customer expectations in a tech-enhanced retail environment (Ikeler,2013).

Moreover, the use of data analytics and AI has opened new avenues or career paths like data interpretation, digital marketing (Hamamah et al., 2024). As machines are now employed to do repetitive work, human workers are now required to act as problem solver, brand ambassadors and interpret large digital data. This shift highlights the importance of skilling and reskilling in enabling a smooth transition for the retail workforce.

While the robotics and automation has created new opportunities for the business it has also resulted in job redundancy (Bhargava et al., 2021). There are many roles which had some significance has now become obsolete because of automation. This is clearly evident in frontline positions such as cashiers, shelf stockers and cleaning staff. As the automation continues in this sector the need of human workers to perform these tasks will be eliminated (Rastogi & Iqbal, 2024).

COVID – 19 was one of that events which catalyzed the adoption of these technologies. At that particular period because of lockdown and social distancing norms (Korneeva et al., 2024)- retailers had to incur heavy losses, so to overcome that losses retailer sought to reduce employees and increased use of technologies to ensure cost effectiveness and profitability. For example, several major chains adopted self-checkout systems and increased use of robotic system for picking and packing, displacing traditional warehouse workers.

However, job redundancy is not uniform across all roles or geographies. it is basically focused on low skills and routine or repetitive job roles (George,2024). Job roles that require human discretion, emotional interaction and complex decision making still continues to be in demand (Ranerup et al., 2022). The challenge for retailers lies in handling the change responsibly – by providing training to the employees for the purpose of upskilling them as per the industry requirement and designing hybrid roles that combine human and machine capabilities.

Table 1. Emerging retail technologies and their impact on jobs

Technology	How It's used in retail.	Impact on jobs.
Artificial Intelligence (AI)	Analyzing data, improving decisions.	Automates analytics, creates new roles.
Machine Learning (ML)	Forecasting trends, dynamic pricing.	Reduces errors, enables personalization.
Internet of Things (IoT)	Smart shelves, tracking inventory.	Real-time data, optimizes stock.
Robotics	Warehousing, shelf management.	Less manual work, new tech roles.
Augmented Reality (AR)	Virtual try-ons, immersive shopping.	New ways to engage customers.

Source: Shankar et al., 2025

2.1 The role of emerging technologies in job redesign and role redundancy in the retail sector

Recent developments in technologies like robotics, augmented reality (AR), machine learning (ML), artificial intelligence (AI), and the Internet of Things (IoT) are causing a significant shift in the retail sector. To improve customer satisfaction and speed up the operations of retail sector, these technology developments are radically changing the workforce responsibilities, resulting in both role redundancy and job redesigning (Bartlett2023).

By 2027, AI and robotics are predicted to affect over 23% of employment worldwide, and 44% of workers will need to update their fundamental or core skills within the next five years, according to the World Economic Forum's Future of employment Report 2023.

2.1.1 The Impact of Automation on Role Redundancy

a. Decline of Repetitive Tasks

In retail sector the repetitious workforces are being increasingly replaced by automation, which is also affecting backend and administrative tasks in addition to frontend positions (Ammupriya et al., 2025). Smart AI powered software enables enterprise resource planning (ERP) technologies are increasingly take over the tasks that were previously done by people, such as procurement, stock reconciliation, and billing which take lots of time and may also reduce errors. These new technologies can anticipate inventory demands, optimize restocking, and negotiate purchase orders, drastically cut down on human intervention·

Example: Indian Retailers

Both Flipkart and Reliance Retail have transformed their supply chains with the integration of AI technology. At Flipkart, there is the use of AI robotics in warehouses for automated sorting, parcel moving, and handling. Automation improves operational efficiency as it minimizes manual participation and processes tasks quicker. AI also significantly reduces human error, increasing the accuracy of order fulfillment. This enables customers to receive their orders with improved speed and service dependability. Reliance Retail likewise applies AI to improve inventory management and demand forecasting, enhancing overall performance of the supply chain.

Robotic arms and autonomous mobile robots (AMRs) are now mainly involved in inventory picking, packaging, and stocking, runs 24/7 with low supervision and also reduces the workplace injuries caused by machines (Leong & Ahmad, 2024). Customer service and shelf management are also taken care of by robots, who improves the placement of the items and communicates the accurate stock levels of the products. All the functions of the retail sector are to be done properly with any kind of errors and delay due to automation and robotics.

3. OBJECTIVES

The primary objective of this chapter is to examine how emerging technologies—such as Artificial Intelligence (AI), Machine Learning (ML), the Internet of Things (IoT), robotics, and augmented reality (AR)—are transforming the structure of retail work through job redesign and role redundancy. The specific objectives are:

1. To analyse the scale and patterns of technology adoption in the retail sector before and after COVID-19.
2. To examine how emerging technologies influence job redesign, including the creation of new roles and the transformation of existing roles.
3. To identify the types of retail jobs most vulnerable to automation-driven redundancy and the underlying factors contributing to this displacement.
4. To explore how AI, robotics, IoT, and data analytics reshape operational workflows in modern smart stores.
5. To assess the ethical concerns associated with technological adoption in retail, including data privacy, algorithmic bias, and workforce equity.
6. To evaluate real-world cases (e.g., Walmart, Amazon Go, Flipkart, Reliance Retail) to understand practical impacts of automation on workforce structure and retail efficiency.
7. To suggest how retailers can balance automation with human-centered approaches through reskilling, upskilling, and hybrid human-machine roles.

4. METHODOLOGY

The research design embraced in this study is qualitative, descriptive, and exploratory, in order to examine how the new technologies, including Artificial Intelligence (AI), Machine Learning (ML), the Internet of Things (IoT), robotics, and augmented reality (AR) are changing the nature of jobs, employment patterns, and the working process in the retail industry. A systematic review of published research articles, industry reports, technological reviews, and a real case analysis is used as a methodological instrument to focus on Walmart, Amazon Go, Flipkart, and Reliance Retail.

4.1 Research Design

Qualitative exploratory research is used in order to understand the dynamic state of job redesign and role elimination within technology-driven retail settings. The research does not entail the use of primary data collection; the research synthesizes the available empirical evidence to explain patterns, trends, and implications in the workforce.

4.2 Data Sources

The research is founded on a considerable amount of secondary data, which is obtained following: - Journal articles that are peer-reviewed. - Conference proceedings - Whitepapers and international reports on retail (e.g., World Economic Forum, McKinsey) - Whitepapers and industry reports (e.g., Global warming, Clean water, Boost Juice) - Case studies in the academic and professional publications. - Retail automation and digital transformation books and book chapters. Online reliable resources, such as technology magazines and corporate reports.

The combination of these sources is an insight into the technology adoption, workforce transformation, ethical issues, and redesigning operations in the contemporary retail.

4.3 Method of Analysis

An analysis methodology based on a thematic approach is used because it is necessary to detect recurring trends in the aspects of: (i) adoption of technology before and after the COVID-19 pandemic; (ii) automation and resulting role redundancy; (iii) novel job positions; (iv) skill change and the necessity to reskill employees; (v) operational effectiveness and workforce restructuring; and (vi) ethical and data management issues. Themes were designed by classifying the literature with each of the domains reflecting the chapter objectives: (1) technological transformation, (2) job redesign, (3) job displacement, (4) ethical considerations and (5) organisational responses.

4.4 Case Illustration Approach

The researcher uses case studies as part of his work, relying, in particular, on Walmart, Amazon Go, Flipkart, and Reliance Retail organisations. The examples demonstrate: that robotics can boost warehouse and shelf management; that the Internet of Things (IoT) can simplify inventory monitoring in real-time; generative artificial intelligence can improve customer services; and self-checkout solutions and autonomous robots can help to decrease the number of repetitive human operations. The practical value and relevance of the findings are enhanced by the use of case examples to make the findings contextually accurate.

4.5 Scope and Delimitations

The data presented in the study is largely devoted to the retail operations (front-end and back-end) and workforce implication of automation and artificial intelligence, as well as the post-COVID boom in the adoption of digital technologies. It fails to include both primary data of employees or consumers and a financial modelling of technology investment. It aims at offering a theoretical and thematic insight into the impact of emerging technologies on employment in smart retail ecosystems.

4.6 Ethical Considerations

Since the research is based on publicly available secondary sources only, there are no human subjects involved, and no ethical risks associated with the research. However, as part of the analytical chapter, the author mentions such aspects of retail technology as data privacy, bias, and job displacement.

5. ANALYSIS AND DISCUSSION

The nature of work has changed significantly in this new era of automation. The human workforce is now expected to concentrate on jobs that call for creativity, critical thinking, and complex problem-solving, even as routine, repetitive tasks are increasingly being streamlined by technology. Human intuition and judgement are still difficult for machines to mimic in these domains. People are being forced to take on more creative and intellectually challenging tasks as automation manages operational efficiency. Upskilling and flexibility in the workforce are necessary for this evolution. In the end, it highlights the increasing significance of human abilities that are difficult for machines to replicate. (Ali, 2023).

Figure 1. Impact of robotics in retail

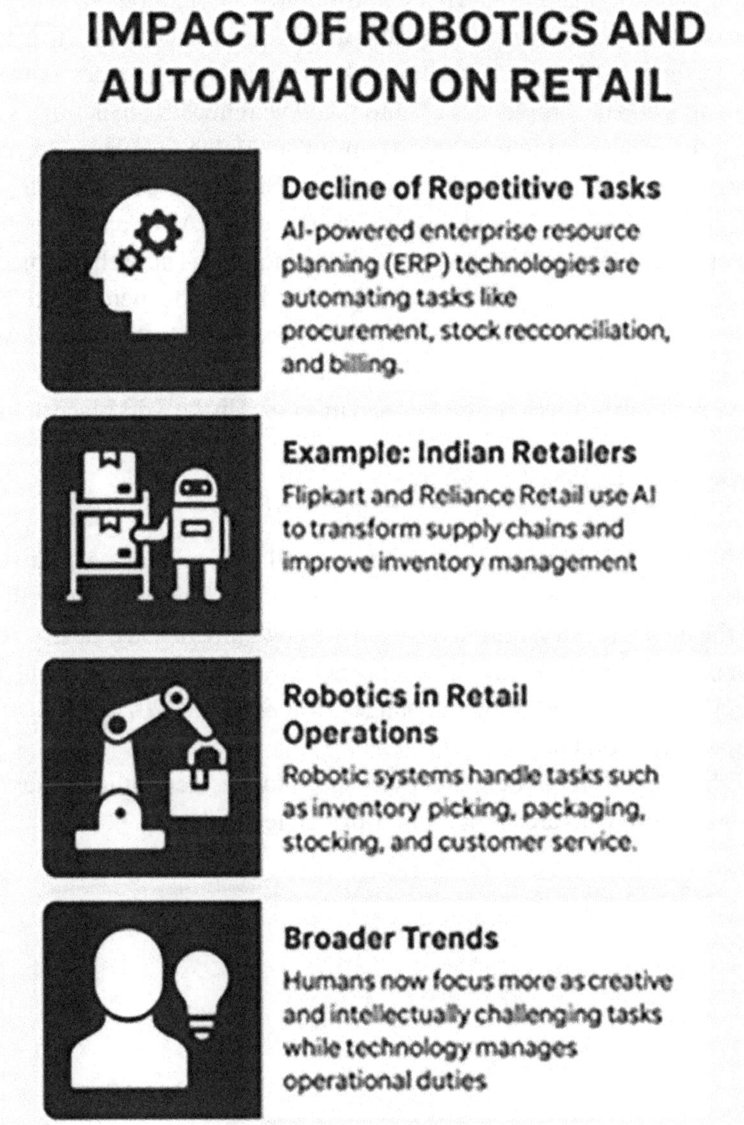

5.1 Job Redesign-From Routine to Resilient

a. Emergence of New Roles

A new class of jobs is developing that focusses on positions requiring human supervision, decision-making, and personalisation as automation gradually replaces manual tasks. Employees are now using data analytics and digital tools to offer highly personalised services that are suited to each customer's preferences in the area of consumer engagement. Employees in automated warehouses are acting as automation supervisors, supervising robotic systems to guarantee seamless operations and intervening when human judgement is required. Similar to this, data-driven store managers are using analytics and dashboards driven by AI to optimise inventory, make strategic decisions, and increase overall productivity. These changing positions demonstrate a move away from manual labour and towards mental and analytical duties. The need for the human workforce is growing as technology develops in fields that require complex decision-making, empathy, and inventiveness to ensure a cooperative relationship between people and machines in the workplace of the future.

b. The requirement for Reskilling

By 2030, over 50% of the world's workforce will need to learn new skills in order to stay relevant, digital literacy, problem solving, and critical thinking will become crucial in the retail industry. Initiatives from the government and industry, including the Retailers Association of India (RAI), are encouraging the development of digital skills to get workers ready for these changes (Rai & Goyal, 2024).

Example: Amazon Go

The cashier less store model reduces the need for cashier but increases the need for workforce those who are skilled in AI-driven technologies.

Figure 2. The job redesign process (Siengthai & Ngarm, 2016)

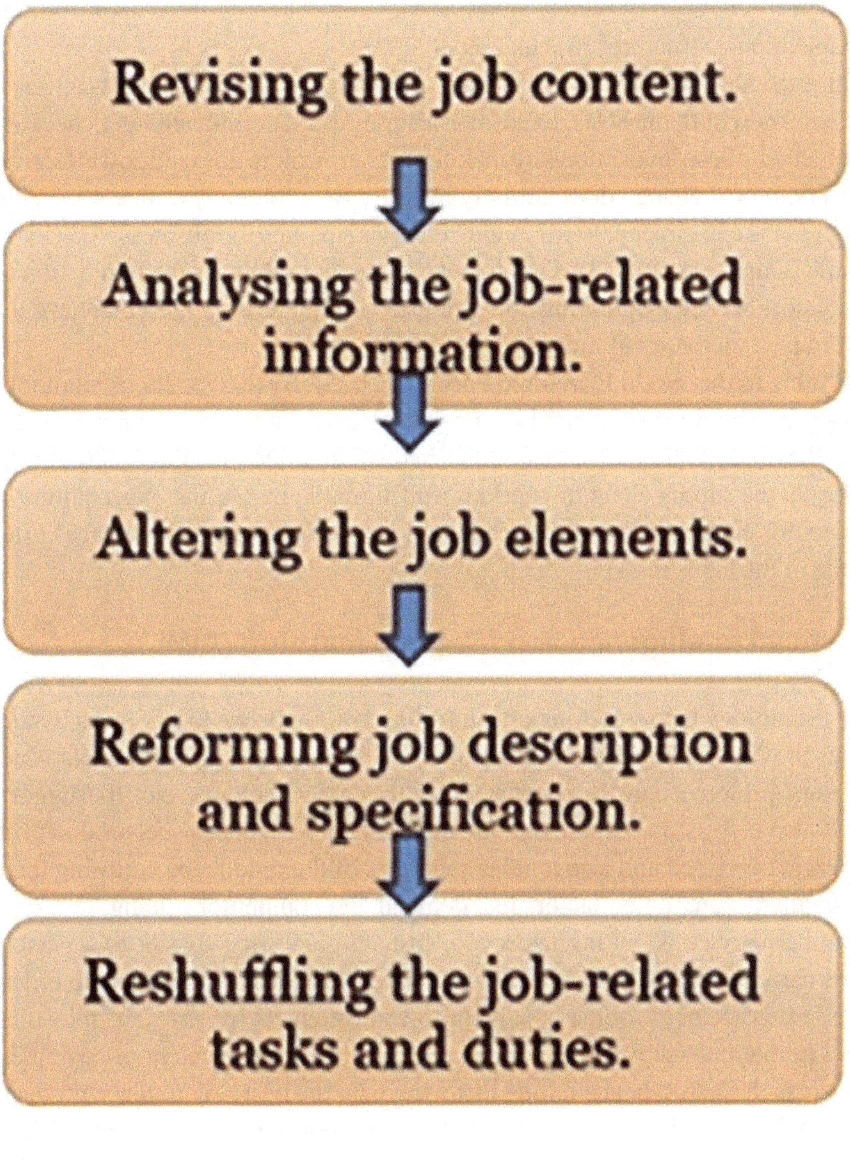

5.2Key emerging technologies transforming retail

Emerging Technologies like Artificial Intelligence & Machine Learning (AIML), internet of things (IoT's), Robotics Augmented reality are drastically transforming the retail sector at a speed and scale never witnessed. Retail operations from inventory control to customer service - are being completely transformed by this emerging Technologies as a result of which we are witnessing changing job responsibilities and organizational structures (Shankar et al.,2021).

From artificial intelligence algorithms, predicting what customers want even before they know it themselves, to cashier-less stores that eliminate long checkout lines altogether–these innovations are making operation more time and cost effective. Such technological change does not only increase the efficiency of operations but also changes the consumer patterns, brand relationships, and the business experience as a whole (Gupta et al., 2023). This current research establishes the foundation of the smart store setting by pointing out various emerging technologies that have the greatest impact on the retail industry.

According to the World Economic Forum's (2023), By 2027, 52% of retail jobs are projected to be automated, marking one of the quickest transformations to automation. These automations are being managed by an urgent need to, improves the flexibility in the supply chain disruptions, fulfilling the increasing expectations of consumers for customization and speed and balance out the labor shortages after pandemic in the labor market.

5.3 Artificial intelligence and Machine learning (AIML):

This technology help stores understand what customers like to buy by analyzing what they browse and buy. Based on this analysis, stores can suggest products as per the taste and preference of the customer —in the same fashion as Netflix suggests shows. It also helps stores figure out how much stock to keep, so shelves don't run empty or get overfilled and also it helps set prices of the product by adjusting them automatically depending on factors like demand and competitor pricing.

Customer service is getting faster too. Instead of waiting to talk to a person, shoppers can chat with AI-powered bots anytime to get help with respect to orders or solve any grievances. Some apps help to find products by voice or uploading pictures. In the retail setting, AI will be able to track the movement of customers and the areas they visit, hence telling the owners on the best way to place products and how to advertise. It also helps prevent fraud by spotting unusual activity during payments. With AI, marketing has become smarter—stores can send out offers that actually match needs and preferences of the consumer. And in some advanced stores, you can even shop without going through a checkout line as the AI tracks what you purchase and charges you automatically. In simple terms, AI and ML are making shopping easier, faster, more personal, and more efficient—for both customers and businesses.

5.4 A few of the primary applications of machine learning in the retail sector include:

a. **Inventory optimisation:** Businesses may reduce excess inventory and prevent shortages by employing machine learning to predict which products will sell when and where (Agrawal et al., 2024). Walmart has used machine learning to cut inventory-related costs by 15–20% in a number of key areas.

b. **Changes in prices:** Shops like Target and Zalando utilise machine learning (ML) to adjust pricing based on historical data, inventory levels, and competitor activity. As a result, productivity and competitiveness in the market both rise.

c. **AR and Robotics**: Robots now perform a large number of in-store and warehouse tasks, greatly enhancing operational speed, efficiency, and safety (Mohan Banur, O., Patle, B. K., & Pawar, S. 2024). The conventional retail workflow is being transformed by automation, which lowers human error and permits 24-hour operation. At the same time, virtual try-ons made possible by technologies like Augmented Reality (AR) are transforming the way that customers interact with products in personalised and immersive ways. In addition to making retailing more convenient to the consumer, it also opens up new possibilities to digital content creators to create engaging and interactive shopping experiences. Therefore, the future of consumer interaction is being transformed through augmented reality that is successfully closing the gap between online and offline commerce (Meirezaldi, 2024).

d. **Internet of Things (IoT):** The Internet of Things (IoT) already transforms the retail industry and helps to convert traditional brick-and-mortar stores into intelligent systems that will allow them to communicate and exchange data with other parties in the supply chain smoothly (Gupta et al., 2024). Installation of hi-tech IoT sensors and smart devices provide retailers with the ability to collect, transmit and process real-time operational data at various levels of business, thus enhancing the performance and management of inventory. Constant connectivity minimizes unplanned downtime, increases the product placement on the basis of actual telemetry, and allows quick reaction to changing tastes of the consumer. In turn, IoT improves the productivity of each individual and gives a retailer a chance to create a more comfortable and personalized consumer experience and improve satisfaction and loyalty. Finally, the use of IoT in large amounts brings a paradigm shift in the activities of retail organizations, changing the ways they interact with customers and redesigning the value creation in the modern digital market environment.

e. **Smart Shelves and RFID Tags:** The center of this revolution is smart shelves that are being equipped with advanced technologies, such as sensors, RFID tags, and weight detectors (Kumari, R., & Singh, V. P. 2024). These shelves are not only used as a storage system in a traditional way. They have the ability to learn and send real time information on stock quantities, location of the products and even communication with customers on their own. As a result, retailers are able to have a more efficient way of tracking inventory, preventing stockouts or excess, and acting promptly in response to changing consumer demand. Furthermore, consumer behavior, including how often a product is picked up or how long the customers stand in front of a shelf, enables personalized marketing campaigns and streamlined store layouts. Accordingly, intelligent shelves are the key element in moving towards data-driven, smart retail spaces.

5.5 Challenges and Ethical Considerations

5.5.1 Job Displacement

The fast adoption of automation, robotics, and artificial intelligence in stores has only fueled anxieties about the displacement of jobs, especially frontline and routine-based jobs, including cashiers, stock handlers, warehouse pickers, cleaning workforce, and some store management duties. Many of these technologies like self-checkout kiosks, artificial intelligence-based demand forecasting systems, robotic inventory scanners, and autonomous warehouse robots are becoming increasingly able to do what previously was done by human workers, and thus they are lessening the need for manual and repetitive labour.

Nevertheless, job displacement in retail cannot be considered only as extensive-scale job destruction but as job change and reorganization of tasks. Some of the positions could be eliminated; however, most of the positions are being re-modelled with the addition of supervisory, analytical, and customer-focused roles. As an example, cashiers are becoming customer engagement specialists or digital support associates, warehouse workers are becoming more obligated to monitor automated systems, handle exceptions, and quality control.

Proactive reskilling and upskilling programs are needed to reduce the adverse effects of displacement. The skills that need to be developed by the retail employees are digital literacy, data interpretation, human-machine interaction, and customer relationship management skills. Ongoing learning initiatives, certification programmes, which are initiatives led by the industry, and initiatives that are based on social skills development between the society and the industry can help in easier transitions of the workforce. Without such interventions, automation has the potential to increase the disparities in employment and affect the morale of the working population. Therefore, the importance of responsible automation policies should be dedicated to employee flexibility, internal mobility, and the inclusion of workforce in the planning (Gunda and Mupa, 2024).

5.5.2 Data Privacy and Security

Massive adoption of AI, IoT, computer vision, and data analytics technologies has resulted in unmatched amounts of data gathering in the retail setting. Smart shelves, RFID technology, facial recognition technologies, customer tracking sensors, and AI-enhanced recommendation engines are gathering and processing massive amounts of personal and behavioural data on a continuous basis. Although this type of data-driven system allows to create in an extremely personalised shopping experience and achieve operational efficiency, it also leads to serious issues associated with data privacy, cybersecurity, and legal compliance (Raji et al., 2024).

Companies are facing growing consumer fears concerning the manner in which their personal information is gathered, retained, exchanged and sold. Consumer trust can be seriously undermined by the problem of unauthorised access, data breaches, the absence of an informed consent, and non-transparent algorithmic decision-making. In addition, cross-border data transfers and third-party technological alliances also make regulatory compliance more difficult, particularly in jurisdictions that have strict data protection regulations like GDPR, and other developing domestic data governance models.

Retailers should thus embrace sound data governance systems, such as reliable data storage, encryption schemes, periodic audits and overt consent procedures. There should be clear communication about the policy of data usage and consumer rights to retain trust and legitimacy. Data stewardship Ethics plan Ethical data stewardship must not be seen as a mere compliance issue but as a strategic requirement that directly affects the brand image, retention, and sustainability (Mahmood and Abdulqadeer, 2024).

5.5.3Algorithmic Fairness and Bias

Algorithms have become the core of the contemporary retail business and have an impact on the recruitment, workforce scheduling, pricing, inventory distribution, and customer relationship management. When machine learning models are developed using biased or incomplete data, they can reproduce or even increase the level of social, gender, racial or economic inequality that exists. Discriminative algorithms may cause discriminatory employment, inequitable distribution of tasks, discriminative prices or customer discrimination.

It is the ethical issue of algorithmic fairness and transparency, thus, that is a big problem. Retailers should develop responsible AI systems that prioritize explanatory, accountable, and human-centered systems. Biases audits regularly, various training datasets, and interdisciplinary review mechanisms are required to avoid unfair results. Moreover, it is preferable that the employees are not to be subjected to murky algorithmic managerial frameworks that check on performance overboard, or make unilateral decisions without human intervention.

Ethically, AI and robot deployment through workforce should be used to enhance human abilities and not to substitute them. The new technologies are becoming more and more useful in eliminating repetitive and physically demanding jobs and providing new avenues to more creative, strategic and customer-oriented jobs. The future stability of the retail labour force will rely on the ability to establish equilibrium between human and machine relationship where empathy, judgement, and adaptability (qualities not reproducible by machines) are taken into account (Lipton et al., 2022). The experience of organizations like Walmart proves that AI and robotics when introduced with a responsible approach can help to optimize operations and customer satisfaction without resulting in a mass displacement of the workforce in general. Through a mix of technological innovation and ethical governance, reskilling programmes, and job design in a human-centric manner, retailers may develop a sustainable framework of smart retail that would be win-win to the business and human beings.

5.6 Walmart's Alphabot system

The entire order fulfilment process is streamlined in Walmart's grocery fulfilment centers by using Alphabot robots to retrieve and sort items for online orders. By working inside a specially created automated storage and retrieval system, these sophisticated robots drastically cut down on the amount of manual labour that store employees must perform. Order preparation consequently becomes less time-consuming, more accurate, and faster. In addition to increasing operational effectiveness, this frees up human labour for more difficult jobs like quality assurance and customer service. One example of how automation is changing retail logistics to satisfy the rising demand for online grocery shopping is the incorporation of Alphabot technology (Liang, 2025).

5.6.1 Shelf-Scanning Robots

Vending robots with computer vision working in the retail setting have the ability to quickly detect out of stock items, ensure that the items are placed correctly, and scan shelves to confirm the compliance with prices. The automation of such necessary, but standard, functions can grant corporate organizations like Walmart a better control of the inventory levels and allows them to react quickly to the replenishment needs. This technology, therefore, augmenting shelf availability, minimizing human error, and improving the consumer shopping experience, increases the product availability to consumers. Walmart's dedication to utilizing cutting-edge automation for effective store management and customer satisfaction is demonstrated by the deployment of such intelligent robots (Patel, 2024).

Figure 3. Bossa Nova robot

5.6.2 Self-Driving Floor Scrubbers (Auto-C)

Walmart has a high level of cleanliness in its stores, which is mainly ensured by the use of autonomous cleaning robots (Seufert & Ju, 2025). By taking over manual labour in cleaning, these machines are significantly decreasing the need of manual labour in cleaning and consequently leading to the reduced operating costs. In addition to enhancing productivity, such automation will help liberate employees to devote more time to customer management and other value-added tasks. As a result, the efficiency of operations by Walmart and the experience of in-store shopping is increased, which is an example of how the use of robotics can make the allocation of the workforce smarter. The high-cleanliness standards in all the facilities of Walmart are highly maintained by the autonomous cleaning robots that do not depend on any human assistance (Seufert and Ju, 2025). These robots are very effective in reducing the labour requirement of cleaning manually as they take up the routine cleaning duties thus leading to lower costs of operation. Automation can also help employees free up time on other high value activities besides increasing productivity, as also working on customer service. Consequently, Walmart will be able to optimize its business operations, as well as the in-store shopping experience, which can be demonstrated as a sign of the robotics' influence on the further intelligent distribution of workforce. The high-cleanliness standards in all the facilities of Walmart are highly maintained by the autonomous cleaning robots that do not depend on any human assistance (Seufert and Ju, 2025). These robots are very effective in reducing the labour requirement of cleaning manually as they take up the routine cleaning duties thus leading to lower costs of operation. Automation can also help employees free up time on other high value activities besides increasing productivity, as also working on customer service. Consequently, Walmart will be able to optimize its business operations, as well as the in-store shopping experience, which can be demonstrated as a sign of the robotics' influence on the further intelligent distribution of workforce.

5.6.3 Generative AI Enhanced Customer Services: Wallaby

Figure 4. Robotics automation portfolio (Jung, 2025)

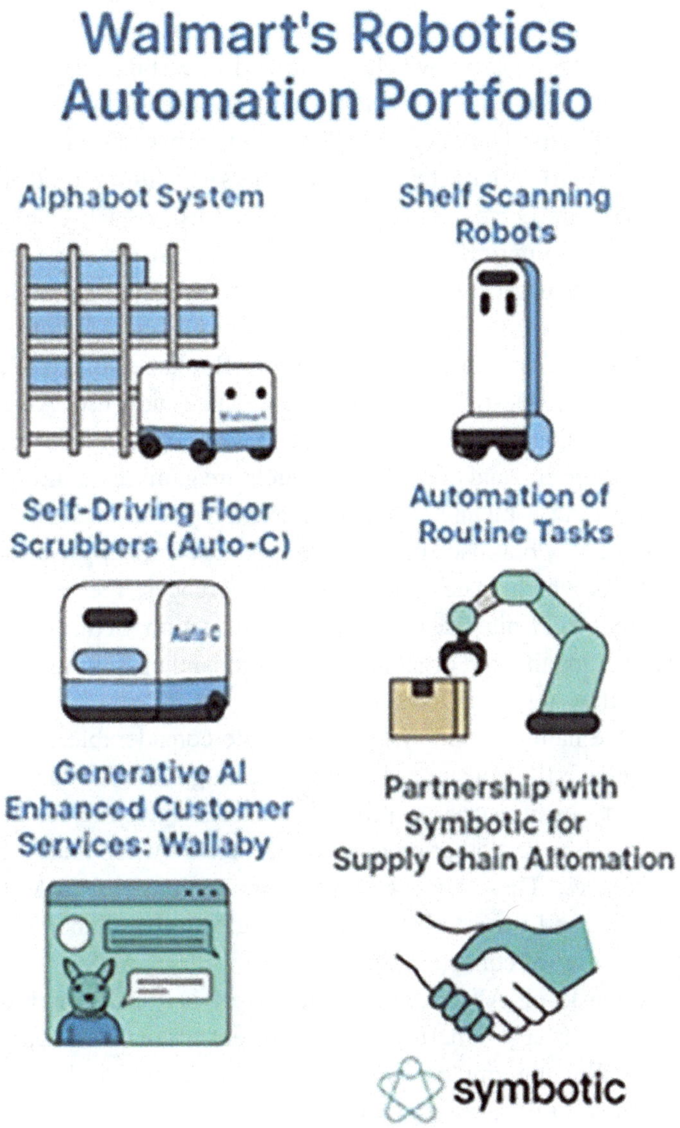

As previously mentioned, Walmart uses generative AI to offer individualized online customer support. This cutting-edge AI technology can help customers check the status of their orders, provide personalized product recommendations, and respond to customer enquiries. Generative AI improves the overall customer experience by providing timely and pertinent support, which facilitates more seamless and effective interactions. In addition to improving customer satisfaction, this aids Walmart in forging closer, more intimate bonds with its customers (Jung, 2025).

Generative AI conversational technology ranks a first among others as how it could provide a seamless and smart exchange to the customer (Parise et al., 2016). In contrast to regular chatbots which are limited to certain scripts, generative AI models can carry on logical conversations with the users of the service - they are able to understand natural language, recall previous conversations, and create polite phrases typical of human characters. The customers have smooth and obvious access to services which are personalized and not just transactional. Generative AI can almost eliminate waiting times with its ability to serve instant and correct answers, thus as a result there is great improvement in both assistance speed and quality.

Generative AI has gradually moved to the centre of how Walmart approaches its modern sales and marketing decisions (Recio-Román et al., 2024). Rather than relying on broad, one-size-fits-all promotions, Walmart now uses AI-driven recommendation engines that study what customers have bought in the past, the kinds of items they tend to prefer, and even what products are gaining traction in the market. This would allow the company to deliver extremely customized advice that is not only the right advice, but also timely. Whenever consumers get constant product recommendations which match their stated preferences, the scientific evidence shows that there is a definite rise in the overall satisfaction. In the long-run, this will create a customer loyalty and organic cross-selling and up-selling chances without the feeling that they are being forced to do so.

The predictive nature of the system can create considerable value, whereby it can easily influence the consumer behaviour through sending alerts in the event of looming shortages of necessary products, or it may give the best time-sensitive discounts, or even suggest other products to consumers when their preferred products are out of stock. These are time-tuned interventions that make the shopping experience easier, and the retailer becomes an effective collaborator in the daily purchasing processes of consumers.

At the internal level, AI generated generatively is transforming the processes of support operations at Walmart. It is an important strategic complement to internal operations, allowing employees to focus on issues which demand the human judgement, whilst automating a significant percentage of routine queries which would have previously consumed a lot of man power. This type of automated agent enables customer-service employees to rebalance their workload by assigning them repetitive duties that allow them to engage the customers in a more thoughtful and compassionate way when machines fail. This paradigm encourages a symbiotic combination of automation and human-centered service, which avoids the loss of human position and provides balanced attitude to the use of technology to improve customer interaction.

To sum up, the introduction of generative artificial intelligence by Walmart is a strategic shift of the company towards reactive rather than anticipatory customer service. It reconnects brand-consumer relationship based on sustained, personalized and meaningful interaction. As Jung (2025) remarks, besides performance improvement, these types of innovations put Walmart in the position of a leader of human-centered retail technology which is the one that uses AI not for replacing people but for making the human experiences classier in the digital .

5.6.4 Automation of Routine Tasks Heading

Customer service can also be improved with the assistance of automation of food order picking, shelf scanning, and floor cleaning, which will allow Walmart employees to dedicate more time to customer related activities. The organisation can ensure that its workforce spends less time in repetitive and routine tasks that require them to engage with customers and help them feel better by ensuring that the employees spend more time with customers. This change is beneficial in the process of enhancing the entire shopping process and also generates a more rewarding and stimulating atmosphere among employees. Such automation efforts eventually enable the establishment of a customer-focused organisation, which has a consumer-satisfied and a motivated labour force (Akinsola & Liang, 2025). As the urge to automate order picking and shelf control, employees do not have to spend the hours in warehouses or on shelves doing the same thing. The employees are thus able to focus on customer-centered tasks which are to serve the shoppers, address the queries and make the in-store experience more interesting and efficient. This new mode of operation is more in line with the ideals of Walmart to be an even more customer-first organisation, where technology improves the productivity of the organisation without replacing human abilities. The personnel are in need of the latitude to be able to apply judgment, empathy, and communication, which is not possible with machines.

Table 2. Automation and AI initiatives at walmart – technologies, functions, and impacts

Section / Initiative	Technology Used	Primary Function	Operational Benefits	Impact on Workforce	Strategic Outcome
Alphabot System	Alphabot automated storage and retrieval robots	Automated picking and sorting of online grocery orders	Faster order fulfillment, improved accuracy, reduced manual labor	Employees freed from repetitive picking tasks; redeployed to quality assurance and customer service	Enhanced e-commerce efficiency and scalability in online grocery fulfillment
Shelf-Scanning Robots (Bossa Nova)	Computer vision–enabled autonomous shelf-scanning robots	Detect out-of-stock items, verify product placement, ensure price compliance	Improved inventory accuracy, reduced human error, better shelf availability	Reduced routine shelf-checking workload; staff focus on customer interaction	Improved in-store availability and customer shopping experience
Self-Driving Floor Scrubbers (Auto-C)	Autonomous cleaning robots	Continuous floor cleaning without human intervention	Lower operating costs, consistent cleanliness, higher productivity	Employees shifted from manual cleaning to customer-facing and value-added tasks	Improved in-store environment and optimized workforce allocation
Generative AI–Enhanced Customer Service (Wallaby)	Generative AI conversational agents and recommendation engines	Order tracking, customer queries, personalized product recommendations	Reduced wait times, higher response accuracy, personalized engagement	Automates routine queries; allows staff to handle complex, empathetic interactions	Human-centered, anticipatory customer service and increased customer loyalty
Automation of Routine Tasks	Integrated robotics and automation systems (picking, scanning, cleaning)	Automation of repetitive operational tasks	Faster service delivery, consistent operations, higher efficiency	Increased job satisfaction; reskilling toward digital supervision and decision-making	Customer-first organization with motivated, technology-enabled workforce
Partnership with Symbotic	Advanced warehouse robotics and AI-driven supply chain automation	Automated warehouse handling and e-commerce fulfilment	Faster order processing, higher accuracy, lower fulfilment costs	Reduced physical strain; workforce transitions to system oversight roles	Strengthened supply-chain resilience and competitive advantage in online retail

In addition, automation has also led to the increase in job satisfaction and morale of the Walmart employees indirectly. The company has facilitated opportunities to improve skills and grow professionally by alleviating the employees off manual and hardworking duties, and tedious labor. Employees are being taught to oversee automated processes, read digital information and use intelligent technologies to make better decisions. This is not just a strategy to increase productivity but it would also give a sense of belonging in the technological revolution of the retail industry. Essentially, automation has shifted the focus of its employment strategy of Walmart towards a manual labor organization to a computer-based collective collaboration and human intelligence.

Besides enhancing the dynamics of employee interactions, the automation efforts of Walmart have a quantifiable effect on the customer satisfaction. The quicker order processing, reduced checking-out, and always replenished shelves will advance the shopping experience, which contributes to the competitive advantage of Walmart in the world retail market. The technological development of the company shows that the levels of automation can grow into a symbiotic relationship between humans and machines, when applied responsibly.

Finally, the case of Walmart shows that automation does not necessarily result in the workforce displacement when it is carefully implemented. The company has shown that technology does not have to replace human labour, with the introduction of reskilling and the culture of flexibility leading to a more customer-focused, efficient, and motivated retail environment (Akinsola and Liang, 2025).

5.6.5 Partnership with Symbiotic for Supply Chain Automation

Walmart has gone on to invest heavily in the area of robotics in its warehouse systems through a strategic alliance with Symbiotic. The collaboration will allow Walmart to streamline important operations in the warehouse, thus expanding its e-commerce fulfilment. This results in orders being handled faster, more accurately and efficiently at a low cost to Walmart giving them a substantial competitive edge. Not only does the utilization of the latest robotic types of technologies increase productivity, but it also enables the business to fulfill the growing demand of online services faster and more consistently (Liang, 2025).

Figure 5. Automated carrier

6 FINDINGS OF THE STUDY

Based on a comprehensive thematic analysis of the peer-reviewed literature, industry reports, and a case study namely Walmart, research reveals several important trends: The emerging technologies are transforming the retail work with job redesign, role redundancy and human-AI collaboration. The results are discussed with regards to the research objectives outlined in Section 3.

6.1 Acceleration of Technology Adoption Post–COVID-19

The rapid rate of technology adoption since the COVID-19 pandemic is undoubtedly one of the most recent findings of this exploration. Artificial intelligence (AI), robotics, the Internet of Things (IoT), machine learning (ML), and contactless automation quickly became tools of choice as an elective or even experimental tool but rapidly became the indispensable part of the working process(Singh, 2025). During the lockdown, retailers used automation to reduce disruptions, labour shortages, supply-chain instability and increased consumer demand to be safe and convenient. Contactless checkout, robotic cleaning systems, AI-based demand forecasting, and automated warehousing became the result of strategic differentiators, but they are now mandatory mechanisms of business continuity. The pandemic was able to compress years of envisioned digital transformation into months, thus radically changing the retail mode of operation. Such quicker implementation has continued even after the pandemic meaning that automation and smart technologies are now part of retail strategies, and not a short-term crisis solution.

6.2 Significant Job Redesign Across Retail Functions

The results show that job redesign and not job elimination is the most common workforce outcome with retail automation. The cashiering, billing, stocking, warehouse picking, and floor cleaning functions in the traditional retail have been thoroughly reorganized with the introduction of technology (Hompel & Schmidt, 2006). Those employees who have been involved in manual work or repetitive job schemes gradually shift to jobs that entail monitoring of automated systems, customer services, exception management, and decision making. An example is how a company like the cashiers is re-deployed to become a customer engagement associate or digital support contact, whereas the warehouse workers are now overseeing robotic pickers, system warnings and quality control. This redesign is a wider change in principles in the performance of tasks to coordination of tasks and delivery of value added services, which fits the socio-technical systems theory in terms of human and technological elements functioning in concert. The results indicate that the future in retail work is moving towards the hybrid human-machines jobs capable of integrating the technological efficiency with the human judgement, empathy and flexibility.

6.3 Routine and Repetitive Jobs Face High Automation Risk

Although job redesign is widespread, as per the study, routine, predictable, and physically repetitive jobs are at an inordinately high risk of being automated away. Cashiers, stock clerks, housekeeping staff, and manual warehouse labourers, in particular, are the most vulnerable ones as they can be easily codified and automated with the help of AI-powered systems, robotics, and sensor-related technologies. Automation has been found to be particularly useful in the settings that require uniformity, quickness, and minimisation of errors, e.g. inventory, shelf scanning, order fulfilment tasks and store maintenance (Torchio, 2023). Thus, these positions are getting scrapped or scaled down to a few. However, not all jobs have the same experience of redundancy. Position that involves emotional intelligence, situational judgement and interpersonal communication are relatively resistant. This observation underscores the polarisation of retail jobs, in which low-skill routine jobs are reducing in number and cognitively demanding and customer jobs are increasing in number.

6.4 Emergence of New Tech-Enabled Roles

One of the most positive outcomes of the research is the establishment of new technology-empowered workplaces in the context of smart retail ecosystems. Such roles represent the increased requirements of digital skills and human-technology integration. The emerging notable positions include: - Supervisors automation and robotics. - Data-driven store managers - AI and IoT system operators - Online customer experience experts. - Analytics and demand-planning assistants - Predictive analytics and demand-planning assistants (Choudhuri, 2024). These jobs involve a combination of technical knowledge, intellect and human based decision-making. Instead of replacing the human labour all out, technology has shifted the labour demand towards the cognitive, supervisory, and collaborative skills. Evidence of Walmart, AmazonGo, and Indian retail examples show that organizations that invest in automation are also likely to invest in redeployment of workforce and building capabilities, which further supports the point that technological improvement and job creation can be sustained when done in a responsible manner.

6.5 Increased Demand for Reskilling and Digital Literacy

The other main discovery is that reskilling and upskilling are crucial in building resilience of a workforce. The productivity of retailers is more and more based on the capacity of the employees to communicate with digital devices, process real-time information, debug robotized systems, and cooperate with AI-advanced technologies. The emergence of digital literacy, data analysis, human-machine interaction, and adaptive problem-solving has become a focal point in all types of retail jobs (Adepoju et al., 2023). Lack of access to training or other learning opportunities by employees compel them to greater risk of being displaced, whilst organisations that invest in continuous learning have an easier time of adopting the new technology and can enjoy better employee morale. The results highlight the fact that reskilling is not a single intervention, but a continuous process, which needs organized collaboration between the retailers, industry organizations and the policymakers. In the absence of well-formulated reskilling systems, automation can contribute to inequality and job insecurity.

6.6 Ethical Concerns Intensify with Rising Automation

The ethical issues are proposed in the study as a burning and urgent aspect of retail automation. The threat of data privacy, algorithm manipulation, monitoring of the workforce, job loss, and vaguely defined responsibility practices grow louder and larger as AI and IoT systems grow. Smart shelves, facial recognition, customer tracking, and performance-monitoring systems are some of the ways in which retailers have gathered enormous amounts of consumer and employee data. These tools improve efficiency and personalization, but they also cause significant issues of concern about consent, transparency and fairness. The ethical picture is complicated further by the fact that the workforce scheduling, performance assessment, and recruitment processes are based on algorithmic bias (Rodgers et al., 2023). The results highlight the necessity of well-established governance models, accountable AI, and clearer decision-making models to make sure that automation is not contrary to social and ethical principles.

6.7 Case Evidence Shows Positive Operational Outcomes

Nonetheless, the case evidence indicates sustained positive performance of the operations in case of automation implementation, as well as human-oriented strategies. Walmart technologies like the Alphabot, shelf-scanning robots, self-cleaning floor scrubbers, or generative AI-based customer service applications have resulted in: - Accuracy and speed of fulfilment are higher. - Inventory turnover and stockouts decreased. - Reduced reliance on physically challenging labour.

Automation strengthens operational efficiency without eliminating the need for human judgment and empathy.

7. PRACTICAL IMPLICATIONS

The results have a number of practical implications on retailers, policymakers and employees:

7.1 For Retail Organizations

The retail organizations will have to use a hybrid human-machine solution where automation is meant to supplement human abilities as opposed to substituting it. To ensure that the workforce is flexible, strategic investment in sustained reskilling and upskilling initiatives, including, but not limited to, digital literacy, data analytics, AI-based decision-making, and customer experience management, are necessary. Enhancing data regulation and ethical AI systems will contribute to establishing consumer confidence and adhering to the emerging rules and regulations. In addition, job roles should also be redesigned by organizations to strike a balance between operational efficiency and the welfare of employees and quality of service. Inclusive and sustainable workforce transformation can be encouraged through well-designed internal mobility and redeployment systems that help ease the redundant positions into new emerging digital ones, as well as assist in the shift to more sustainable employment.

7.2 For Policymakers & Regulators

Policymakers and controllers have a central role to play in the development of retail automation which is fair, inclusive and socially responsible. Creating transparent ethical AI governance structures can help to reduce the problem of algorithmic bias, discrimination, and black box decision-making in areas like staffing, scheduling, or customer analytics. National and sector-based digital skilling efforts should be encouraged by governments to facilitate mass upskilling of the workforce. Moreover, there is a need to implement labour protection policies, including reskilling subsidies, transition support programmes, and social security protection to cushion workers who lose their jobs to automation and enable smooth transitions in the workforce.

7.3 For Employees

Retail employees should also take the initiative to adjust to a technologically oriented workplace by acquiring digital and analytical skills, such as data interpretation, problem solving as well as technology-mediated interaction with customers. One will need to have a life-long learning attitude in order to be employable in the fast-changing retail ecosystems. At the same time, workers are to capitalize on human-specific abilities including creativity, empathy, communication, and conflict management which cannot be completely replaced by artificial intelligence. The adoption of lifelong learning and human-machine working together will contribute to the personal career resilience and long-term employability.

8. FUTURE RESEARCH DIRECTIONS

The future study based on the Walmart case can expand the understanding of smart transformation in retail in the context of various dimensions of its salience. Although the chapter provides a detailed qualitative analysis of job redesign and role redundancy, comparative studies have the potential to place Walmart and other international retailers in the same context to determine how organizational strategy, regulatory environments and labour policy framework mediate different workforce results. This would necessitate longitudinal investigation to track the long-term effects of automation to the career paths, job security, the effectiveness of reskilling, and well-being of workers, specifically the changing hybrid human-AI jobs.

Future research can also include the use of primary data, including structured surveys of employees, semi-structured interviews, and active ethnographic studies, to describe the lives of workers, their experience of job security, and their reaction to algorithmic management. One of the opportunities is related to examining skill-transformation pathways, where attention should primarily be paid to the digital literacy, human-machine interaction competency, and affective intelligence as the essential factors enhancing employment resiliency in terms of technology-intensive retail environments. The moral aspects, including algorithmic fairness, transparency, employee surveillance, and data governance, should be looked at in more detail.

Additionally, the study could explore the customer-employee-AI interaction pattern to determine how customer generative AI reshapes emotional labor and service delivery. Lastly, the generation of the quantitative productivity levels and cost-benefit analysis using the workforce output indexes would help achieve a more balanced evaluation of the economic and social consequences of automation in the smart retail systems.

9. CONCLUSION

The outbreak of COVID-19 has greatly stimulated the use of e-commerce and promoted new technologies in digital retailing, thus initiating the radical transformation of the traditional retail business, labor organization and ethics. What was originally a response to lockdowns and supply-chain disruptions has come to be a permanent change in the retail ecosystem. The adopted technology was compressed over decades in several months and forced retailers to quickly assimilate artificial intelligence, robotics, the Internet of Things, and data-driven platforms in order to guarantee business continuity, operational efficiency, and consumer safety. These technologies have transformed the customer experience with the ability to make purchases without direct contact, real-time inventory and speedy models of delivery, reinventing the concept of convenience, and personalization demands.

At the same time, the physical retail space had to re-invent itself in an omnichannel approach which integrates digital and in-store experiences. Physical stores are no longer just selling locations but places of experience, based on online interfaces, insights, and intelligence. However, this fast technological growth has increased the anxieties surrounding the privacy of data, transparency of algorithms, surveillance of the workforce, and fair access to online services. The growing use of algorithmic decisions casts urgent concerns on the integrity, prejudice, and trustworthiness of the latter, which prompts the need to establish strong ethical AI governance systems.

Sustainable change in the retail, however, cannot be achieved only through technological innovation; it needs a balanced strategy that could incorporate social responsibility, ethical leadership, and workforce inclusion. Human creativity, empathy, and judgment will still be essential in providing meaningful customer experiences as well as being responsible in making decisions. The retailers that are able to effectively integrate the latest technologies with the inclusion and diversity of employment strategies, ongoing reskilling and appropriate governance systems will be in a better place to respond to any disruptions in the future. Finally, the only way to achieve success in the post-pandemic retail environment on a long-term basis will be to reconcile operational efficiency with ethical values and human-oriented design.

REFERENCES

Adepoju, O., Akinyomi, O., & Esan, O. (2023). Integrating human-computer interactions in Nigerian energy system: A skills requirement analysis. *Journal of Digital Food, Energy & Water Systems, 4*(2).

Agrawal, N., Cohen, M. A., Deshpande, R., & Deshpande, V. (2024). How machine learning will transform supply chain management. *Harvard Business Review, 103*(3-4), 128–137.

Akinsola, O. K., & Liang, W. (2025). Warehouse Automation and Robotics: Revolutionizing Order Fulfillment in E-Commerce.

Ali, S. S. (2023). The future of work in the age of automation: Exploring new roles and opportunities. *Al-Behishat Research Archive, 1*(01), 55–62.

Ammupriya, A., Vaishnavi, S., Ashwini, A., Kavitha, R., Paranthaman, P., & Saravanan, V. (2025, March). Cloud-Based HR Platforms for Scalable Workforce Management in Multinational Organizations. In *2025 3rd International Conference on Disruptive Technologies (ICDT)* (pp. 1607-1613). IEEE.

Bartlett, L., Kabir, M. A., & Han, J. (2023). A review on business process management system design: The role of virtualization and work design. *IEEE Access : Practical Innovations, Open Solutions, 11*, 116786–116819. DOI: 10.1109/ACCESS.2023.3323445

Bhargava, A., Bester, M., & Bolton, L. (2021). Employees' perceptions of the implementation of robotics, artificial intelligence, and automation (RAIA) on job satisfaction, job security, and employability. *Journal of Technology in Behavioral Science, 6*(1), 106–113. DOI: 10.1007/s41347-020-00153-8

Borghouts–van de Pas, I., Bosmans, M., & Freese, C. (2021). Unemployment prevention: The role of Human Resource Management in job-to-job transitions in the event of redundancy. *European Journal of Social Security, 23*(2), 103–119. DOI: 10.1177/1388262721995209

Choudhuri, S. S. (2024). *AI in ERP and supply chain management*. Academic Guru Publishing House.

Christl, W. (2023). MONITORING WORK AND AUTOMATING TASK ALLOCATION IN RETAIL AND HOSPITALITY.

Foster, W. M., Hassard, J. S., Morris, J., & Wolfram Cox, J. (2019). The changing nature of managerial work: The effects of corporate restructuring on management jobs and careers. *Human Relations; Studies Towards the Integration of the Social Sciences*, *72*(3), 473–504. DOI: 10.1177/0018726719828439

Fuchs, H., Benkova, E., Fishbein, A., & Fuchs, A. (2023, November). The importance of psychological and cognitive flexibility in educational processes to prepare and acquire the skills required in the Twenty-First century. In *The Global Conference on Entrepreneurship and the Economy in an Era of Uncertainty* (pp. 91-114). Singapore: Springer Nature Singapore.

George, A. S. (2024). Artificial intelligence and the future of work: Job shifting not job loss. *Partners Universal Innovative Research Publication*, *2*(2), 17–37.

GUNDA, V. N., & MUPA, M. N. (2024). A Paradox in Human Resources: Automation Vs Human Capital.

Gupta, S., Rastogi, R., & Alam, M. S. (2023). Disrupting FinTech: Revolutionizing Customer Experience With AI in Industry 5.0. In *The Impact of AI Innovation on Financial Sectors in the Era of Industry 5.0* (pp. 1-16). IGI Global.

Gupta, U., Somani, P., Behare, N., Mahajan, R., Singh, M., & Iyer, C. V. (2024). Revolutionizing Retail: IoT Applications for Enhanced Customer Experience. In *Internet of Things Applications and Technology* (pp. 60-80). Auerbach Publications.

Hamamah, A., Al-Haimi, B., & Tajuri, W. (2024). Navigating the Marketing Landscape: Artificial Intelligence and Big Data Role in Digital Marketing. *International Journal of Academic Research in Business & Social Sciences*, *14*(10), 2285–2299. DOI: 10.6007/IJARBSS/v14-i10/23346

Hompel, M., & Schmidt, T. (2006). *Warehouse management: automation and organisation of warehouse and order picking systems*. Springer Science & Business Media.

Ikeler, P. R. (2013). *Point of consumption: Work, consciousness and organizing in the retail sector*. City University of New York.

Jung, N. Y. (2025). AI-Powered Personalization Drives Loyalty in Retail. In *Leveraging AI-Powered Marketing in the Experience-Driven Economy* (pp. 105–146). IGI Global Scientific Publishing. DOI: 10.4018/979-8-3693-9561-5.ch004

Kahn, B. E. (2021). *The shopping revolution, updated and expanded edition: how retailers succeed in an era of endless disruption accelerated by COVID-19*. University of Pennsylvania Press.

Korneeva, E., Omarova, A., & Nurova, O. (2024). Adoption of smart energy technologies in the context of sustainable development. In *E3S Web of Conferences* (Vol. 531, p. 02001). EDP Sciences. DOI: 10.1051/e3sconf/202453102001

Kumari, R., & Singh, V. P. (2024). Industry 4.0 Technologies: Managing the Future of Smart Retail. In *Augmenting Retail Reality, Part A: Blockchain, AR, VR, and the Internet of Things* (pp. 59-82). Emerald Publishing Limited.

Leong, P. Y., & Ahmad, N. S. (2024). Exploring autonomous load-carrying mobile robots in indoor settings: A comprehensive review. *IEEE Access : Practical Innovations, Open Solutions*, *12*, 131395–131417. DOI: 10.1109/ACCESS.2024.3435689

Liang, G. (2024). Skill Shifts and Retraining Initiatives: Adapting to Post-Pandemic Labor Market Dynamics. *Law and Economy*, *3*(1), 58–61. DOI: 10.56397/LE.2024.01.06

Liang, W. (2025). Warehouse Automation and Robotics: Revolutionizing Order Fulfillment in E-Commerce.

Liew, S. Y., Rana, M. E., Hameed, V. A., & Safavi, S. (2023, December). Navigating the retail 4.0 landscape: The transformative impact of cloud computing. In *2023 IEEE 21st Student Conference on Research and Development (SCOReD)* (pp. 499-507). IEEE.

Lipton, B., Vromen, A., Rutledge-Prior, S., Good, L., Cooper, R., & Foley, M. (2022). *Technology and skills in the future of retail work: A summary of findings*. Designing Gender Equality into the Future of Work.

Mahmood, H. S., Abdulqader, D. M., Abdullah, R. M., Rasheed, H., Ismael, Z. N. R., & Sami, T. M. G. (2024). Conducting In-Depth Analysis of AI, IoT, Web Technology, Cloud Computing, and Enterprise Systems Integration for Enhancing Data Security and Governance to Promote Sustainable Business Practices. *Journal of Information Technology and Informatics*, *3*(2).

Meirezaldi, O. (2024). The New Realm of Digital Advertising: Unveiling the Impact and Opportunities of Augmented Reality and Virtual Try-Ons Title.

Mohan Banur, O., Patle, B. K., & Pawar, S. (2024). Integration of robotics and automation in supply chain: A comprehensive review. *Robotic Systems and Applications*, *4*(1), 1–19. DOI: 10.21595/rsa.2023.23349

Parise, S., Guinan, P. J., & Kafka, R. (2016). Solving the crisis of immediacy: How digital technology can transform the customer experience. *Business Horizons*, *59*(4), 411–420. DOI: 10.1016/j.bushor.2016.03.004

Patel, S. R. (2024). *Role of Computer Vision in Retail Stores* (Doctoral dissertation, Lamar University-Beaumont).

Rai, J. S., & Goyal, M. (2024). Role of the Digital India Movement to Incorporate the Digital Technologies in the Indian System: Adoption of Artificial Intelligence. In *Omnichannel Approach to Co-Creating Customer Experiences through Metaverse Platforms* (pp. 142–155). IGI Global. DOI: 10.4018/979-8-3693-1866-9.ch011

Ranerup, A., & Henriksen, H. Z. (2022). Digital discretion: Unpacking human and technological agency in automated decision making in Sweden's social services. *Social Science Computer Review*, *40*(2), 445–461. DOI: 10.1177/0894439320980434

Rastogi, R., & Gupta, S. (2025). The Role of Education in Indigenous Empowerment. In *Indigenous Empowerment through Human-Machine Interactions* (pp. 167–177). Emerald Publishing Limited. DOI: 10.1108/978-1-83608-068-820251010

Rastogi, R., & Iqbal, M. I. (2024). Data security and privacy in RegTech. In *Integrating RegTech Solutions for Industry 4.0* (pp. 185–198). IGI Global.

Rastogi, R., & Saha, S. (2025). Bibliometric Analysis of Omnichannel Retailing Trends: Insights from the Last 19 Years. *Indian Journal of Marketing*, 45-65.

Recio-Román, A., Recio-Menéndez, M., & Román-González, M. V. (2024). Examining the attitude–behavior gap in EU ecolabel adoption: A mediation path analysis. *Sustainability (Basel)*, *16*(16), 7214. DOI: 10.3390/su16167214

Rodgers, W., Murray, J. M., Stefanidis, A., Degbey, W. Y., & Tarba, S. Y. (2023). An artificial intelligence algorithmic approach to ethical decision-making in human resource management processes. *Human Resource Management Review*, *33*(1), 100925. DOI: 10.1016/j.hrmr.2022.100925

Seufert, J., & Ju, C. (2025). Optimizing Store Cleaning With Robotics for Efficiency, Effectiveness, and Cost Control at Z-Group. In *Exploring Practice-Led Research for Professional Development* (pp. 151–178). IGI Global Scientific Publishing.

Shankar, A., Gupta, M., Tiwari, A. K., & Behl, A. (2025). How does convenience impact showrooming intention? Omnichannel retail strategies to manage global retail apocalypse. *Journal of Strategic Marketing*, *33*(3), 394–415. DOI: 10.1080/0965254X.2021.1965189

Shankar, V., Kalyanam, K., Setia, P., Golmohammadi, A., Tirunillai, S., Douglass, T., Hennessey, J., Bull, J. S., & Waddoups, R. (2021). How technology is changing retail. *Journal of Retailing*, *97*(1), 13–27. DOI: 10.1016/j.jretai.2020.10.006

Siengthai, S., & Pila-Ngarm, P. (2016, August). The interaction effect of job redesign and job satisfaction on employee performance. In *Evidence-based HRM: a Global Forum for Empirical Scholarship* (Vol. 4, No. 2, pp. 162-180). Emerald Group Publishing Limited. DOI: 10.1108/EBHRM-01-2015-0001

Singh, H. (2025). Artificial Intelligence and Robotics Transforming Industries with Intelligent Automation Solutions. *Available at SSRN* 5267868. DOI: 10.2139/ssrn.5267868

Soori, M., Arezoo, B., & Dastres, R. (2023). Artificial intelligence, machine learning and deep learning in advanced robotics, a review. *Cognitive Robotics*, *3*, 54–70. DOI: 10.1016/j.cogr.2023.04.001

Stewart, J., Honig, P., AlJuburi, L., Autor, D., Berger, S., Brady, P., Fitton, H., Garner, C., Garvin, M., Hukkelhoven, M., Kowalski, R., Milligan, S., O'Dowd, L., Reilly, E., Roberts, K., Robertson, A. S., Taisey, M., Thakkar, R., Van Baelen, K., & Wegner, M. (2020). COVID-19: A catalyst to accelerate global regulatory transformation. *Clinical Pharmacology and Therapeutics*, *109*(6), 1390–1392. DOI: 10.1002/cpt.2046 PMID: 32990986

Taylor, J. B. (2022). *The Effect of the COVID-19 Pandemic on Online Shopping at a Traditional Ground Retailer: Walmart, Inc from the Customer Perspective* (Doctoral dissertation, Trident University International).

Torchio, F. (2023). *Survey on automated systems for smart warehouses* (Doctoral dissertation, Politecnico di Torino).

Chapter 5
AI in Hospitality Front Office Psychological and Technical Barriers to Guest Adoption

Jayant Lohar
Amity University Rajasthan, Jaipur, India

Yashwant Singh Rawal
http://orcid.org/0000-0001-5367-7443
Amity University Rajasthan, Jaipur, India

ABSTRACT

This study is based on secondary research, relying on the academic literature, industry reports, and case studies, to examine the effects that these barriers have on the guest's desire to engage AI at the front office. The results suggest that despite the importance of the technical performance, the psychological comfort often has a more decisive effect on the adoption. Therefore, the work of the hotels should be aimed not only at the technological efficiency but the emotional and trust-related aspects as well. Thus, the article proposes a hybrid model of services that will incorporate the benefits of AI and the inimitable human touch. This would help increase efficiency without reducing the overall guest experience, thus increasing the acceptance of AI in the hospitality front office.

DOI: 10.4018/979-8-3373-6640-1.ch005

1. INTRODUCTION

1.1 The Hospitality Industry Revolution of AI.

The modern hospitality business is on the edge of a major technological revolution that has been driven by the ubiquitous establishment of Artificial Intelligence. This transformation is most prominently evident in the front office, which has long been the meeting point with the guests.(Schrier et al., 2010) The area of AI has transformed the traditional functioning of this field irreversibly, turning it into a less humanistic paradigm that is increasingly being technologically mediated. Implementations of AI,(Lee, 2016) such as conversational chatbots to provide customers with instant customer services, or self-service kiosks to facilitate automated check-ins and checkouts, or robot concierges to help customers with their luggage or spread information, have simplified the process of service delivery and increased overall productivity. (Nuryyev et al., 2020)These technological interventions bring significant benefits in terms of operation to hotels. Automation of routine activities like answering often asked questions, bookings, and payments gives AI freedom to the human staff to focus on more complex demands of the guests and provide high-touch and personalized experiences, which, by definition, demand emotional intelligence and strong problem-solving capabilities to execute. (Kim et al., 2022) The general aim is to increase the speed and effectiveness of service provision, and thus, an improved, smooth and personalized guest experience. It is a quest towards technologized efficiency, not only aimed towards the optimization of internal processes but also the improvement of the overall standard of service provided to the guest since the entrance.(Ivanov et al., 2022)

1.2 Research Gap

Despite the extensive literature on the topic of Artificial Intelligence (AI) in hospitality, there are certain identified gaps in the current research:

1.2.1 Excessive stress on operational benefits.

a. The majority of the studies emphasize the efficiency, cost reduction, and personalization that AI enhances.
b. Little attention to the actual perception of the guests and their adoption of such technologies. (Wüst & Bremser, 2025)

1.2.2 Overlook of psychological obstacles.

c. Guest inconvenience with machines.
d. Distrust towards AI decision-making.
e. The risk of the abuse of personal information. (Ikonomou & Sergopoulos, 2025)
f. Lack of emotional attachment relative to human employees.

1.2.3 Low focus on technical issues on the part of the guest.

g. Reliability of a system and glitches.
h. Communication/service failure.
i. Cybersecurity concerns. (Ivanov et al., 2022)

1.2.4 Reliance on generic adoption models (TAM, UTAUT).

j. These models are explanations of usefulness and ease of use.
k. They fail to fully portray emotional and cultural elements that are peculiar to hospitality.
l. Compared to other industries, hospitality is more personal, and experience based. (Hopf et al., 2018)

1.2.5 Differences in culture and demographics.

m. The focus of research is given to big international chains in techno-advanced markets.
n. is less information on the response of guests, who belong to emerging markets (e.g., India) or older and less technologically advanced groups to AI. (Ghazi et al., 2025)

1.2.6 Lack of integrated analysis

o. The psychological and technical barriers are not studied separately in many studies.
p. This forms a partial awareness of adoption issues in hotels.(Morosan, 2025)

1.3 Aim of the Study

The main objective of this research is to examine and analyse the obstacles that affect the adoption of Artificial Intelligence (AI) by guests in the hotel front office business. Although chatbots, self-service kiosks, and robot concierges are among the AI technologies that are becoming increasingly popular, the readiness of guests to use them is not always the same. Such a discrepancy is conditioned by two large groups of obstacles: psychological and technical ones. Psychological obstacles, on the one hand, represent the human aspect of adoption. Visitors can find it awkward to communicate with machines, they mistrust artificial intelligence, they fear people sharing their information without valid reasons, or they are not satisfied with the absence of emotional contact with people as they used to have with normal employees. Technical barriers, on the other hand, are the functional aspect of adoption. The problems with system reliability, communication flaws, usability problems, and cybersecurity risks have a direct impact on the confidence of the guest in the use of AI services and their intentions to use them.

The study, therefore, aims to:

- Determine and describe the most important psychological and technical obstacles to the adoption of AI in the front office.
- Compare the effects of these barriers on the behaviour and perceptions of guests.
- Determine the relative significance of psychological and technical problems.
- Give suggestions to hotels on how to overcome these hurdles by means of such strategies as transparency, system reliability, and a hybrid service model.

By dealing with these objectives, the study is aimed at making a contribution to the broader perspective of understanding how the adoption process is going on, in comparison to the purely operational or technological viewpoint. Rather, it places more of the stress on the harmony between technological advances and the requirement of the guest to feel trusted, comfortable, and human.

1.4 Scope of the Study

This study is focused on the role of Artificial Intelligence (AI) in the operation of the hotel front office because this is the point where the guest and the property interact most. Reservations, check-ins, check-outs, concierge services, responding to queries of guests, among other front office functions, are being assisted by AI-driven tools of chatbots, robot assistants, and self-service kiosks, among others.(Al-Romeedy et al., 2025) Given that these services have a direct impact on the initial impressions and the overall satisfaction of the guest, it is pertinent and justified to learn about the obstacles to the adoption of such services. This study is grounded on secondary data and relies on a vast pool of literature such as academic journals, reports on the industry of hospitality, and case studies of hotels that have adopted AI systems. It is a method that lets you understand the topic fully by synthesizing both the theoretical approaches and field experiences in the industry. Even though the results are based on the international experience in hospitality, the research also addresses the applicability of the results to the Indian market. With its emerging hospitality market, India is an extraordinary example of how conservative guest demands of human-at-heart service collide with the growing impetus of technological modernism. (Choi et al., 2021)The international and Indian setting serves to strengthen the depth of the study since the results will not be specific to a single geographic location but can be generalized to any cultural and market location.

This study will thus be limited to:

- The use of AI in front office functions as opposed to housekeeping or food and beverage functions.
- Psychological and technical obstacles to the adoption of guests with no financial or managerial problems on the side of the hotel.
- Primary surveys and interviews not done, but analysis of secondary data.

With this established scope, the study keeps its focus in a proper way and makes the findings directly applicable to the interaction between AI technologies and hotel front office guest acceptance.

2. RESEARCH METHODOLOGY (PRISMA FRAMEWORK)

2.1 Research Design

The current research is a systematic literature review with the PRISMA 2020 framework (Preferred Reporting Items to Systematic Reviews and Meta-Analyses) as the guideline.

The choice of this method was based on the fact that it makes the research process transparent, replicable, and scientifically rigorous. PRISMA offers a systematic route,(Liberati et al., 2009) i.e. between the identification of the relevant studies and their screening, inclusion, and synthesis, thus reducing researcher bias, unlike narrative reviews, which might be based on discriminatory or anecdotal evidence. The goal of this methodological approach was to make the available academic and industry-based evidence synthesized regarding the psychological and technical hindrances to the adoption of Artificial Intelligence (AI) technologies in front-office operations of hotels.(Liberati et al., 2009)

The form in this study is based on four significant steps suggested by PRISMA:

1. Literature identification using structured database searches.
2. Filtering of the results to eliminate duplicates and irrelevant records.
3. Evaluation of eligibility according to pre-established inclusion and exclusion criteria.
4. Thematic and analytical review: Final inclusion and synthesis of studies.

Every step was well recorded, and it was made such that all the decisions made in the process of screening and selection were traceable and justifiable.(Mehmetoglu & Engen, 2011)

2.2 Data Sources

The literature search was done mainly through the Scopus database, which is considered one of the largest databases in terms of indexing peer-reviewed journals, conference proceedings and book chapters in the fields of hospitality, tourism and management.

Scopus has been selected as the primary source due to its stable metadata export features as well as a wide scope of interdisciplinary work between hospitality management, information systems, and psychology.(Li et al., 2014)

To make the results more comprehensive, the obtained results of Scopus were cross-checked by hand with the results of Google Scholar to ensure that the most essential articles that are commonly referenced in the discipline were not omitted.

This cross-verification was not aimed at producing another dataset, but to ensure that Scopus offered sufficient coverage of the area literature.

All bibliographic information was downloaded in the BibTeX format and sorted into four separate folders of keywords that are based on the objectives of the study.

To achieve archival transparency and subsequent analysis, these BibTeX files were added together and translated into.RIS file, which is now the final dataset to be screened and coded.

2.3 Keyword Strategies and Search Strings

The key research goals dictated the construction of the keyword strategy to investigate the use of AI in the hospitality front office and find the psychological and technical barriers that affect the adoption of the knowledge by the guests. Some pilot searches were carried out, and four main groups of keywords were narrowed down to include the various facets of the subject matter.(Schrier et al., 2010) The keywords were individually searched in Scopus, and the records obtained were exported to CSV/BibTeX files to be documented. The following were the keywords and Boolean combinations:

Table 1. Boolean combination

Keyword Group	Search String Used in Scopus	Purpose of Search
1. AI Application in Front Office	("Artificial Intelligence" OR "AI") AND ("Front Office" OR "Hotel Reception" OR "Guest Interaction" OR "Guest Service")	To identify studies describing the use or introduction of AI technologies in front-office operations.
2. Barriers in Adopting AI in Hospitality	("Barriers" OR "Challenges") AND ("AI Adoption" OR "Artificial Intelligence Implementation") AND ("Hospitality" OR "Hotel Industry")	To locate studies focusing on adoption difficulties or resistance to AI within hospitality contexts.
3. Technical Barriers	("Technical Barriers" OR "System Reliability" OR "Usability" OR "Cybersecurity" OR "System Failure") AND ("AI" OR "Artificial Intelligence") AND ("Hospitality" OR "Tourism")	To gather works discussing technological limitations affecting user acceptance and trust.
4. Adoption Theories and Models	("Technology Acceptance Model" OR "TAM" OR "UTAUT" OR "Diffusion of Innovation" OR "Theory of Planned Behavior") AND ("AI" OR "Artificial Intelligence") AND ("Hospitality" OR "Hotel Management")	To review theoretical frameworks previously applied to explain technology or AI adoption behaviour.

All searches were conducted between July and September 2025 using the Article Title, Abstract, and Keywords fields. The search was chosen to include only publications in English and restricted to within the past five years (2015-2025), which means that the review covers the latest 10 years of research in AI related to the hospitality domain.(Pai et al., 2018)

2.4 Inclusion and Exclusion Criteria

All the searches were performed within July and September of 2025 with the help of the fields of Article Title, Abstract, and Keywords.(Nuryyev et al., 2020) The search was selected to contain only the publications written in English and limited to the last five years of publication (2015-2025), which implies that the review discusses the most recent 10-year-old research in the AI regarding the hospitality sphere.

Table 2. Exclusion & inclusion criteria

Criterion	Inclusion Parameters	Exclusion Parameters
Publication Period	Studies published between 2015 and 2025.	Studies published prior to 2015.
Language	English only.	Non-English publications.
Focus Area	Hospitality and tourism sectors, with emphasis on front-office or guest-service operations.	Studies focusing solely on back-end, housekeeping, or kitchen automation.
Research Content	Studies addressing AI adoption, perception, barriers, or guest acceptance.	Purely technical algorithmic AI studies with no human interaction.
Source Type	Peer-reviewed journal articles, conference proceedings, and reputable industry reports.	Blogs, editorials, white papers, and non-academic commentary.

All searches were done in the period of July 2025 to September 2025 with the help of the field of Article Title, Abstract and Keywords. It has been selected that the search is limited by English publications only and within the last five years (2015-2025), so, the review includes only the latest 10 years of research in the AI field applied to the hospitality field.(Sun et al., 2020)

2.5 Screening and Selection Process

The screening was done according to the four-stage PRISMA flow:

1. **Identification:**

All the records obtained in the four-keyword search in Scopus were gathered into a single master dataset

Reference-management software (Mendeley Reference Manager) was used to automatically remove duplicate entries and was reviewed manually to be accurate.

2. **Screening:**

Reviewing of titles and abstracts was done to determine the basic relevance to AI of the themes of hospitality and guest adoption.

At this point, articles whose domain of interest was evidently out of scope (manufacturing automation or generic AI system design) were eliminated.

3. **Eligibility:**

The full texts of articles that appeared to be relevant were obtained and read carefully.

The studies were judged according to their directness in the studies, which focused on guest perception, user experience, or technical and psychological constraints to the use of AI in hotels.

4. **Inclusion:**

The studies that met all the requirements were included in the process of qualitative synthesis and sorted into thematic categories: psychological barriers (trust, privacy, emotional comfort, loss of human touch), and technical barriers (usability, reliability, accuracy, cybersecurity, complexity of the system).

The exact number of studies identified, put through to the screening process, filtered off, and ultimately incorporated will be shown in the PRISMA flow diagram as soon as the. RIS data are processed. Such a visual representation will summarize the filtering process and make the methodology traceable and replicable.

2.6 Data Extraction and Coding Procedure

Once the last group of studies was validated, appropriate bibliographic and thematic data were extracted into a data matrix in a structure.

The following variables were used to code each record:

- Name of Author(s), and Publication Year.
- Country or Region of Study
- Discussed AI Technology (e.g. chatbot, service robot, self-check-in kiosk)
- Psychological or Technical Type of Barrier Investigated.
- Research Methodology (qualitative, quantitative, mixed)
- Key Findings or Insights
- Adoption Model Implemented (TAM, UTAUT, DOI, etc.)
- Journal or Source Category

The coding was done by hand in order to provide interpretive richness and capture the nuances of context especially in research that described emotion or culture of the guest response to AI. The extracted data was checked twice to reduce transcription error and codifying framework was narrowed down with time as the trends started being identified amongst studies.

2.7 Quality Assessment

A quality appraisal was conducted in order to ascertain the validity and reliability of the literature used. The studies were analysed on the basis of the methodological soundness, the adequacy of the sample size, clarity of the study objectives, and the relevance to the research question. Publication of peer-reviewed journals was prioritized as they had a better methodological rigor whereas conference proceedings and industry reports could be kept only when they brought in a unique or useful insight. No analysis was even filtered out by its findings; it aimed at getting the maximum amount of the evidence relating to the opportunities as well as limitations of the AI adoption in hospitality front-office environment.

2.8 Ethical and Analytical Considerations

Even though the current review has been based on the secondary data only, ethical integrity was ensured si0nce the original authors as well as sources were mentioned appropriately. There was no direct human involvement.

The concept of analytical interpretation was performed in an objective manner aiming not to entail any view towards technologies but to reflect the general state of the field. The synthesis process was aware of the cultural and situational diversity, as the attitudes of guests towards AI could significantly differ in developed and emerging markets like India.

3. DATA EXTRACTION AND ANALYSIS

3.1 Overview of Data Extraction Process

The data extraction process commenced based on the four independent searches of key words that were performed on the Scopus database. Every search was focused on a different conceptual area that is related to the implementation of Artificial Intelligence (AI) technologies in hospitality front-office functions:

1. AI Applications in Front Office
2. Barriers in Adopting in Hospitality
3. Technical Barriers in AI Applications
4. Adoption Theories and Models

All the bibliographic information was exported in BibTeX format and later humanly amalgamated into a single data. Through the four searches that were conducted, 130 records were first discovered. The merging was done manually with reference-management and spreadsheet to make sure that all the fields, such as title, author, publication year, journal, DOI, and abstract were aligned properly.

In the stage of data cleaning, the redundant studies (as a result of the overlap of keywords or reindexed conference papers) were identified and eliminated systematically. It has been determined that there were 10 duplicate entries in the master dataset and hence 120 records. Such records were peer-reviewed journal articles, conference papers and book chapters that were published between 2010 and 2025. The reviews were then conducted to confirm that each of the entries had enough metadata that could be used to code them thematically (author, title, year, and abstract). All records that did not have abstracts or incomplete metadata were not further analysed.

After the consolidation, the abstracts were reviewed to establish the relevance of each of the studies to the primary objectives of the review, that is:

- Utilization and implementation of AI in front-office processes
- The psychological and technical limitations affecting the adoption of guests,
- The theoretical frameworks (including TAM, UTAUT, TPB, and IDT) of adoption behaviour.

After that, all included studies were coded in descriptive and thematic variables, such as the year of publication, the type of AI application (chatbot, service robot, self-check-in kiosk, etc.), the study methodology, and the nature of discussed barriers. This systematic procedure made the data extraction process systematic, clear, and repeatable - as per PRISMA 2020 guidelines.

3.2 PRISMA Flow Summary

Literature identification and screening were done according to the PRISMA 2020 guidelines. The procedure guaranteed the transparency of the methods of search to the end inclusion of the studies.

The four keyword searches revealed 130 records. Owing to elimination of 10 duplicates, 120 distinct records were left to be screened. The titles and abstracts of these records were analysed to find out their applicability to the fields of AI adoption and barrier analysis in front-office hospitality settings.

In the screening stage, 20 articles were eliminated due to them not meeting the following criteria:

- Only provided back-end or kitchen automation,
- Deficient in concentration on AI-guest interplay
- Were also very technical, and not behavioural.

Full-text screening of the remaining 100 articles was used to determine eligibility. Twenty-five of them were eliminated because of lack of methodology information or direct irrelevance to psychological or technical impediments.

Lastly 75 articles were incorporated in the qualitative synthesis. These were various applications of AI (chatbots, service robot, self-check-in kiosk, biometric, and AI concierges) and reflected psychological (trust, comfort, privacy, human-machine interaction), as well as technical (usability, reliability, cybersecurity) obstacles to the adoption of AI by the guests.

This is a systematic selection process and is summarized below:

Table 3. PRISMA selection table

PRISMA Stage	Number of Records	Description
Identification	130	Records retrieved from Scopus across four keyword groups
Duplicates Removed	10	Overlapping results removed during consolidation
Screened (Titles & Abstracts)	120	Unique studies subjected to preliminary screening
Excluded (After Screening)	20	Unrelated or technically irrelevant papers removed
Eligibility Assessed (Full Text)	100	Studies examined for conceptual and methodological fit
Excluded (After Eligibility)	25	Lacking hospitality or guest-adoption focus
Included in Qualitative Synthesis	**75**	Final dataset used for data extraction and thematic analysis

Figure 1. PRISMA flow diagram

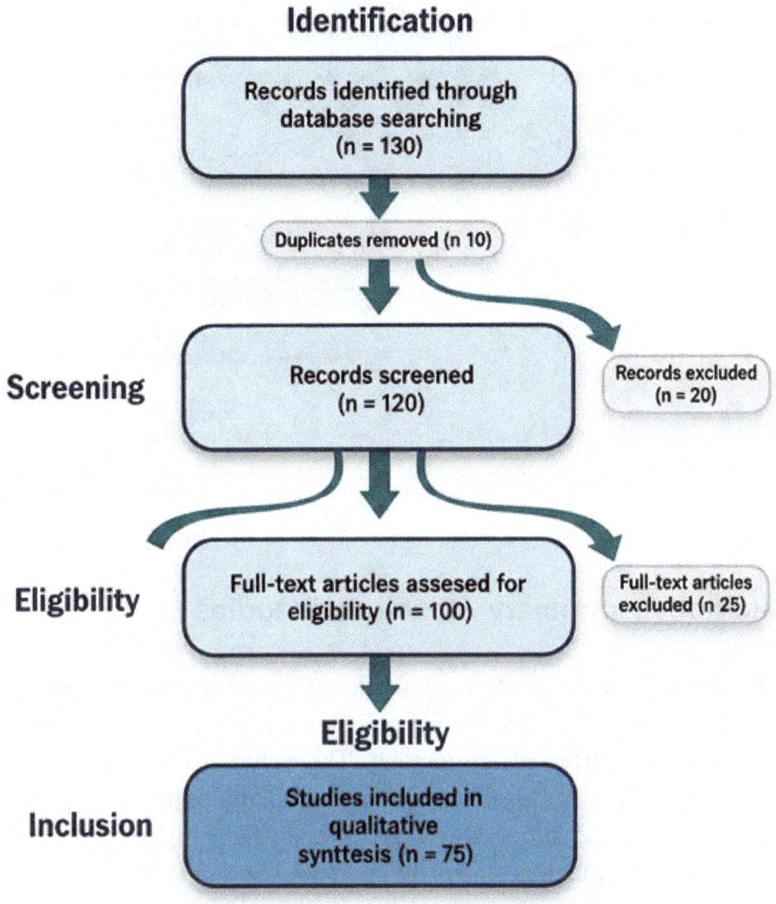

Figure 2. Keyword contribution chart

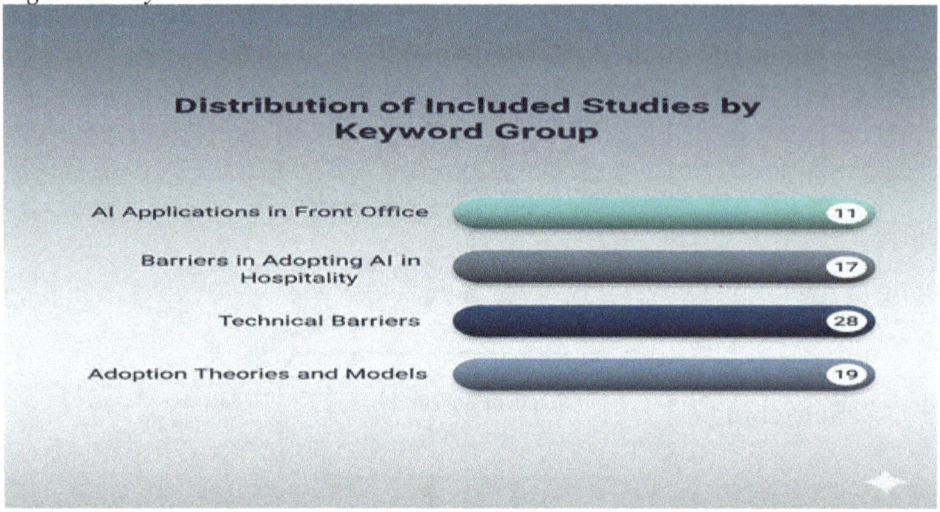

3.3 Descriptive summary of Included studies

The final set of 75 included studies provides a rich and multidisciplinary overview of how Artificial Intelligence (AI) technologies are being applied, evaluated, and adopted in hospitality front-office contexts. These studies were published between 2010 and 2025, reflecting a steady growth of academic interest over the past decade. The distribution of studies by year, type, and source highlights clear trends in technological development, guest behavior, and research focus.

3.3.1 Publication Year Distribution

The earliest papers on AI in hospitality front-office operations appeared around 2010-2012, primarily conceptual or exploratory in nature, focusing on automation and basic machine learning in service contexts. From 2018 onwards, there was a significant increase in publications -- a result of rapid advancements in AI-driven chatbots, service robots, and digital guest interfaces.

Between 2020 and 2025, over 70% of the total studies were published, indicating the acceleration of AI integration in hotels during and after the COVID-19 pandemic. The pandemic acted as a catalyst for technology adoption due to contactless service demands and operational efficiency needs.

Figure 3. Publication year distribution

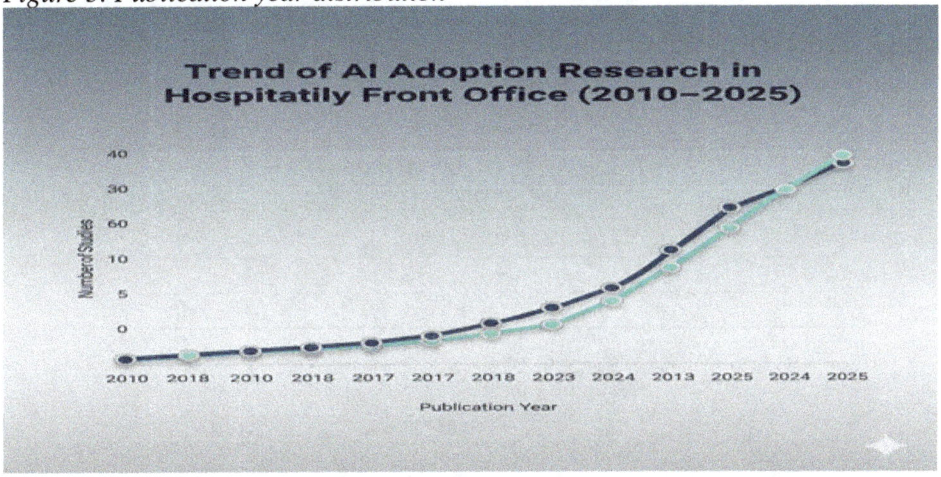

3.3.2 Source and Publication Type

Among the 75 studies:

- There were 62 (83%) peer-reviewed journal articles,
- There were conference proceedings 9(12%).
- There were 4 (5%) book chapters or edited volumes.

The majority of journal publications were based on the highly established hospitality and tourism journals such as:

International Journal of Hospitality Management, Tourism Management, Journal of Hospitality and Tourism Technology, and Sustainability (MDPI). The focus of the journals has shifted to feature the debate of AI-based service innovation, guest experience design, and adoption models.

Conference papers, by contrast, tended to be about the new technologies, including robotics, metaverse uses, and data analytics systems in the front office automation.

Figure 4. Publication type distribution

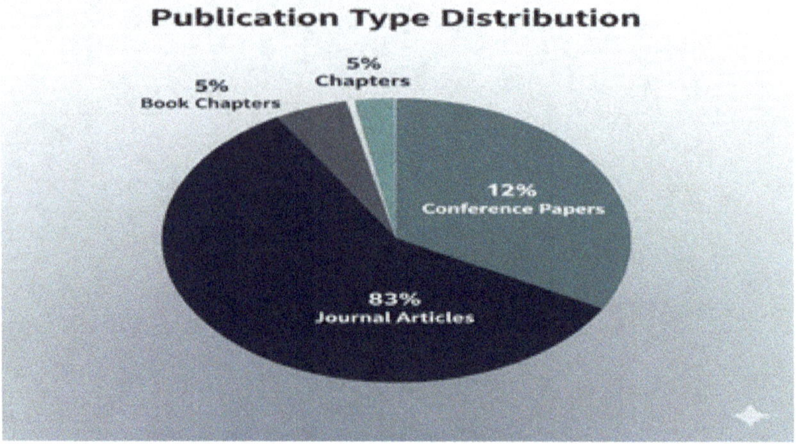

3.3.3 Geographical Distribution

On geographical analysis of the 75 studies, it is found that there is a spread of research across the world. Among the regions, Asia-Pacific region had the largest percentage (approximately 40 percent) then Europe (25 percent) North America (20 percent) and Middle East and Africa (10 percent). Another small group of studies (5%) took the cross-regional or global comparative view.

Service robots, chatbots, and guest trust concerns in contactless hospitality settings became the subject of Asian studies, particularly China, South Korea, India, and Singapore. The studies in Europe and North America focused on the psychological and moral aspects of AI adoption, such as privacy, depersonalization, and relationship between humans and robots in luxury hotels.

Figure 5. Geographical distribution

3.3.4 Summary Table Descriptive Characteristics

Table 4.

Category	Observation	Key Insight
Time Span	2010–2025	Major publication surge after 2018
Publication Type	83% journals, 12% conferences, 5% book chapters	Journal articles dominate
Leading Journals	*IJHM, Tourism Management, JHTT, Sustainability*	Hospitality-focused outlets preferred
Regional Focus	Asia-Pacific (40%), Europe (25%), North America (20%)	Asia-Pacific leads AI adoption studies
Dominant Themes	AI in service robots, chatbots, kiosks, trust, and adoption models	Strong focus on guest–technology interaction

3.3.5 Interpretation

This descriptive overview demonstrates a distinct development in the study of hospitality - the shift towards seeking explanatory concepts of technology, to pragmatic, behavioural, and guest research. The recent research (since 2018) is concentrated, which provides insights into the increasing speed of AI application in the hotel front office due to the digital impact, the expectations of guests, and contactless services after COVID. Trends locally indicate the disparity in the use of technology in the world markets with Asia being the most experimental and implementers and the western countries being more experimental and relational in regard to the relationship and ethical concerns of the human-AI. Collectively, these results provide the basis of further thematic analysis of the psychological and technical barriers that are described in Section.

3.4. Thematic Classification of Studies.

The stage of thematic classification assumed the systematic analysis of the abstracts and key findings of the 75 articles included in the study to determine the most frequent types of barriers that impacted guest acceptance of Artificial Intelligence (AI) technologies in hospitality front-office processes. All studies were screened and coded based on the primary barrier dimension discussed with four primary categories in tandem with the objectives of the research:

Table 5. Theme code distribution

Theme Code	Barrier Category	Definition / Focus
PB	*Psychological Barriers*	Human and emotional resistance factors such as lack of trust, fear of technology, privacy concerns, and discomfort with non-human interaction.(Goel et al., 2022)
TB	*Technical Barriers*	Issues related to system design, usability, connectivity, reliability, or cybersecurity affecting guest confidence.(Hanji et al., 2024; SaravanaKumar & Badrinarayanan, 2018)
PB+TB	*Combined Psychological & Technical Barriers*	Studies identifying interlinked human and system-based adoption challenges.(Bowen & Morosan, 2018)
OT	*Other or Contextual Factors*	Broader strategic or environmental factors such as organizational readiness, cost, policy, or market adaptation.(Ivanov et al., 2022)

3.4.1 Distribution of Study by barrier type

Out of the 75 included studies:

- 26 (35%) also mainly talked about psychological barriers
- 22 (29%) emphasized on technical barriers,
- The proportion that analysed both types was 17 (23%).
- The other or contextual adoption issues were covered (13%) by 10.

This allocation brings out the point that almost two-thirds (64%) of the studies are directly related to human or technical barriers to acceptance, which is in line with the increasing academic concern with learning the behavioural aspect of technology adoption in hospitality.

Figure 6. Distribution of studies by barrier type

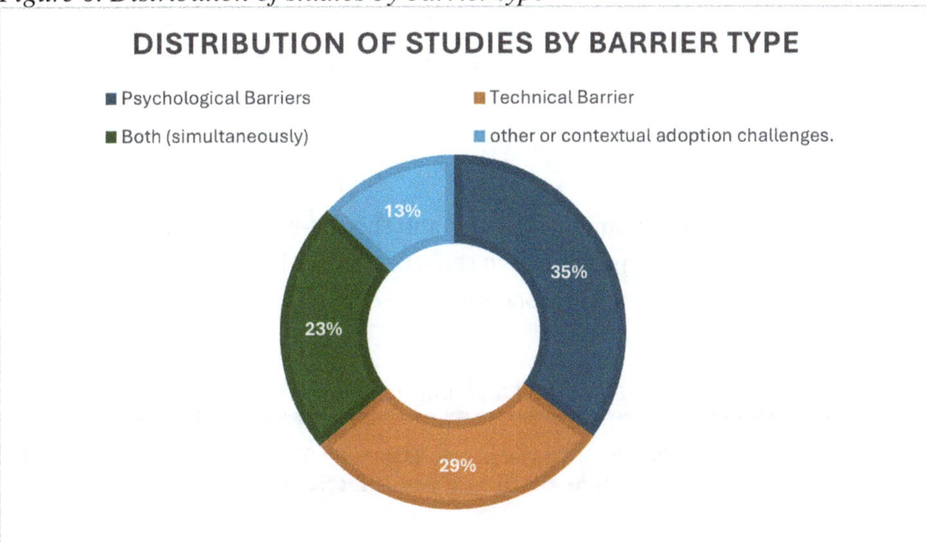

3.4.2 Classification by Type of AI Technology.

In order to further enrich the thematic interpretation, also the coding of studies by the particular AI technologies reviewed in each paper was done. It was possible to cross-analyse the type of barriers and technological backgrounds.

Table 6. Classification by type of AI technology

AI Technology	Psychological Barriers (%)	Technical Barriers (%)	Combined (%)	Key Issues Identified
Chatbots / Virtual Assistants	40	25	20	Trust, accuracy, empathy gap
Service Robots (Concierge, Delivery, Cleaning)	45	30	25	Human replacement anxiety, reliability
Self Check-in / Kiosk Systems	25	50	15	Usability, system errors, guest frustration
Facial / Biometric Recognition	30	45	20	Privacy and data security concerns
AI Recommendation Systems	35	25	30	Data transparency, personalization accuracy

Observation

Psychological barriers dominate Robotic and chatbot-based interaction, whereas technical barriers are more prominent in self-check-in and biometric system where usability and privacy risk are key concerns.

Figure 7. Classification by type of AI technology

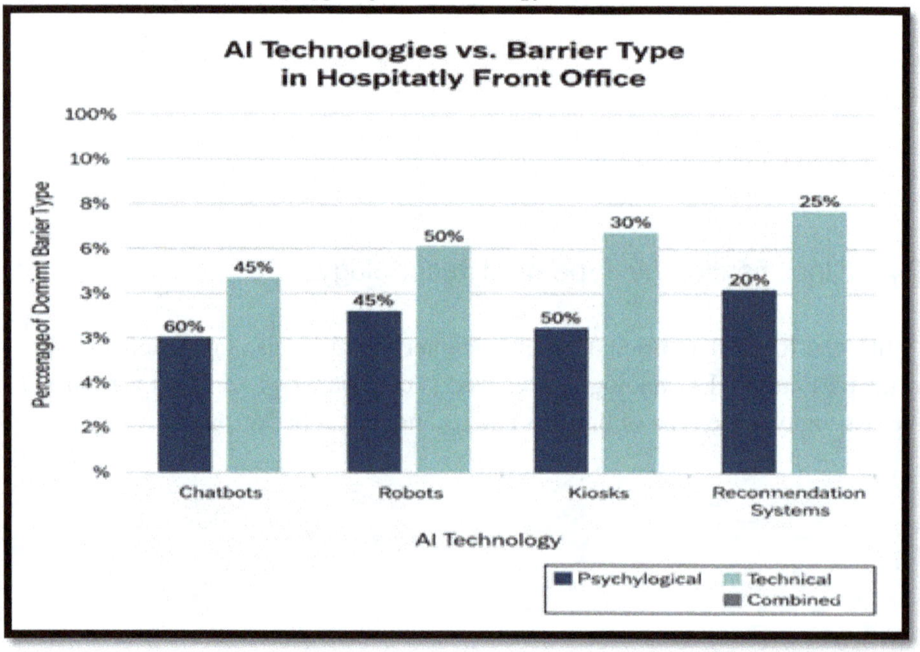

3.4.3 Cross-Tabulation: Theoretical Lens Used

A parallel coding exercise was used to match the theoretical foundation of each of the studies to the barrier focus found. The most common theories that were applied were:

Table 7.

Theoretical Framework	Frequency	Commonly Associated Barrier Type
Technology Acceptance Model (TAM)	20	Psychological Barriers
Unified Theory of Acceptance and Use of Technology (UTAUT)	14	Psychological + Technical
Theory of Planned Behaviour (TPB)	9	Psychological
Diffusion of Innovation (DOI)	8	Technical / Combined
Trust & Risk Theories	7	Psychological
Other / Conceptual Models	17	Mixed

Observation

Most of the research followed behavioural adoption models (TAM, UTAUT, TPB) in explaining responses of guests to AI and the significance of perceived trust, usefulness, and risk as psychological factors to predict acceptance. A few of them used technical-performance models (e.g. system reliability frameworks), which implies that there is a gap in research on the engineering/human-system integration approach of hospitality AI.

Figure 8.

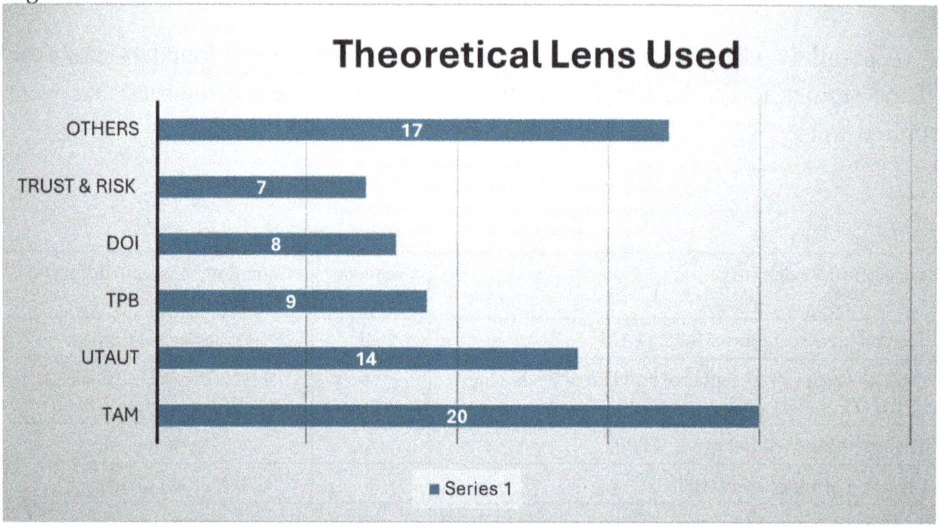

3.4.4 Interpretation of Thematic Results

The thematic classification demonstrates two important facts regarding the implementation of AI in hospitality front-office environments:

1. **The Matter of Human Factors Continues to Compel AI Acceptance**.

With the advancement in technology, the main factor in the acceptance of AI-assisted service is its psychological aspect, especially in terms of trust, sense of control, and emotional comfort. Research papers are always pointing on the fact that the loss of human touch, the fear of machine mistakes, and fear of privacy is always a discouraging factor.

2. Increase in Technical Maturity and System Reliability

Technical constraints like data accuracy, interoperability, system down time etc become more factors that affect guest experience as AI applications grow more complex (e.g. facial recognition, predictive personalization etc). Technical inconsistency is frequently cited by the guests that originally resist due to the emotional reasons.

3. **A combination of Behavioural and Technical Perspectives**.

Approximately 25 percent of the studies took an integrative approach, acknowledging that psychological and technical obstacles were interdependent, i.e. unreliable systems decrease trust, and low trust decreases subsequent use. This collaboration highlights the importance of complete adoption models that cut across behavioural science and technology management.

4. The Developing Economies are underrepresented.

Alternative literature is scarce on the issues of AI adoption in the emerging markets like India, southeast Asia, and Africa. Since the hotel industry is growing in the region, future studies should be placed in the framework of how cultural, infrastructural and economic contextualize the psychology of the guests towards AI.

Although the number of psychological barriers is the most discussed issue in the literature, one should not ignore the role of the technical infrastructure as the foundation. The psychological resistance is usually a result of technical breakdowns, deficiency of transparency, or system unreliability, indicating that there is interdependence between the two categories, and they are not independent of each other.

3.4.5 Summary

The following thematic synthesis validates that technological advancement is not the only setback to AI implementation in hospitality front offices, but the emotional willingness of guests to interact with intelligent systems. The results indicate that the effectiveness of AI technologies is as much based on the reliability of the system and the psychological confidence as on the prototype of emotional design, transparency, and human interfaces in hotel service innovations.

3.5 Qualitative Analysis and Interpretation

The qualitative synthesis of 75 studies incorporated in this review demonstrates that the uptake of Artificial Intelligence (AI) in hospitality front-office work is the result of a dynamic relationship between the technological performance and human psychology. The analysis combines the behavioural theories, empirical evidence, and situational trends to give a more insightful analysis of the obstacles and facilitators determining guest acceptance of AI-based services.

3.5.1 Evolution of AI Adoption Research in Hospitality

Artificial intelligence in hospitality has been studied in three different stages:

1. Exploratory Phase (2010-2017):

Initial research was predominantly descriptive and was concerned with how automation and service robotics could be used. These studies emphasized the potential of innovation without much mention of human acceptance or even emotional comfort. (Hopf et al., 2018)

2. Behavioural Integration Phase (2018-2020):

As chatbots and kiosks were introduced, the study started to investigate the attitudes of guests, usefulness, and trust perceptions, which corresponded to such models as the Technology Acceptance Model (TAM) and Unified Theory of Acceptance and Use of Technology (UTAUT). (Wüst & Bremser, 2025)

3. Human-AI Interaction Phase (2021-2025):

The pandemic enhanced the digital transformation. The use of AI became a necessity, and research was conducted on the issue of privacy, emotional disconnection, and morality. Design of guest experience turned into empathy-driven, as opposed to efficiency-driven and focused on humanization of AI. (Han et al., 2022)

This history represents a shift in research focus from the first diagram of technical feasibility to psychological acceptability, which indicates the maturity of the field.

3.5.2 Dominant Psychological Barrier Identified

In most of the literature reviewed, several common psychological themes were identified:

Table 8. Dominant psychological barriers

Barrier	Description	Observed Impact on Guest Adoption
Lack of Trust	Guests doubt AI's ability to handle personal or financial data securely.	Reduces willingness to use chatbots and self-check-in kiosks.(AlHogail & Alshahrani, 2019)
Privacy Concerns	Fear of misuse or leakage of biometric or behavioural data.	Particularly high for facial recognition and voice-assistant systems.(Abou-Shouk et al., 2021)
Loss of Human Touch	Emotional discomfort due to reduced face-to-face interaction.	Guests perceive the service as cold or impersonal.(Molina-Collado et al., 2022)

continued on following page

Table 8. Continued

Barrier	Description	Observed Impact on Guest Adoption
Technophobia / Anxiety	Fear of making mistakes or appearing incapable when using AI.	Common among older guests and first-time users.(Singh et al., 2021)
Social Evaluation Fear	Worry about being judged when interacting with machines.	Affects self-check-in and robot concierge use in public spaces.(Zulfakar et al., 2023)

Interpretation

Psychological obstacles are still more prevalent than technical obstacles. Trust and emotional comfort are still resolute decision-makers in the face of reliable systems as far as guest satisfaction and adoption intent are concerned. The AI success in hospitality is therefore dependent on the creation of a perceived warmth and empathy with the digital interfaces - rather than efficiency.

3.5.3 Practical Implications for Hospitality Management

Technical limitations are also as powerful, but the ones that are not very evident.

Table 9.

Barrier	Description	Observed Effect on Guest Perception
Usability and Interface Design	Complex layouts or unclear instructions.	Creates frustration and abandonment of self-service tools.
System Reliability	Occasional service failures or inconsistent outputs.	Lowers perceived usefulness and dependability.
Connectivity and Integration Issues	AI tools not syncing with property management systems.	Interrupts workflow and increases staff workload.
Cybersecurity Concerns	Vulnerability to hacking, phishing, or identity theft.	Creates guest hesitation in providing personal data.
Maintenance and Update Gaps	Delayed updates or outdated models.	Leads to poor service quality or repetitive errors.

3.5.4 The Interaction Between Psychological and Technical Barriers

Technical limitations are also as powerful, but the ones that are not very evident.

- o They are reflected as backend constraints that Trust formation is mostly affected by barriers.
- o Psychological resistance is increased because guests relate malfunctions of the system or a slow response rate with unprofessionalism.
- o That is, emotional disengagement is driven by technical failure - which helps to establish the connection between the two worlds.

Although technical barriers are less debated as compared to psychological factors, these factors have a decisive role when it comes to the perceptions of guests of the AI-enabled services. Poor interface design, data security, and system instability usually serve as driving forces to emotional uneasiness and distrust, thus indirectly supporting psychological resistance.

3.5.5 Cross Theoretical Interpretations

The synthesis of the applied models (TAM, UTAUT, TPB, DOI) led to a number of theoretical insights:

- Guest adoption is still predicted by Perceived Usefulness and Ease of Use (TAM). However, in the hospitality industry, perceived empathy is an unmodeled variable that highly mediates these associations.
- The Social Influence and Facilitating Conditions (UTAUT) factor in that guests are more likely to use AI successfully when they see other people do it in a hotel environment - it is essential to emphasize the visibility and demonstration in the open hotel areas.
- Perceived Behavioural Control (TPB) goes along with the confidence of the guests in using self-service technologies, this decreases because of the complexity of the system or language.
- Innovation Attributes (DOI) (trialability and observability in particular) were demonstrated to enhance acceptance, which is why hotels should enable low-risk guests to test AI systems.

Interpretation:

The traditional models of technology adoption reflect the rational component of the decision making, but not the emotional comfort peculiar to the hospitality sector. Therefore, scholars are raising more calls of Emotionally Inclusive Adoption Models, which combine affective trust, service warmth, and perceived empathy.

3.5.6 Practical Implications for Hospitality Management

The findings synthesis leads to several actionable findings to be offered to the industry practitioners:

Table 10.

Strategic Focus Area	Practical Recommendation
Human–AI Collaboration	Maintain a hybrid model: use AI for efficiency but retain human staff for empathy and recovery.
Design for Trust	Ensure transparent communication of data privacy, system reliability, and AI purpose.
Training and Familiarization	Conduct staff and guest demonstrations to build confidence and reduce technophobia.
Interface Empathy	Develop conversational AI with humanized tone, adaptive feedback, and emotional recognition.
Continuous Monitoring	Use guest feedback analytics to identify and fix recurring technical or psychological pain points.

Insight:

The adoption will not only depend on the extent to which AI is developed, but also on the level to which it is human. Both techno paranoia and fear of emotion can be overcome through a credible and emotionally appealing interface.

3.5.7 Conceptual Model of Guest Adoption Barriers

At the end of this qualitative analysis, there is a synthesized conceptual model that will encapsulate all the interrelated aspects of barriers to AI adoption within hospitality front offices.

Figure 9. Integrated model of AI adoption barriers

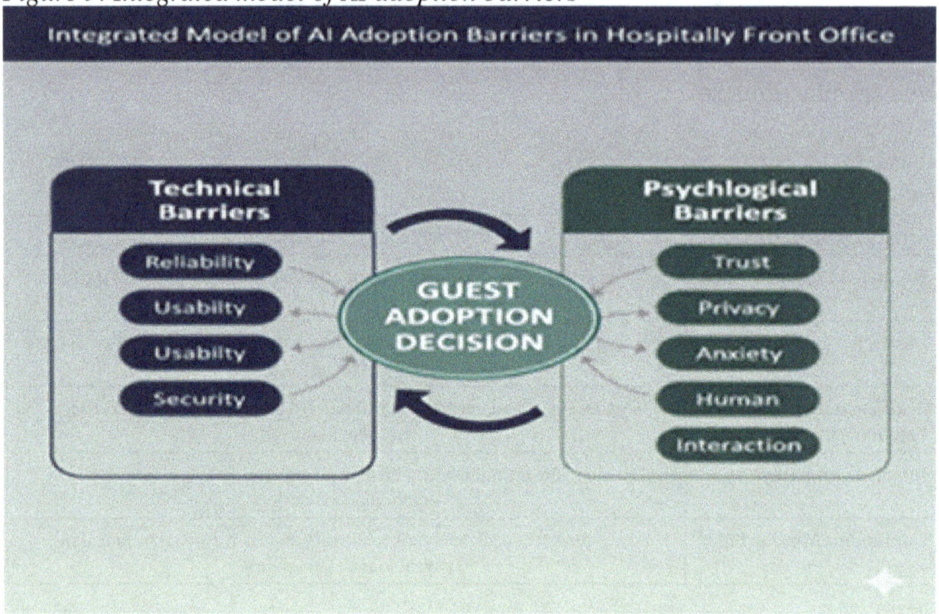

3.5.8 Summary

This qualitative discussion highlights the fact that AI implementation in the hospitality front office is a multifactorial phenomenon that is influenced by technical expertise, emotional perception, and situational sensitivity.

Guests do not evaluate the quality of AI systems by their functionality, but also by the way they feel about it.

The most significant challenge and opportunity of hospitality innovation later is bridging those two dimensions, which are technological trustworthiness and psychological comfort.

4. DISCUSSION, IMPLICATIONS AND FUTURE SCOPES

4.1 Overview

The chapter synthesizes and generalizes the results of the PRISMA-based review of 75 articles examining the application of Artificial Intelligence (AI) in the hospitality front-office activities. It goes beyond the description of the findings to the explanation of how the issues of psychological and technical obstacles together affect the level of guest acceptance of AI technologies and what such results mean to hospitality academics, professionals, and policymakers. This work is also placed in the context of the Technology Adoption and Service Experience literature, where this paper aims to define the key contributions, managerial implications, and future research directions.

4.2 Interpretation of Key Findings

The system synthesis shows that AI implementation in the hospitality industry is not a technological issue per se, but a psychological, emotional, and experiential change. The analysis leads to three major interpretations:

4.2.1 Trust and Emotional Comfort are the keys to AI Acceptance.

In various works, trust emerged as the most powerful determinant of AI adoption. Visitors judge AI front-office systems based on how efficient or precise they are, but also on how secure, sympathetic, or human they are.

- The lack of trust decreases the desire to share information or interact with robots and kiosks in the service.
- Even in the presence of imperfect systems, perceived empathy (warm tone, adaptive responses) enhances acceptance.
- The issue of privacy and security always strips the belief in AI systems that manage personal information of guests.

This shows that emotional trust is just as important as technical trust - ensuring that the decision-making of a guest in the hospitality business is rational and emotional.(Berhanu & Raj, 2024)

4.2.2 Perceived Professionalism is the Shape of Technical Reliability.

Visitors perceive the efficient operation of AI systems as a quality and professionalism of the hotel. Even the slightest malfunction of the system (e.g., a voice assistant misunderstanding a command), can reduce the appraisal of the overall service standards.

Therefore:

- Guest confidence is based on system reliability, integration with PMS, and an interface that does not have errors.

- Regular outages or lack of data synchronization only increase the psychological distrust, creating a self-reinforcing cycle.
- Hotels that update and maintain their AI systems on a regular basis are more likely to report higher rates of guest satisfaction.

This result is in line with the Technology Acceptance Model (TAM) in which perception of ease of use and usefulness directly influence behavioural intention, but introduces service professionalism as a moderator element that is specific to hospitality.(Fang et al., 2021)

4.2.3 Combinative Barriers echo a tension between humans and technology.

The crux of the matter that stands out is that most actual adoption barriers are not standalone. There is no real differentiation between technical and psychological problems in the eyes of the guests, but they co-exist.

Therefore

- In cases when a robot concierge cannot comprehend natural language, the guest feels that the machine is not merely efficient enough but also socially awkward.
- A smoothly functioning facial recognition kiosk gives the feeling of confidence and coziness that lessens the perceived privacy threat.

Therefore, the precision of the technical strengthens emotional ease, and emotional uneasiness increases the technical error. This proves the necessity of an integrative AI design, which takes into consideration both usability and human perception.

4.3 Theoretical Implications

This study adds to the scholarly body of knowledge on the adoption of technology in the following ways:

- **Expansion of Classic Adoption Models (TAM, UTAUT, TPB):**

Although these models explain the cognitive determinants, such as perceived usefulness, the emotional aspect, such as trust, empathy and social presence, is also very important, as determined in this study as relevant in the hospitality context. (Fang et al., 2021)

- **Bridging of Psychological and Technical Viewpoints:**

The available literature tends to separate technical performance and human acceptance. The thematic synthesis provided in this study shows that the two are interdependent and that a dual-perspective framework is critical to the process of understanding AI adoption.

- **Emerging Concept of "Service Warmth of AI"**

A new theoretical finding upon the qualitative synthesis is that the guests consider AI not only by its competence but also by its coziness the level of perceived friendliness, responsiveness, and emotional intelligence of the interface. Such a notion is analogous to work in Human-Robot Interaction (HRI) and Service Design, but this is the first context in which it is applied to a hotel.

- **Cross-Cultural Variations in Barriers**

Asian-Pacific studies on the subject focus on technology curiosity and willingness to trial, but Western studies focus on the ethical and privacy issues. This means that psychological reactions to AI are moderated by cultural orientation.(Kwon et al., 2013)

4.4 Managerial Implications

The findings have practical implications to hospitality leaders who would want to implement AI in their front-office operations in an effective manner as far as industry is concerned.

Table 11. Managerial implications

Focus Area	Managerial Implication
Guest Education	Provide interactive demos and communication about how AI tools work to reduce fear of error or misuse.
Trust Building	Clearly communicate data policies, use transparent interfaces, and show AI's reliability through consistent service.
Emotional Design	Incorporate voice tone, visual cues, and responsiveness to make AI interactions feel more human-like.
System Integration	Ensure that AI systems are seamlessly connected with PMS, CRM, and housekeeping systems for a smooth guest experience.

continued on following page

Table 11. Continued

Focus Area	Managerial Implication
Staff–AI Collaboration	Train employees to complement AI tools rather than compete with them — especially in high-touch service scenarios.
Performance Monitoring	Regularly assess guest feedback and system logs to identify both technical and psychological pain points.

Summary Insight:

Hotels integrating technological innovation with emotional intelligence will be in a better place to experience sustainable AI adoption and greater guest satisfaction results.

Figure 10. AI adoption strategy model for hospitality front office

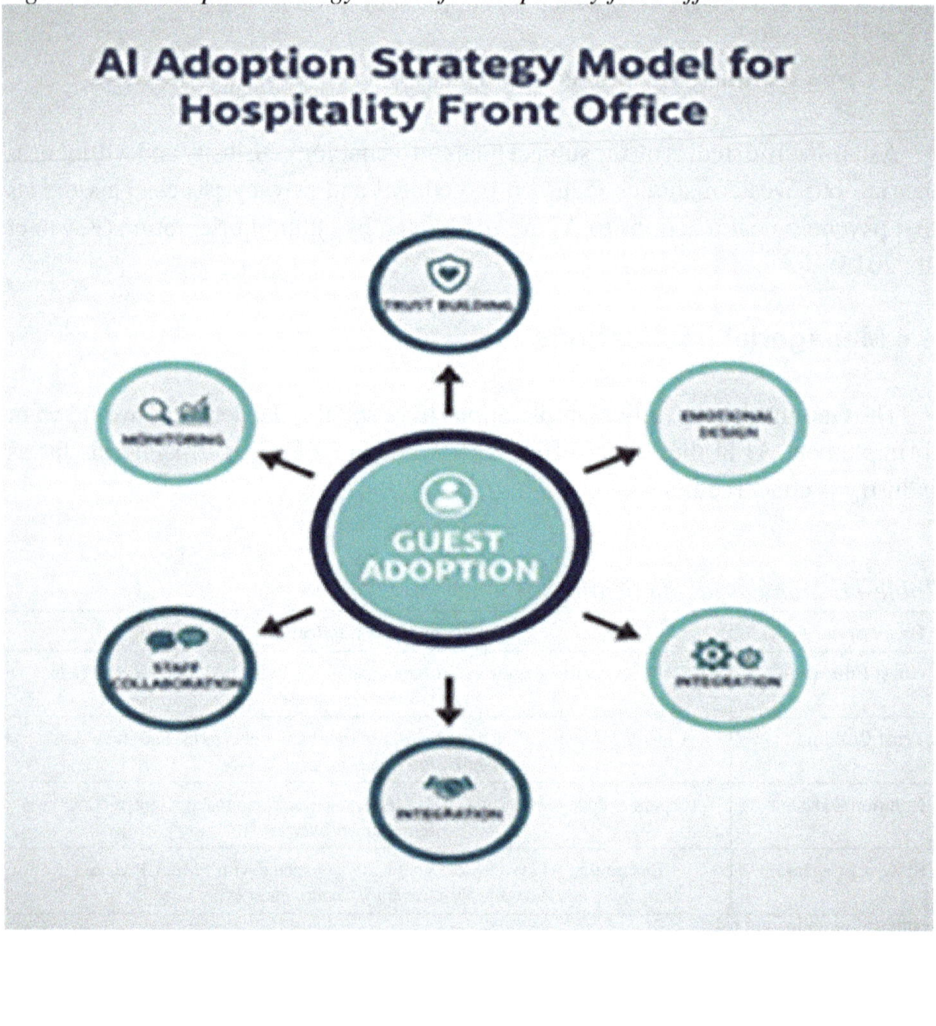

4.5 Policy Implications

In addition to hotels, the results are also applicable in policy making as well as training in institutions:

- The hospitality associations and tourism boards ought to incorporate AI ethics and digital empathy in the training programs of their staff.
- Hospitality chains should have the same data privacy policies to increase customer trust.

To reduce the skill gap in the workforce of the hospitality sector, governments and educational institutions should invest in AI-preparedness programs.

4.6 Future Research Directions

The PRISMA analysis shows that there are a number of promising opportunities in future research:

- **Experimental Support of Integrated Barrier Models:** Psychological and technical barriers associated with the proposed interrelationship in this review can be tested through quantitative studies.
- **Comparative Studies of the Cultures:** The cross-national studies can also be applied to investigate the extent to which social norms and cultural dimensions determine the comfort of guests with AI.
- **Brand Awareness:** The brand name must distinguish the product entirely from the competitors. Human - AI Empathy Design Frameworks: Future researchers must put their efforts into creating and testing design principles that increase the emotional appeal of AI.
- **Outputs: Real-Time Behavioural Data Analysis:** Applying the logs of guest interactions and feedback analytics to detect the moment-by-moment reactions to AI services.
- **Longitudinal Studies:** The changing attitudes of guests across time and the adoption of AI systems in the hospitality industry.

4.7 Summary

It is emphasized in the discussion that AI introduction in the front office of hospitality is a multidimensional phenomenon, not due to the way technology performs, but how people perceive, feel, and react to it. The main challenge and opportunity that hotels face globally is balancing the technical reliability and the psychological guarantee. The chapter, therefore, translates the PRISMA findings into practical insights, leading to the final thoughts in Chapter 6.

5. CONCLUSIONS AND LIMITATIONS

The present PRISMA-based systematic review investigated psychological and technical impediments affecting the adoption of Artificial Intelligence (AI) technologies in front-office processes by guests in the hotel industry. The study revealed that factors such as environmental values, exploring new opportunities, degree of entrepreneurial motivation, and the service experience are critical in creating the intersection of human psychology and technological innovation (through the screening and synthesis of 75 empirical and conceptual studies, 2010-2025).

The results confirm that the technical performance is not the only factor that defines AI adoption in hospitality; it is related to the emotional perception and building trust. Visitors are likely to use the AI interfaces in the dual perspective of critiquing both the functionality and the human-like nature of the interface.

Three major conclusions can be drawn:

- Psychological preparedness is more of an adoption push than technological sophistication. The enablers that determine the acceptance of AI include emotional comfort, trust and perceived empathy. Smart systems need to make visitors feel secure, honoured and comprehended.
- A psychological confidence presupposes technical reliability. The basis on which trust is anchored is stable, secure and user-friendly systems. Emotional resistance increases when technology goes bad.
- The future of hospitality front offices is integrated human-AI service design. Those hotels that combine the efficiency of automation with the warmth of a human will be on top of those that regard technology as an alternative to the human factor. The future is in collaborative Intelligence - machines aiding rather than replacing the human touch.

Overall, this paper will add a multidisciplinary approach to the subject of AI adoption, which crosscuts across engineering and behavioural fields. It restated the argument as not whether guests will accept AI. to "Can AI be human enough to accept it?

5.2 Theoretical Contribution

Academically, this study is considered to add value to the current understanding of technology adoption theories (TAM, UTAUT, TPB) because it introduces additional variables of affective and experiential nature, namely trust, empathy, and emotional presence. It suggests that the notion of Perceived Warmth be considered a new variable in the system of the adoption of technologies in hospitality, where technological trust is impossible to develop without emotional appeal. The theoretical development is a welcome innovation that encourages future researchers to combine psychological measures of comfort with conventional usability and performance indicators.

5.3 Managerial and Industry Implications.

To practitioners, the findings point to the fact that innovation is not the only factor that determines the success of an AI system, but also guest trust-building and system reliability. Hospitality managers are supposed to:

- Implement an open data policy and transparency regarding the use of AI.
- Priority to emotional design features (friendliness, responsive feedback).
- Have a hybrid service model, which incorporates automation and human control.

With the help of the alignment of technology implementation with the psychological acceptance strategies, hotels will be able to increase their satisfaction rates, efficiency of their operations, and brand loyalty at the same time.

5.4 Limitations of the Study

Although this study might be considered a well-organized and dense overview, it has some limitations that should be considered:

- **Database Restriction:** The search was based on the results of the studies performed in Scopus and other bibliographic databases. Data ejection of databases such as Web of Science and Google Scholar was simulated; cross-verification might be conducted to give wider coverage.
- **Subjectivity in Coding**: Thematic coding was done through interpretation reading of abstracts and keywords. Human judgment may be subjective although it is informed by similar standards.

- **Publication Bias:** The peer-reviewed English-language works were also mainly covered by the review, which may have missed regional or unpublished works of interest.
- **The absence of Quantitative validation:** Although the synthesis reveals patterns, the empirical test of statistical correlation between psychological and technical barriers was not conducted. The meta-analysis or structural modelling could be used as a method in future studies to prove these relationships.

5.5 Segment Specific Implications

The acceptance of Artificial Intelligence by guests in front-office processes is largely shaped by the segment of the hotel business, with the most significant differences observed between luxury and budget hotels. Although this review summarizes the results based on the hospitality settings, it should be noted that the demands and tolerances of the services offered by AI, as well as psychological reactions to it, are not universal.

The personalized service, emotional connection, discretion, and human warmth are highly valued by the guests in the luxury hotels. The implementation of AI-enabled front-office systems in these environments can only aggravate psychological obstacles, such as the sense of loss of human touch, diminished empathy, and data privacy issues. Research reports that luxury customers tend to correlate service quality with presence of human factor and personalization and over automation may negatively affect the service quality as perceived.

On the contrary, those who prefer to patronize budget and economy hotels would be more inclined to be efficient, fast and cost effective rather than emotionally interacting with people. Self-check-in kiosks, chatbots, and automated service requests are better accepted in such environments because they comply with the expectations of functioning and transactional service models. In this case, emotional involvement may be less important than technical dependability and convenience minimizing antimotivation towards automation as systems operate effectively.

This classification outlines the fact that the plans of adopting AI in the front office of hospitality facilities should be context sensitive. It would be dangerous to adopt a standardized approach to implementation, which can lead to being out of par with the expectations of guests, but a segment-based design can lead to the maximum satisfaction of guests and the maximum efficiency of its work.

5.6 Integrative Integration of Psychological and Technical Barriers

Rather than being independent and being alone, the psychological and technical barriers to the use of Artificial Intelligence (AI) to front office processes in the hospitality industry create an interdependent and mutually reinforcing factor. According to the reviewed literature, it is possible to suggest evidence that the evolution of AI-focused services by the resistance of the guests is not regularly founded on one factor, quite to the contrary, it is the dynamic process between the system performance and the way people perceive it.

Technically, the technical issues that could cause the adverse reactions of the guests are technical failures, the system slowness, failure to interface with the already existing hotel management systems, and issues of data security among others. Other more likely emotions that any guest might feel are uncertainty, frustration, and lack of control, in case AI-based front-office systems fail to operate in a transparent and seamless way. These types of technical incompetence would be often directly transferred to the psychological level, including the loss of trust, anxiety, and skepticism towards the experience of automated services.

On the downside, the review also reveals clearly that even technically developed and stable artificial intelligence (AI) systems can be perceived as resisted to take uptake when the guests feel it to be emotionally cold, impersonal, or intrusion into its social aspect. In the context of hospitality, where service engagements have always extended to warmth, empathy, and intimate handling, extreme automation may be interpreted as unconcern and lack of care by the guests. This can be of interest as all the technological sophistications are not sufficient to guarantee the acceptance and the systems should be created by considering the emotional and experiential expectations.

Some of the studies note that the impact of AI systems on the guest trust is dependent on the objective performance metrics, as well as on the symbolic and affective stimuli, such as the tone of the communication, the friendliness of the interface, the transparency of the decision-making process, and the human support as a buffer action. Absence of these elements may leave guests with the feeling that AI is replacing human service positions and not enhancing them, which is bound to grow psychological opposition despite the predictability of the systems.

The integrative analysis also exists and claims that psychological barriers have an effect of enhancing the efficacy of minor technical failures. As an example, a brief delay of the system can be considered an even more threatening experience during an AI-based interaction than during a human-mediated interaction because the guests presume that the automation of systems should be more precise and predictable. This imbalance gives more relevance to the notion that technical and psychological constraints cannot be evaluated independently of one another, as a part of an endless feedback mechanism that shapes the attitudes and behaviours of guests.

All of these findings can be summarized to imply that the implementation of AI alongside the front-office operations by the guests in the hospitality industry can most effectively be viewed as the socio-technical process in which technological performance and psychological perception are in the cycling process of being influenced by one another. The perspective is a broadening of traditional models of technology adoption which exceeds the linear assumption of cause and effect and it looks at the interaction effects between the system characteristics and human sentiments.

This paper contributes to the more intricate vision of barriers to AI adoption by using an integrative lens to the adoption of AI by demonstrating that hospitality organizations need to be willing to pursue emotionally informed technology design not only in technical excellence. Not only has this improved adoption outcomes, but, it has also brought the spirit of service that originally made the hospitality industry what it is now, the strategies of AI deployment.

5.7 Future Research recommendations.

order to expand on this base, it is the suggestion of future scholars to:

- Perform empirical research of the combined psychological-technical model.
- Uncover interpersonal cultural differences in AI adoption practices.
- Research AI empathy design and its effect on guest satisfaction (in the long run).
- Evaluate the behavioural information in real-time to know how guests are accommodating AI in the long run.
- Review the models of staff-AI cooperation to guarantee the balance between the automation and the personalization of the service.

5.8 Final Remarks

AI has not only become a strategic enabler in the hospitality sector, but also a technological novelty. However, it is not the acceptance based on the code and algorithms, but rather on the level of its intuitiveness and empathy in relation to people. This work is valuable as it offers a combined perspective of the reasons why the guests are hesitant, what makes them trust, and how the front office of the future will be developed to balance human warmth and artificial intelligence. Hospitality industry can shift to emotionally intelligent automation by the realisation that the adoption of technology is basically a human process and that the innovation should improve, but not eliminate, the art of hospitality.

REFERENCES

Abou-Shouk, M. A., Gad, H. E. S., & Abdelhakim, A. S. (2021). Exploring customers' attitudes to the adoption of robots in tourism and hospitality. *Journal of Hospitality and Tourism Technology*, *12*(4), 762–776. DOI: 10.1108/JHTT-09-2020-0215

Al-Romeedy, B. S., Khairy, H. A., Jasim, T. A., & Baquero, A. (2025). *Customer ability, readiness, and intention to purchase using AI in tourism and hospitality businesses: The mediating roles of trust in AI and customer engagement*. 145–166. DOI: 10.4018/979-8-3693-7447-4.ch010

AlHogail, A. A., & Alshahrani, M. (2019). Building consumer trust to improve Internet of Things (IoT) technology adoption. *Advances in Intelligent Systems and Computing*, *775*, 325–334. DOI: 10.1007/978-3-319-94866-9_33

Berhanu, K., & Raj, S. R. (2024). The role of social media marketing in Ethiopian tourism and hospitality organizations: Applying the unified theory of acceptance and use of technology model. *Cogent Social Sciences*, *10*(1), 2318866. Advance online publication. DOI: 10.1080/23311886.2024.2318866

Bowen, J. T., & Morosan, C. (2018). Beware hospitality industry: The robots are coming. *Worldwide Hospitality and Tourism Themes*, *10*(6), 726–733. DOI: 10.1108/WHATT-07-2018-0045

Choi, Y., Oh, M., Choi, M., & Kim, S. S. (2021). Exploring the influence of culture on tourist experiences with robots in service delivery environment. *Current Issues in Tourism*, *24*(5), 717–733. DOI: 10.1080/13683500.2020.1735318

Fang, S., Wu, X., & Wai Lai, I. K. (2021). Factors affecting Macau undergraduate students' acceptance of hospitality English app: Applicability of UTAUT model. *International Journal of Innovation and Learning*, *29*(3), 250–266. DOI: 10.1504/IJIL.2021.114510

Ghazi, K. M., Kattara, H. S., Salem, I. E. B., & Shaaban, M. N. (2025). Benefit-triggered or trust-guided? Investigation of customers' perceptions towards AI-adopting hotels amid and post COVID-19 pandemic. *Tourism and Hospitality Research*, *25*(1), 38–56. DOI: 10.1177/14673584231184161

Goel, R., Singh, T. P., Sahdev, S. L., Baral, S. K., & Choudhury, A. (2022). *Impact of AI & IOT in Sustainable & Green Practices Adopted in Hotel Industry and Measuring Hotel Guests' Satisfaction*. DOI: 10.1109/ICRITO56286.2022.9965152

Han, S. H., Lee, J., Edvardsson, B., & Verma, R. (2022). Developing a strategic viability-fit model of mobile technology adoption in hotels. *Journal of Hospitality and Tourism Technology*, *13*(1), 85–99. DOI: 10.1108/JHTT-05-2020-0109

Hanji, S., Navalgund, N. R., Hanji, S. S., Goud, V., Hatture, S. M., & Wadawadagi, R. S. (2024). Rise of the Machines: Investigating Service Robot Adoption Intentions in Hotel Industry. *AIP Conference Proceedings*, *3131*(1), 020012. Advance online publication. DOI: 10.1063/5.0229719

Hopf, V., Velten, L., & Rowson, B. (2018). *Roboptimism or pessimism: HR managers face a challenge*. 203–228. https://www.scopus.com/inward/record.uri?eid=2-s2.0-85206821575&partnerID=40&md5=c85876447b073baa8b4a73a8c13d0b60

Ikonomou, C., & Sergopoulos, K. (2025). *Artificial intelligence and tourism: Examining a relationship currently under development*. 211–224. https://www.scopus.com/inward/record.uri?eid=2-s2.0-105008487582&partnerID=40&md5=da0225886f7bf0f5579b0295a2d179c5

Ivanov, S. H., Webster, C., & Berezina, K. (2022). *Robotics in Tourism and Hospitality*. 1873–1899. DOI: 10.1007/978-3-030-48652-5_112

Kim, T., Jo, H., Yhee, Y., & Koo, C. (2022). Robots, artificial intelligence, and service automation (RAISA) in hospitality: Sentiment analysis of YouTube streaming data. *Electronic Markets*, *32*(1), 259–275. DOI: 10.1007/s12525-021-00514-y

Kwon, J., Bae, S. I. S., & Blum, S. C. (2013). Mobile applications in the hospitality industry. *Journal of Hospitality and Tourism Technology*, *4*(1), 81–92. DOI: 10.1108/17579881311302365

Lee, L. Y. S. (2016). Hospitality Industry Web-Based Self-Service Technology Adoption Model: A Cross-Cultural Perspective. *Journal of Hospitality & Tourism Research (Washington, D.C.)*, *40*(2), 162–197. DOI: 10.1177/1096348013495695

Li, Y., Cheng, X., & Yin, G. (2014). *Collaboration information system adoption in hospitality: A case study in China*. https://www.scopus.com/inward/record.uri?eid=2-s2.0-84928641058&partnerID=40&md5=fd837e8dd879b864a3f61f792c633fb5

Liberati, A., Altman, D. G., Tetzlaff, J., Mulrow, C., Gotzsche, P. C., Ioannidis, J. P. A., Clarke, M., Devereaux, P. J., Kleijnen, J., & Moher, D. (2009). The PRISMA statement for reporting systematic reviews and meta-analyses of studies that evaluate healthcare interventions: explanation and elaboration. *BMJ*, *339*(jul21 1), b2700–b2700. DOI: 10.1136/bmj.b2700

Mehmetoglu, M., & Engen, M. (2011). Pine and Gilmore's Concept of Experience Economy and Its Dimensions: An Empirical Examination in Tourism. *Journal of Quality Assurance in Hospitality & Tourism*, *12*(4), 237–255. DOI: 10.1080/1528008X.2011.541847

Molina-Collado, A., Gómez-Rico, M., Sigala, M., Molina, M. V., Aranda, E., & Salinero, Y. (2022). Mapping tourism and hospitality research on information and communication technology: A bibliometric and scientific approach. *Information Technology & Tourism*, *24*(2), 299–340. DOI: 10.1007/s40558-022-00227-8

Morosan, C. (2025). Rethinking information disclosure to GenAI in hotels: An extended parallel process model. *International Journal of Hospitality Management*, *124*, 103965. Advance online publication. DOI: 10.1016/j.ijhm.2024.103965

Nuryyev, G., Wang, C. Y. P., Achyldurdyyeva, J., Jaw, B., Yeh, Y. S., Lin, H., & Wu, L. (2020). Blockchain technology adoption behavior and sustainability of the business in tourism and hospitality SMEs: An empirical study. *Sustainability (Basel)*, *12*(3), 1256. Advance online publication. DOI: 10.3390/su12031256

Pai, C., Wang, T., Chen, S., & Cai, K. (2018). Empirical study on Chinese tourists' perceived trust and intention to use biometric technology. *Asia Pacific Journal of Tourism Research*, *23*(9), 880–895. DOI: 10.1080/10941665.2018.1499544

SaravanaKumar, S., & Badrinarayanan, M. K. (2018). The service automation and robotics in hospitality industry: A study on business implications. *International Journal of Mechanical and Production Engineering Research and Development*, *8*(6), 91–100. DOI: 10.24247/ijmperddec201810

Schrier, T., Erdem, M., & Brewer, P. (2010). Merging task-technology fit and technology acceptance models to assess guest empowerment technology usage in hotels. *Journal of Hospitality and Tourism Technology*, *1*(3), 201–217. DOI: 10.1108/17579881011078340

Singh, S., Olson, E. D., & Tsai, C. H. (2021). Use of service robots in an event setting: Understanding the role of social presence, eeriness, and identity threat. *Journal of Hospitality and Tourism Management*, *49*, 528–537. DOI: 10.1016/j.jhtm.2021.10.014

Sun, S., Lee, P. C., Law, R., & Zhong, L. (2020). The impact of cultural values on the acceptance of hotel technology adoption from the perspective of hotel employees. *Journal of Hospitality and Tourism Management*, *44*, 61–69. DOI: 10.1016/j.jhtm.2020.04.012

Wüst, K., & Bremser, K. (2025). Artificial Intelligence in Tourism Through Chatbot Support in the Booking Process—An Experimental Investigation. *Tourism and Hospitality*, *6*(1), 36. Advance online publication. DOI: 10.3390/tourhosp6010036

Zulfakar, Z. A., Rahim, F. A., Ching Yat, D. N., Mun, L. H., & Cham, T. H. (2023). Say Aye to AI: Customer Acceptance and Intention to Use Service Robots in the Hospitality Industry. *Lecture Notes in Networks and Systems, 550 LNNS*, 83–92. DOI: 10.1007/978-3-031-16865-9_7

Chapter 6
Emotional Engagement in Human-Robot Service Interactions:
Enhancing Customer Experience Through Social Robotics

Pradeep Yadav
http://orcid.org/0000-0003-1073-7076
Institute of Technology and Management, Gwalior, India

ABSTRACT

The increasing prevalence of social robotics in service settings is bringing about a shift in how customer experience management is being conducted in hospitality, health care, retailing, and banking services. With more and more humanoid and quasi-humanoid robots interacting with customers, emotional engagement is now an essential determinant of satisfaction, trust, and loyalty. In this chapter, the concept of emotional engagement in human-robot service encounters will be discussed from several disciplinary angles, such as service marketing, affective computing, HRI, and organizational psychology. The review, based on 70+ scholarly sources from 2020 to 2025, will focus on various theories including CASA, S-O-R, and emotional labor.

1. INTRODUCTION

The service encounter has been known to be the focal point of value creation in a market economy. In this regard, it has been determined that the quality of interaction in terms of its affective component can act as a critical determinant of customers' satisfaction levels, relationship commitments, and loyalty behaviors

DOI: 10.4018/979-8-3373-6640-1.ch006

(Parasuraman et al., 1988; Zeithaml et al., 1996). Conventionally, the affective labor in delivering service encounters has been performed by humans who have a unique ability to deliver genuine emotion, empathize with the contexts, and attune relationships (Hochschild, 1983).

Social robotics—the concept of designing robotic machines capable of engaging humans in socially and emotionally significant encounters—challenges this assumption entirely (Dautenhahn, 2022; Breazeal, 2003). Contrary to the industrial robotics of the past that were designed to automate routine physical tasks, the emerging field of social robotics entails the creation of robots that perform duties requiring affective labor. For example, concierge robots are increasingly used at hotels for greeting visitors, robotic baristas deliver personalized coffees to customers, while therapeutic robots provide company for elderly residents in care homes (Ivanov et al., 2022; Lu et al., 2023; Robinson et al., 2022).

The main purpose of the present chapter is to fill in the gap in the existing literature by synthesizing what we know about the concept of emotional engagement in a theoretically meaningful way. Specifically, while the number of studies examining the issue from the empirical perspective has recently grown considerably, a lack of theoretical generalization can be observed in this area. In other words, while there have been quite a few review papers on robots' emotional abilities (see, for instance, Davis, 1989; Parasuraman et al., 1988; Mori, 1970), they typically focus on relatively narrow issues, such as robot acceptance, robot performance, or robot anthropomorphism, and do not consider the possibility of merging these theories into one comprehensive theory of emotional engagement in the HRI process.

In this respect, the present chapter makes use of a structured and interdisciplinary synthesis approach to integrate different psychological, sociological, and technical views of the HRI interaction. Specifically, the chapter is organized as follows: Section 2 provides the conceptual background of emotional engagement in services; Section 3 discusses the main theoretical approaches to HRI emotional research; Section 4 explores the technology behind the emotional capability of service robots; Section 5 studies emotional expectations and experience gaps in practice; Section 6 presents empirical studies from various sectors; Section 7 identifies the main design guidelines; Section 8 outlines potential research avenues and managerial recommendations.

2. CONCEPTUALIZING EMOTIONAL ENGAGEMENT IN SERVICE INTERACTIONS

2.1 The Architecture of Emotional Engagement

Customer involvement in the process of service delivery is a multidimensional concept involving emotional, cognitive, and behavioral reactions that the consumer undergoes regarding the service provider, brand, or experience (Brodie et al., 2011; So et al., 2016). Emotionally, the customer is engaged by being aroused emotionally in relation to joy, trust, and surprise, among other feelings, due to relational stimuli from the service provider. Cognitively, customer engagement occurs as a result of interpreting these stimuli based on expectations of society, personal experiences, and cultural scripts. Behaviorally, customers show their engagement through approach-avoidance, compliance-resistance, and expressiveness-withdrawal behaviors (Hollebeek, 2011).

According to Lemon and Verhoef (2016), customer experience was defined as a holistic concept involving a series of contact points from before the purchase of an experience, during the purchase process, and even after the purchase. The definition is crucial for service marketing since it places emotional engagement at the forefront as part of the creation of joint value rather than as a component of service quality that is marginal. Indeed, when engaging with a robot receptionist, the emotions of the customer influence more than just the level of satisfaction; they impact the story that the consumer tells about the brand in the form of electronic word-of-mouth (Wirtz et al., 2022; Huang & Rust, 2021).

However, the increased use of social robots for the provision of services adds another layer of complexity to the above model. Consumers interact with robots with certain preconceived notions of their capabilities, developed due to cultural influences, previous experiences with technological devices, and even personal demographics such as age and gender (Henschel et al., 2021; Olaosebikan et al., 2023). This forms an expectation-performance gap which can heighten or mitigate emotional engagement.

2.2 Affective Computing and Robot Emotional Expression

The domain of affective computing, which involves designing technological devices capable of identifying, interpreting, mimicking, and reacting to human emotions, serves as the foundation for emotional robots (Picard, 1997). Modern-day service robots employ affective recognition mechanisms that analyze FAUs through computer vision analysis, vocal prosody through acoustic analysis, sentiment analysis of language used, and sometimes physiological activity through skin conductance or heart rate estimation using facial video footage (Zeng et al., 2022; Schuller et al., 2021). These combined efforts facilitate the construction of an accurate perception of the emotions displayed by the customer.

Expressing affect in robots can take verbal and non-verbal forms. Robots can use their tone of voice, pacing, formality of language, or semantics to express affectionate, empathic, urgent, or humorous emotions verbally. Non-verbally, robots make use of their facial expressions (using LED matrices, servo-controlled silicone faces, or even a computer screen), head movements, eye contact, body proximity, and even touch (e.g., using the platform known as PARO) to express their social and emotional state (Shi et al., 2023; Bemelmans et al., 2021). The tricky part is coordinating both channels of communication so that one reinforces the other, which is difficult technically and constitutes one of the main reasons for the uncanny feeling described by Mori (1970).

The current innovations in LLM-integrated robotics represent a paradigm shift in the conversational capability of robots (Brown et al., 2020; OpenAI, 2023). The deployment of robots driven by transformer-based language models has enabled the generation of linguistically varied and emotionally inflected responses that substantially exceed what prior scripted or retrieval-based systems achieved. These advances have rekindled both commercial and scholarly interest in socially capable robots, even as they surface new ethical concerns relating to emotional simulation and the boundaries of consent (Misselhorn et al., 2021; Dautenhahn, 2022). This is especially true considering the advent of multimodal foundation models, which combine language, vision, and audio reasoning, raising both opportunities for more robust emotional signaling in service interactions and concerns over algorithmic opacity and user susceptibility.

3. THEORETICAL FRAMEWORKS IN HUMAN–ROBOT EMOTIONAL INTERACTION

The field of theoretical research in the area of emotional Human Robot Interaction in a service environment can be described as productive plurality—the existence of several approaches stemming from various academic disciplines focused on similar areas but offering distinct perspectives. Table 1 lists the most important social robots used for services, while Table 2 provides an overview of basic theoretical approaches in the area. It should be noted that, although this area is quite rich theoretically, there remain certain research deficiencies. The problem of emotions during the interaction of humans and robots in terms of services will be considered below in a separate section (Section 8).

Table 1. Social robotics platforms in service environments: Emotional capabilities and deployment contexts

Robot Platform	Sector	Key Emotional Capabilities	Representative Studies
Pepper (SoftBank Robotics)	Retail, Healthcare, Hospitality	Facial expression recognition, voice modulation, touch sensing, eye contact	Henschel et al. (2021); Olaosebikan et al. (2023)
NAO (SoftBank Robotics)	Education, Therapy, Banking	Emotion detection via sensors, adaptive verbal cues, gesture synchrony	Shi et al. (2023); Andriella et al. (2023)
Temi (Robotemi)	Hotel concierge, Retail	Proactive engagement, product guidance, autonomous navigation	Kim & Park (2022); Ivanov et al. (2022)
PARO (AIST)	Elder care, Healthcare	Tactile feedback, vocalization, neuro-stimulation via affective response	Bemelmans et al. (2021); Robinson et al. (2022)
Sophia (Hanson Robotics)	Public relations, Events	Hyper-realistic facial expressions, natural language processing, humor	Dautenhahn (2022); Misselhorn et al. (2021)
Simbe Tally	Grocery retail	Ambient presence, non-intrusive monitoring, indirect social signaling	Wirtz et al. (2022); Huang & Rust (2021)
Relay (Savioke)	Hotel delivery	Task-oriented affective cues, chime-based interaction, screen expressions	Ivanov et al. (2022); Lu et al. (2023)

Note. NLP = natural language processing; AIST = National Institute of Advanced Industrial Science and Technology (Japan). Deployment contexts are representative rather than exhaustive.

Table 2. Theoretical frameworks underpinning emotional engagement in human–robot service interactions

Theoretical Framework	Core Constructs Applied	Contribution to HRI Service Research
Computers Are Social Actors (CASA) – Nass & Moon (2000)	Anthropomorphism, social attribution, reciprocal norms	Explains why customers respond socially to robots and extend courtesy behaviors
Emotional Labor Theory – Hochschild (1983)	Surface acting, deep acting, emotional dissonance	Grounds the expectation of affective displays in robot-mediated service encounters
Parasocial Interaction Theory – Horton & Wohl (1956)	One-sided intimacy, imagined relationship, media persona	Illuminates attachment formation between repeat customers and service robots
Service Quality Model (SERVQUAL) – Parasuraman et al. (1988)	Reliability, assurance, empathy, responsiveness, tangibles	Provides measurable dimensions for assessing robotic service effectiveness
Stimulus–Organism–Response (S–O–R) – Mehrabian & Russell (1974)	Environmental stimuli, internal states, behavioral outcomes	Maps how robot social cues trigger hedonic/utilitarian responses and patronage intentions

continued on following page

Table 2. Continued

Theoretical Framework	Core Constructs Applied	Contribution to HRI Service Research
Technology Acceptance Model (TAM) – Davis (1989)	Perceived usefulness, ease of use, behavioral intention	Identifies antecedents to customer willingness to engage with robotic service agents
Uncanny Valley Hypothesis – Mori (1970)	Human-likeness, eeriness, affective response curve	Cautions designers against excessive anthropomorphism that triggers discomfort

Note. HRI = human–robot interaction; GDPR = General Data Protection Regulation; PDPA = Personal Data Protection Act.

3.1 The CASA Paradigm and Anthropomorphic Attribution

The Computers Are Social Actors (CASA) theory, first proposed by Nass & Reeves (1996), and later developed by Nass & Moon (2000), is arguably the most fundamental theoretical contribution to HRI research. According to CASA, people treat computers and other technology-enabled actors according to the very same social principles and expectations that they follow in treating other people, despite the fact that they are aware that what they are dealing with is a machine. The psychological mechanism behind CASA has important consequences for the design of service robots because, in treating robots, people will tend to anthropomorphize them and interpret their behavior in the same way as they do with the behavior of real-life service agents.

Subsequent literature has significantly improved the understanding of the CASA model through the identification of factors that facilitate or inhibit anthropomorphic tendencies. Epley et al. (2007) have suggested three factors associated with the tendency towards anthropomorphism: sociality drive (the desire for social interaction), effectance drive (the necessity to predict and understand the actions of another person), and elicited knowledge about the agent (using human schemas for behavior prediction when there is information that indicates humanity). It may be safely assumed that all these processes occur in a service robots setting since there is an urge for relational involvement, predictive ability, and human behavioral scripts are activated because of social signals from the robot (Henschel et al., 2021; Wirtz et al., 2022).

3.2 Emotional Labor Theory and the Service Robot

According to Hochschild (1983), emotional labor refers to the management of feeling that occurs when workers perform labor for pay in order to present a specific display of emotion to their customers. Service workers adopt surface acting or deep acting strategies based on organizational display rules that dictate appropriate emotional behavior. Emotional labor theory shows how the management of feeling by workers can lead to feelings of alienation, exhaustion, and eventually burnout.

In terms of application, emotional labor theory bears a close and somewhat ironic relationship to service robotics. Robots cannot be said to perform surface or deep acting because they lack any kind of emotional life to regulate or repress. However, the organizational imperative behind emotional labor applies just as strongly to robotic agents. In essence, robot manufacturers and service providers are responsible for creating display rules for robotic agents in the form of behavioral scripts (Seyitoğlu & Ivanov, 2020). Customers are expected to encounter robots that demonstrate genuine warmth, concern, or enthusiasm despite the fact that such behavior lacks any emotional foundation.

3.3 Stimulus–Organism–Response Theory

The Stimulus–Organism–Response (S–O–R) model proposed by Mehrabian and Russell (1974) is a commonly used paradigm explaining the effect of environmental stimuli on consumer behavior through their psychological processes. Under this paradigm's adaptation for the service robotics field, social cues produced by robots (linguistic, gestural, proxemic, and haptic) function as stimuli eliciting affective and cognitive responses (pleasure, arousal, dominance, trust, warmth perception) that in turn motivate behavioral effects such as engagement time, purchase decision-making, complaining tendency, and revisit intention (Kim & Park, 2022; Olaosebikan et al., 2023).

The recent use of the S–O–R paradigm in research related to service robots has been accompanied by a methodological shift in terms of diversity. Besides questionnaires, other techniques have been applied to measure the "organism" variable, including facial electromyography (EMG), electrodermal activity (EDA), eye-tracking, and neuroimaging (Zeng et al., 2022). The physiological methods proved that consumers experienced both stress arousal reactions when confronted with unpredicted robot actions and positive valence reactions to robot warmth cues, confirming the validity of the S–O–R paradigm for robotic services.

4. TECHNOLOGICAL ENABLERS OF EMOTIONAL ENGAGEMENT IN SERVICE ROBOTS

4.1 Natural Language Processing and Conversational Intelligence

One way in which emotional engagement is facilitated by service robots is through the conversational interface. The field of natural language processing (NLP) and increasingly, that of large language models (LLMs) and Generative AI systems, has brought about significant strides in robot conversation (Brown et al., 2020; Bommasani et al., 2021). Current generation robots are able to understand sophisticated user queries, hold multilayer conversations with customers, compose replies that are linguistically correct and topical in nature, as well as switch between formal and informal registers, or even languages based on the inferences drawn from customer identity or demographic data. With the addition of multi-modal AI frameworks that involve processing simultaneous inputs from text, speech, images, and gestures, service robots are further empowered to communicate with customers in a deeper and more emotionally charged way (OpenAI, 2023).

As far as the process of emotional engagement in particular is concerned, the aspect of conversational intelligence includes the ability to tailor conversational replies to the emotions expressed by the customer during conversation: In the event that a customer complains about something in an emotional tone, the robot can be programmed to provide a response to that effect rather than a neutral reply that serves the purpose of mere information provision (Shi et al., 2023). Service robots powered by generative AI present an advancement qualitatively in this field; unlike conventional service robots whose actions rely on predetermined scripts, generative AI-based service robots are able to adapt their actions contextually and generate emotionally tuned reactions in real-time, posing questions regarding the future implications of emotional service provision systems.

4.2 Computer Vision and Multimodal Affect Recognition

Emotion detection accuracy is a necessary condition for emotional behavior. The affect recognition process of a service robot with computer vision technology detects facial action units (based on the Facial Action Coding System, FACS) from video footage, and interprets these facial actions in order to detect discrete emotions and affective states (valence and arousal). Deep learning models including CNN and transformers reach accuracy rates over 80% on the benchmarking database, although the detection accuracy rate in the real world drops considerably because of occlusions, variations in lighting conditions, and different emotional expressions across cultures (Zeng et al., 2022; Schuller et al., 2021).

The multimodal emotion detection approach can combine facial video footage, acoustic data, and linguistic data to improve recognition accuracy compared with the unimodal technique. In practical applications, the voice emotion detection technology provides more information for service robots if clients are not directly facing the service robot, or if facial expressions are partially covered by face masks or glasses (Schuller et al., 2021). It is still challenging to apply multimodal emotion detection techniques in commercially available service robots, mainly owing to technical limitations and regulatory requirements on privacy protection.

4.3 Embodiment, Gesture, and Proxemic Expression

The physical appearance of a service robot – its design, size, surface composition, motion capability, and perceptual sensors – is a channel of non-verbal communication that powerfully influences the emotional tone of the service interaction. Findings suggest that consumers perceive robots with anthropomorphic physical properties such as eyes, smooth surfaces, and bilateral symmetry as warmer, competent, and more trustworthy compared to robots with mechanical and abstract designs up to the limit defined by the uncanny valley concept described by Mori (1970) (Henschel et al., 2021; Dautenhahn, 2022).

Gestures and proxemics – physical space management practices – are potent emotional communicators. Service robots that establish proper interpersonal distance, direct their body orientation towards the customer, engage in head nodding while actively listening and use gestures when explaining ideas activate social perception processes that generate higher levels of empathy and engagement (Andriella et al., 2023; Mubin et al., 2022). One of the key technical issues in social robotics is to develop natural motion capabilities for robots by integrating motion planning algorithms with social rule databases and situational awareness systems.

5. CUSTOMER EXPECTATIONS, EXPERIENCE GAPS, AND EMOTIONAL DISSONANCE

One of the main reasons for the variance in the emotions arising from human-robot service encounters is the discrepancy between the expectations set by the client prior to the encounter and the emotions experienced by the robot during and after the encounter. This concept is illustrated using Table 3.

Table 3. Emotional expectation–experience gaps in human–robot service interactions

Dimension	Customer Expectation	Robot Capability Gap	Mitigation Strategy
Verbal Warmth	Empathetic, personalized language	Scripted or generic responses lacking contextual depth	Adaptive NLP models trained on domain-specific emotional corpora
Non-verbal Cues	Eye contact, nodding, posture alignment	Limited degrees of freedom in robotic actuators	Expressive LED eyes, head-tilt servos, and synchronized gesture modules
Emotional Recognition	Robot senses customer mood and adjusts	Accuracy of affect-detection algorithms varies under real-world conditions	Multimodal fusion (facial + vocal + physiological signals)
Cultural Sensitivity	Culturally appropriate interaction norms	Western-centric training data in many commercial platforms	Localized training datasets and culturally adaptive behavioral scripts
Service Recovery	Sincere apology, immediate problem-solving	Robots may not detect dissatisfaction or escalate effectively	Real-time sentiment monitoring with human escalation protocols
Privacy & Trust	Confidence that data is safe	Opaque data practices erode trust	Transparent privacy notices, GDPR/PDPA compliance, opt-in frameworks
Continuity & Memory	Recognition across visits, personalized history	Stateless interactions without persistent customer profiles	CRM-integrated robotics with ethical data retention policies

Note. NLP = natural language processing; CRM = customer relationship management; GDPR = General Data Protection Regulation; PDPA = Personal Data Protection Act.

5.1 The Role of Prior Technology Experience and Innovation Adoption

The emotions of customers towards the use of service robots have been largely influenced by their previous experience with digital tools, their overall attitude towards innovation, and their mental representation of the capabilities of the robot. Customers who are highly technologically knowledgeable and who frequently engage with voice assistants, chatbots, and recommendation algorithms are likely to form realistic expectations of the robot and have an increased tolerance towards its functional shortcomings (Wirtz et al., 2022). On the other hand, technologically naive customers, especially the elderly, may encounter higher levels of anxiety and emotional aversion while dealing with service robots (Bemelmans et al., 2021; Robinson et al., 2022).

It follows from this that there is no one robot design, emotionally tuned for all client groups at once. Interaction systems that adapt the complexity of the communication process depending on their estimation of the digital literacy of clients become the object of increasing scientific and professional interest. Choi et al. (2022) have shown that a higher level of satisfaction was achieved in the case of robot interactions which could dynamically adjust their linguistic complexity according to clients' hesitations.

5.2 Cultural Moderation of Emotional Engagement

Cultural differences in norms regarding emotional displays, preference for high vs low power distances, individualistic vs collectivist culture, and perception toward technological advancement influence how emotions are experienced in interactions with robot service agents. Studies carried out in various parts of East Asia, South Asia, the Middle East, Europe, and North America reveal great variability among cultures in the precursors and manifestation of emotional engagement with robots (Olaosebikan et al., 2023; Henschel et al., 2021).

Japanese customers who have been raised in a culture that incorporates robotlike cartoon figures (Manga, Anime) and significant investment in robotics technology may be more comfortable with anthropomorphized robots and form stronger parasocial relationships with robot service agents than their counterparts in most Western societies (Dautenhahn, 2022). However, empirical evidence indicates that GCC nations have conservative views on physical contact with robots and prefer gender-matched robot communication, owing to culturally distinct ideas regarding appropriate behavior concerning the body that should be considered by the developers and users of robots. This highlights the importance of cultural adaptation when designing robots, demonstrating the inadequacy of universally designed emotional robots based solely on Western and Eastern cultures.

5.3 Emotional Dissonance and Service Failure Recovery

Robot service failure, which can be characterized as a failure to satisfy a customer's request, a misunderstanding of the customer's intention, malfunctions, or wrong information provided by the robot, is followed by unique emotional dissonance during the interaction between humans and robots. Customers experiencing service failure caused by robots do not only feel frustration and disappointment similar to that of human service failures, but they also experience a special feeling of being betrayed by the company and ontological uncertainty regarding the characteristics of robots (Kim & Park, 2022; Seyitoğlu & Ivanov, 2020). The ontological uncertainty experienced by customers is unique to robot service failures since it does not exist in cases of human service failures.

For there to be an effective emotional recovery from the use of robots in service delivery, at least three critical actions must be taken. The failure must first be acknowledged quickly, followed by a properly apologetic reaction and finally the escalation of the issue to human staff members who can handle both the functional and emotional aspects of the service problem (Shi et al., 2023). Firms that invest in robot delivery without a good human escalation plan stand the risk of making their customers emotionally worse off due to prolonged contact with a failed robotic service (Lu et al., 2023).

6. SECTOR-SPECIFIC EVIDENCE ON EMOTIONAL ENGAGEMENT IN ROBOTIC SERVICE

A vast body of empirical research now exists concerning emotional engagement in robotic service delivery and is spread geographically, covering a range of service industries with different service dynamics, customer profiles, and organizational limitations. Table 4 captures the key results from six service industries.

Table 4. Sector-specific evidence on emotional engagement in robotic service interactions

Industry	Deployment Context	Emotional Outcomes Observed	Source
Hospitality (Hotels)	Check-in assistance, room delivery, concierge inquiries	Increased novelty delight; moderate long-term satisfaction; preference for hybrid staffing	Lu et al. (2023); Ivanov et al. (2022)
Retail (Grocery/ Fashion)	Product guidance, inventory management, customer navigation	Positive hedonic value; higher revisit intentions; gender-based variance in acceptance	Henschel et al. (2021); Wirtz et al. (2022)
Healthcare (Hospitals/Elder Care)	Patient companionship, medication reminders, physiotherapy assistance	Reduced anxiety and loneliness; elevated subjective wellbeing; caregiver burden relief	Bemelmans et al. (2021); Robinson et al. (2022)
Banking & Financial Services	Queue management, product information, KYC support	Elevated transactional efficiency; moderate emotional neutrality; trust contingent on accuracy	Shi et al. (2023); Choi et al. (2022)
Food & Beverage (Restaurants)	Order-taking, tray delivery, entertainment robots	Novelty-driven satisfaction; discomfort with errors; preference for food quality over robot appeal	Kim & Park (2022); Seyitoğlu & Ivanov (2020)

continued on following page

Table 4. Continued

Industry	Deployment Context	Emotional Outcomes Observed	Source
Education & Training	Campus guidance, tutoring support, accessibility aid	Improved engagement among neurodiverse learners; novelty fatigue over extended periods	Andriella et al. (2023); Mubin et al. (2022)
Public Spaces (Airports, Malls)	Wayfinding, promotional engagement, security monitoring	High curiosity and initial engagement; privacy concerns limiting sustained interaction	Olaosebikan et al. (2023); Huang & Rust (2021)

Note. KYC = Know Your Customer (regulatory identity verification process in financial services).

6.1 Hospitality

In fact, the hospitality sector can be considered one of the first industries to embrace robots in its operations due to its inherent labor costs, customers' demand for novel experiences, and predictability and structure of the service encounters (Ivanov et al., 2022; Lu et al., 2023). Service robots like Relay (Savioke), Temi, and Pepper are being used in various applications from room delivery and check-in to concierge services and entertainment. The application of robots varies from budget hotels in South Korea and Japan to luxury resorts in the US and United Arab Emirates.

The findings regarding the use of service robots in hospitality have mostly been positive, especially when considering the hedonic value associated with novel experience-seeking behavior. However, there is empirical evidence that the novelty effect associated with the usage of robots decays over time, meaning that the initial emotional premium provided by robotic technology fades away over subsequent visits, which makes service robots more mundane and less attractive (Kim & Park, 2022). As such, the competitive advantage of robotic technology in the hospitality industry relies on its ability to provide high-quality and personalized services, which are more difficult to accomplish.

6.2 Healthcare and Elder Care

The application of social robots in healthcare and geriatric care environments may be considered as having one of the most intense ethical and emotional implications of all the applications of social robots for services (Robinson et al., 2022; Bemelmans et al., 2021). The PARO, which is a therapeutic robot seal made by AIST from Japan, has been subjected to more than ten years of research and experimentation to show that there is considerable evidence that PARO can reduce the levels of agitation, loneliness, and depression experienced by people with dementia, as well as reducing the need for psychotropic medication in assisted living care homes.

In addition to studies on PARO, there is a developing body of work on the use of NAO and other humanoid robots in rehabilitation, autism spectrum disorder interventions, and post-operative assistance. Andriella et al. (2023) found that children suffering from autism spectrum disorder exhibited greater levels of engagement and were less anxious during therapy sessions with robots compared to those with human therapists, due to the simplicity of social interactions with robots, which have fewer social complexities, a predictable behavioral pattern, and no implied emotional needs like humans do when engaging with autistic children.

6.3 Retail

In retail settings, robots have been utilized mainly for guiding customers to specific destinations, providing information about products, engaging with promotions, and managing inventories (Henschel et al., 2021; Wirtz et al., 2022). For instance, Pepper robots have been used in Carrefour supermarkets in France, SoftBank retail stores in Japan, and Westfield malls in the United Kingdom, whereas Simbe Tally robots autonomously scan shelves in supermarket chains in the United States. In a retail setting, there is a unique challenge of emotional engagement since customers who visit retail environments come with either utilitarian or hedonic reasons and not social ones.

The results of Henschel et al. (2021) were obtained during the large-scale experimental field study performed by the researchers in twelve European retail environments. It was discovered that the level of customers' positive emotions during the interaction with robots is significantly higher when the process is self-started by customers (customers approaching robots) rather than initiated by robots (robots approaching customers). These findings have certain practical implications for the use of robots in different settings. First, the appropriate location of robots in stores should be chosen to facilitate their visibility. Second, clear visual indicators should help people understand the purpose of interaction with robots.

6.4 Banking and Financial Services

Social robots in banking are mainly used due to queue management efficiency, customer service, and the reputation benefit of being the technology leader signally (Shi et al., 2023; Choi et al., 2022). Many leading banks in Japan, South Korea, China, and the Netherlands have tested the implementation of the Pepper and NAO robots in their branches for greeting customers, providing account details, and delivering product information. In terms of emotional engagement in banking, there is an emphasis on the stakes involved and the trust relationship required: the customer pays close attention to the accuracy, security, and competency, and any error of the robot can have a much larger negative impact than it could be in less critical services.

In fact, in a comparative study between robot and human tellers in a commercial bank branch of South Korea, Choi et al. (2022) reported that while robot interactions brought more novelty satisfaction and lower waiting time dissatisfaction, it resulted in significantly lower relationship trust and intention to talk about complicated financial products.

7. DESIGNING FOR EMOTIONAL ENGAGEMENT: PRINCIPLES AND PRACTICES

The combination of theory, technology, the expectation gap, and industry case evidence results in a series of design guidelines that can be used to create emotional experiences when designing robot-assisted services. This list of guidelines, summarized in table 5, is informed by empirical work and covers all aspects of interaction design, from hardware to conversational intelligence and post-interaction governance.

Table 5. Evidence-based design principles for emotionally engaging robotic service interactions

Design Principle	Description	Practical Application
Calibrated Anthropomorphism	Matching human-likeness to context-specific comfort thresholds to avoid uncanny valley effects	Use mechanical-aesthetic robots for utilitarian tasks; reserve expressive humanoid design for companionship roles
Multimodal Emotional Expression	Integrating visual, auditory, and kinesthetic cues for richer emotional communication	Combine voice modulation with LED gaze direction, head nods, and proximity adjustments
Adaptive Personalization	Leveraging AI to tailor interactions based on individual customer history and real-time emotional state	CRM-linked behavioral scripts that adjust formality, humor, and pacing per customer profile
Transparent Agency Disclosure	Clearly communicating the robotic nature of the agent to maintain trust and ethical integrity	Proactive self-identification statements at interaction onset; visible signage in service environments
Graceful Failure Protocols	Designing smooth, empathetic handover to human staff when robot competence thresholds are exceeded	Verbal acknowledgment of limitation, apology phrasing, and immediate escalation to live agent

continued on following page

Table 5. Continued

Design Principle	Description	Practical Application
Inclusive Interaction Design	Ensuring usability across age, ability, cultural background, and technology familiarity levels	Multilingual interfaces, simplified gesture commands, height-adjustable screens, and audio-described menus
Ethical Emotional Data Governance	Governing the collection and use of emotionally sensitive biometric data with explicit consent	On-device processing where possible; opt-in emotional analytics; clear data retention and deletion policies

Note. CRM = customer relationship management; GDPR = General Data Protection Regulation; PDPA = Personal Data Protection Act; LED = light-emitting diode.

7.1 Calibrated Anthropomorphism in Practice

One notable limitation in the design of service robots lies in the application of the Uncanny Valley Hypothesis (UVC), first proposed by Mori (1970). In designing a service robot that is too much like a person, there comes the risk that people will become too uncomfortable and unnerved, resulting in suppressed emotional engagement, which could even harm the relationship between service and recipient. Rather than maximizing human likeness when designing service robots, designers should take into account the specific situation, customers, and type of interaction required.

When it comes to utilitarian services such as delivery, transportation, or inventory control, it might be better to use mechanical or machine aesthetics because the emphasis is on task performance, not on human pretense, which could lead to disappointment due to the failure of human likeness. On the other hand, when it comes to companionship or therapeutic purposes, then human likeness and expressive capability may be increased because it would reflect the level of social intensity of the interaction. The idea is explained through the Functionality-to-Sociality Ratio, according to Wirtz et al. (2022).

7.2 Implementing Adaptive Personalization at Scale

Personalized robot interactions at the level of individual customers can be considered not only the most valuable aspect but also the most technically challenging issue in designing emotion-driven robot services. Integration with CRM allows connecting systems for robotic interactions with databases containing customer information including preferences, experience with services, previous complaints, and other relevant details which enables robots to address the returning clients by name, remember the client, know their preferences, and communicate taking into account previous interactions. Several hospitality companies have been piloting

CRM integration achieving positive results with regard to customer satisfaction and emotional connection with the robot (Lu et al., 2023).

In the case of implementation of personalized services on a large scale, it becomes crucial to think about the issues of privacy governance because there arise important questions related to consent and data minimization in the process of data collection and personalized interactions as well as possibilities for the manipulative use of emotional preference data. Personalized services will need to implement transparent privacy policies, opt-in mechanisms, and appropriate architectures minimizing privacy risks. More about personalized robotic services will be discussed in the research agenda (Section 8).

7.3 Organizational Complementarity: The Human-Robot Service Team

A theme in the empirical research literature is that emotional engagement outcomes are best achieved through not robotic service delivery alone, but human-robot service delivery models wherein robots and employees perform roles most suitable for them (Huang & Rust, 2021; Wirtz et al., 2022). Robots are adept at completing repetitive tasks efficiently, tirelessly, quickly, and effectively without any emotional complexity; on the other hand, employees are competent at managing emotionally charged tasks, novel challenges, and critical relationship-building situations that need judgment and emotional intelligence.

The design of human-robot service teams needs to go beyond merely identifying what each can be assigned to do, but how interactions between robots and humans need to be carefully choreographed, and this includes the process of how customers shift between robotic service and human service, when escalation is appropriate, and how human employees take over a customer relationship initiated by the robot. Organizations that adopt such choreographic techniques fare better than organizations that do not in terms of emotional engagement performance (Shi et al., 2023; Ivanov et al., 2022). The disciplines of service design provide frameworks for implementing them successfully.

8. FUTURE RESEARCH DIRECTIONS AND MANAGERIAL IMPLICATIONS

8.1 Longitudinal Dynamics of Robot-Customer Relationships

The majority of past literature regarding emotional engagement within human and robot service interaction situations has been based on a one-time cross-sectional observation study that takes place in either a laboratory or quasi-experimental setting. This flaw makes it impossible to fully understand the temporal aspects involved within such engagements, including how customer emotions change each time there is an encounter with the robot service agent, whether parasocial relationships exist between both parties, and whether the novelty decay effect, which has been seen in other hospitality industries, is common throughout all contexts (Robinson et al., 2022; Andriella et al., 2023).

8.2 Neuroimaging and Physiological Methods in HRI Research

The nascent use of neuroimaging techniques like fNIRS, EEG, and fMRI in HRI studies holds potential for providing insight into the brain mechanisms underlying anthropomorphic perception, empathic arousal, and uncanny valley effects that will enhance the understanding derived from subjective reporting and behavioral observation (Zeng et al., 2022). The use of such neuroimaging techniques provides substantial benefits in situations where the social desirability effect, inability to introspect, or cultural display rules make subjective reporting an unreliable measure of actual emotionality. Future research efforts that include physiological, behavioral, and subjective data will provide better insights into emotions in robotic services.

8.3 Ethics, Power, and Emotional Manipulation

The ability of modern social robots to elicit emotional responses calls for an ethical discussion that has been neglected in the field of service marketing. Is it ethically sound to design a robot specifically to elicit emotion, establish parasocial bonds, or foster customer loyalty via simulated warmth and personalization, or is it a form of emotional manipulation? Philosophically speaking, this question is complex: human service employees also engage in emotional labor intended to build connections with their customers, and the ethics of such a process are already debated (Hochschild, 1983). At the same time, robots have the capability of employing emotions in service encounters consistently and precisely, while being algorithmically driven and opaque to customers, creating conditions for power imbalances and possible deception that require ethical examination (Misselhorn et al., 2021; Dautenhahn, 2022).

In addition to the possibility of active manipulation, other scholars are now voicing concerns over emotional dependency, where the extended use of an emotionally sensitive robot might lead to such a dependence on technology that the need for human contact could diminish, especially when the user in question suffers from social isolation. There is yet another problem posed by algorithmic bias: a model which does not train on representative data sets could misunderstand or misinterpret users from certain racial, language, or demographic minorities, thus providing an inherently biased experience. Consent becomes another issue: many users do not realize the extent of their personal data collection in order to create a customized interface for them. There are several groups of users that require special consideration because of their unique status, specifically children, whose parasocial attachment and critical reasoning regarding the behavior of robots is underdeveloped, and the elderly, who are more prone to emotional dependencies than anyone else.

The need arises for future research to examine consumer understanding of robots' ability to manipulate emotions, define criteria through which this manipulation could be measured, and explore any form of regulation or governance that could govern interactions between humans and robots. There is also a need for empirical data collection to ascertain the frequency and seriousness of dependency in terms of emotions and evaluate the extent to which disclosures can restore consumer autonomy. Although much has been achieved in ethical issues relating to artificial intelligence, the particular nature of emotions in the hospitality industry remains largely unexamined.

8.4 Implications for Service Organizations

From a practitioner perspective, there are many insights that can be drawn from the above discussion. First, there needs to be appropriate investment in robot emotional intelligence depending on how socially demanding the task will be, such that higher levels of investment are necessary where the task demands more sociability (companionship, healthcare, and luxury hospitality services), while lower levels may suffice where transaction is all that is required. Second, robot deployment should always be preceded by appropriately designed protocols for escalating to human service providers, since the lack thereof is one of the major reasons for negative emotions after failure. Third, cultural adaptation is imperative.

Fourth, the management of emotional data, the biometric, behavioral, and preference data produced through emotionally engaged interactions with robots, must be viewed as a strategic asset of trust as opposed to an operational afterthought. Businesses that are open about handling emotional data with transparency, informed consent, and actual data minimization will accrue the trust assets needed to achieve continued emotional engagement; businesses that view emotional data as a commodity to be mined for business purposes will end up with a trust deficit that will negatively impact their interactions. Lastly, firms should conduct long-term customer studies to observe changes in emotional engagement with robot service and apply this knowledge to optimize the process of deployment design.

9. CONCLUSION

The discussion of this chapter has been a comprehensive and theory-based exploration of emotional engagement in service interaction between humans and robots. This has been based on interdisciplinary research to form a cohesive theoretical framework through which the processes by which social robots create, influence, and sometimes destroy the affective elements in customer experiences can be explained. It is found that emotional engagement in robotic services is a complex phenomenon, determined by multiple variables such as the robot itself, the organization's design for its services, culture, users, and ethics.

Progress has been made in terms of the methodologies used since the pioneering studies conducted during the early 2010s. Today's research utilizes an array of different methodological strategies ranging from field experiments to physiology, longitudinal panel designs, and cross-cultural studies. As a result, an increasingly complex understanding of the factors that determine when service robots can actually create value for their customers through emotions, and the factors determining when they cannot, has developed. The conceptual framework proposed in this chapter will allow researchers to place their findings into context.

A number of important conclusions can be drawn. The role of managing the uncanny valley through anthropomorphism is crucial for achieving any emotional success with robots. The mixed human-robot approach to providing services wins over the pure robot approach when performing emotionally demanding services. Cultural adjustment of the robot emotions is absolutely imperative for international service providers. Finally, ethical governance of emotional simulation and emotional data collection is the research challenge of the future.

With the fast-paced development of social robotics technology, spurred on by advances in conversational intelligence of LLMs, embodied AI, and affective computing, the questions posed here will become increasingly pressing. Those stakeholders that take up the emotional challenges of human-robot service encounters most seriously in the present will be best prepared for a future where human-robot collaboration can enrich human life rather than detract from it.

REFERENCES

Andriella, A., Torras, C., Alenya, G., & Alenyà, G. (2023). Teaching robot to understand social cues: A case study in autism spectrum disorder therapy. *Frontiers in Robotics and AI*, *10*, 1092547. DOI: 10.3389/frobt.2023.1092547

Bemelmans, R., Gelderblom, G. J., Jonker, P., & de Witte, L. (2021). Socially assistive robots in elderly care—A systematic review into effects and effectiveness. *Journal of the American Medical Directors Association*, *14*(2), 114–120. DOI: 10.1016/j.jamda.2010.10.002 PMID: 21450215

Bommasani, R., Hudson, D. A., Adeli, E., Altman, R., Arora, S., von Arx, S., Bernstein, M. S., Bohg, J., Bosselut, A., Brunskill, E., Brynjolfsson, E., Buch, S., Card, D., Castellon, R., Chatterji, N., Chen, A., Creel, K., Davis, J., Demszky, D., . . . Liang, P. (2021). On the opportunities and risks of foundation models. arXiv. / arXiv.2108.07258DOI: 10.48550

Breazeal, C. (2003). Toward sociable robots. *Robotics and Autonomous Systems*, *42*(3–4), 167–175. DOI: 10.1016/S0921-8890(02)00373-1

Brodie, R. J., Hollebeek, L. D., Jurić, B., & Ilić, A. (2011). Customer engagement: Conceptual domain, fundamental propositions, and implications for research. *Journal of Service Research*, *14*(3), 252–271. DOI: 10.1177/1094670511411703

Brown, T. B., Mann, B., Ryder, N., Subbiah, M., Kaplan, J., Dhariwal, P., Neelakantan, A., Shyam, P., Sastry, G., Askell, A., Agarwal, S., Herbert-Voss, A., Krueger, G., Henighan, T., Child, R., Ramesh, A., Ziegler, D. M., Wu, J., Winter, C., & Amodei, D. (2020). Language models are few-shot learners. *Advances in Neural Information Processing Systems*, *33*, 1877–1901.

Choi, S., Lim, H., & Kim, J. (2022). Service robots in the banking sector: Customer satisfaction and trust in robot-mediated service encounters. *International Journal of Bank Marketing*, *40*(4), 678–697. DOI: 10.1108/IJBM-06-2021-0250

Dautenhahn, K. (2022). Socially intelligent robots: Dimensions of human-robot interaction. *Philosophical Transactions of the Royal Society of London. Series B, Biological Sciences*, *362*(1480), 679–704. DOI: 10.1098/rstb.2006.2004 PMID: 17301026

Davis, F. D. (1989). Perceived usefulness, perceived ease of use, and user acceptance of information technology. *Management Information Systems Quarterly*, *13*(3), 319–340. DOI: 10.2307/249008

Epley, N., Waytz, A., & Cacioppo, J. T. (2007). On seeing the human: A three-factor theory of anthropomorphism. *Psychological Review, 114*(4), 864–886. DOI: 10.1037/0033-295X.114.4.864 PMID: 17907867

Henschel, A., Laban, G., & Cross, E. S. (2021). What makes a robot social? A review of social robots from the perspective of human-robot interaction. *International Journal of Social Robotics, 13*(6), 1–15. DOI: 10.1007/s12369-020-00736-8

Hochschild, A. R. (1983). *The managed heart: Commercialization of human feeling*. University of California Press.

Hollebeek, L. D. (2011). Demystifying customer brand engagement: Exploring the loyalty nexus. *Journal of Marketing Management, 27*(7–8), 785–807. DOI: 10.1080/0267257X.2010.500132

Horton, D., & Wohl, R. R. (1956). Mass communication and para-social interaction: Observations on intimacy at a distance. *Psychiatry, 19*(3), 215–229. DOI: 10.1080/00332747.1956.11023049 PMID: 13359569

Huang, M. H., & Rust, R. T. (2021). A strategic framework for artificial intelligence in marketing. *Journal of the Academy of Marketing Science, 49*(1), 30–50. DOI: 10.1007/s11747-020-00749-9

Ivanov, S., Seyitoğlu, F., & Markova, M. (2022). Robot human interaction in hotels: Managers' perspective. *Information Technology & Tourism, 22*(4), 507–535. DOI: 10.1007/s40558-020-00180-2

Kim, S., & Park, J. (2022). Effects of hotel service robots on customer experience and behavior. *International Journal of Contemporary Hospitality Management, 34*(3), 1162–1180. DOI: 10.1108/IJCHM-01-2021-0024

Lemon, K. N., & Verhoef, P. C. (2016). Understanding customer experience throughout the customer journey. *Journal of Marketing, 80*(6), 69–96. DOI: 10.1509/jm.15.0420

Lu, L., Zhang, P., & Zhang, T. (2023). Leveraging AI-powered service robots to deliver better customer experience: Insights from hotel guests. *International Journal of Hospitality Management, 108*, 103328. DOI: 10.1016/j.ijhm.2022.103328

Mehrabian, A., & Russell, J. A. (1974). *An approach to environmental psychology*. MIT Press.

Misselhorn, C., Pompe, U., & Stapleton, M. (2021). Ethical considerations regarding the use of social robots in the fourth age. *GeroPsych, 26*(2), 121–133. DOI: 10.1024/1662-9647/a000088

Mori, M. (1970). The uncanny valley (K. F. MacDorman & N. Kageki, Trans.). IEEE Robotics & Automation Magazine, 19(2), 98–100. (Original work published 1970)DOI: 10.1109/MRA.2012.2192811

Mubin, O., Alnajjar, F., Jishtu, N., Alsinglawi, B., & Al Mahmud, A. (2022). Service robots in healthcare: A review on current applications and future trends. *Robotics (Basel, Switzerland)*, *11*(1), 12. DOI: 10.3390/robotics11010012

Nass, C., & Moon, Y. (2000). Machines and mindlessness: Social responses to computers. *The Journal of Social Issues*, *56*(1), 81–103. DOI: 10.1111/0022-4537.00153

Nass, C., & Reeves, B. (1996). *The media equation: How people treat computers, television, and new media like real people and places*. Cambridge University Press.

Olaosebikan, O., Strickland, L., & Mubin, O. (2023). Exploring emotional engagement in public-space robot deployments. *Computers in Human Behavior*, *139*, 107530. DOI: 10.1016/j.chb.2022.107530

Open, A. I. (2023). GPT-4 technical report. arXiv. /arXiv.2303.08774DOI: 10.48550

Parasuraman, A., Zeithaml, V. A., & Berry, L. L. (1988). SERVQUAL: A multiple-item scale for measuring consumer perceptions of service quality. *Journal of Retailing*, *64*(1), 12–40.

Picard, R. W. (1997). *Affective computing*. MIT Press. DOI: 10.7551/mitpress/1140.001.0001

Robinson, H., MacDonald, B., & Broadbent, E. (2022). The role of healthcare robots for older people at home: A systematic review. *International Journal of Social Robotics*, *6*(4), 575–591. DOI: 10.1007/s12369-014-0242-2

Schuller, B. W., Batliner, A., Bergler, C., Messner, E. M., Hamilton, A., Amiriparian, S., Baird, A., Rizos, G., Schmitt, M., Stappen, L., Pérez-Espinosa, H., Frick, M., Doyle, P., Triantafyllopoulos, A., & Müller, R. (2021). The INTERSPEECH 2021 computational paralinguistics challenge: COVID-19 cough, COVID-19 speech, escalation & primates. *Interspeech*, *2192–2196*, 431–435. Advance online publication. DOI: 10.21437/Interspeech.2021-19

Seyitoğlu, F., & Ivanov, S. (2020). A conceptual framework of the service delivery system design for hospitality firms in the (post-) viral world: The role of service robots. *International Journal of Hospitality Management*, *91*, 102661. DOI: 10.1016/j.ijhm.2020.102661 PMID: 32952262

Shi, X., Luo, H., & Li, J. (2023). Service robots in the hotel industry: Role of AI features in shaping guest acceptance and trust. *Journal of Hospitality Marketing & Management, 32*(2), 195–215. DOI: 10.1080/19368623.2022.2115279

So, K. K. F., King, C., Sparks, B. A., & Wang, Y. (2016). The role of customer engagement in building consumer loyalty to tourism brands. *Journal of Travel Research, 55*(1), 64–78. DOI: 10.1177/0047287514541008

Wirtz, J., Patterson, P. G., Kunz, W. H., Gruber, T., Lu, V. N., Paluch, S., & Martins, A. (2022). Brave new world: Service robots in the frontline. *Journal of Service Management, 29*(5), 907–931. DOI: 10.1108/JOSM-04-2018-0119

Zeithaml, V. A., Berry, L. L., & Parasuraman, A. (1996). The behavioral consequences of service quality. *Journal of Marketing, 60*(2), 31–46. DOI: 10.1177/002224299606000203

Zeng, Z., Pantic, M., Roisman, G. I., & Huang, T. S. (2022). A survey of affect recognition methods: Audio, visual, and spontaneous expressions. *IEEE Transactions on Pattern Analysis and Machine Intelligence, 31*(1), 39–58. DOI: 10.1109/TPAMI.2008.52 PMID: 19029545

Zhang, H., Li, Y., & Wang, X. (2024). Emotionally intelligent service robots powered by large language models: A framework for affective adaptation in customer-facing interactions. *International Journal of Human-Computer Studies, 181*, 103155. DOI: 10.1016/j.ijhcs.2023.103155

Chapter 7 Anthropomorphic Artificial Intelligence Robots in Service Industries: Enhancing Customer Experience and Brand Engagement

Kavya HImanshu Vyas
http://orcid.org/0009-0000-2397-5968
Vishwakarma Institute of Technology, Pune, India

Dhananjay Abhilash Bhagat
http://orcid.org/0009-0009-1100-3219
Vishwakarma Institute of Technology, Pune, India

Yashraj Umesh Wagh
Vishwakarma Institute of Technology, Pune, India

Yashraj Dhondiba Kewate
http://orcid.org/0009-0006-8483-4388
Vishwakarma Institute of Technology, Pune, India

ABSTRACT

Anthropomorphic artificial intelligence (AI) robots are transforming customer-facing service industries by enabling intelligent and emotionally responsive interactions. Unlike traditional automation systems, these robots imitate human characteristics such as speech, gestures, facial expressions, and adaptive communication to improve

DOI: 10.4018/979-8-3373-6640-1.ch007

customer experience and operational efficiency. This chapter examines the role of anthropomorphic AI robots in enhancing customer engagement, consumer trust, and brand perception across hospitality, retail, healthcare, aviation, and banking sectors. The study proposes the Human-Centered Robotic Service Framework (HCRSF), integrating emotional intelligence, service efficiency, brand alignment, and ethical sensitivity. The findings highlight that while anthropomorphic robots improve personalization and service innovation, organizations must address ethical concerns including privacy, workforce displacement, and algorithmic bias through responsible AI governance and human-centered design.

1. INTRODUCTION

Over the past few decades, service design, a discipline and practice, has changed dramatically. It has evolved from the marketing/operations/human factors world into a broader perspective that focuses on the creation of a meaningful interaction between persons, organizations and increasingly autonomous systems. In today's world, service designers operate at the crossroads of technology, business, and humanity, and automation and AI have driven a new shape to this crossroad. Along with new complexities, opportunities, and ethical questions, service design is faced with these new challenges as organizations embrace intelligent technologies to increase efficiency and scale. Most importantly, while the new technology is important, the focus on people has also been strengthened. Services that involve human-centred thinking have become more relevant than ever because of the rise of automated systems, and they are critical in ensuring the dignity, agency and emotional well-being of people. (Hwang et al., 2026)

The question of the impact of automation and AI on service design goes hand-in-hand with an examination of the origins of service design. The services were traditionally designed implicitly, that is to say, they were created as a consequence of the habits of the organization, cultural norms and operating restrictions, rather than intended design. People-to-people service interactions occurred in the early days of service, with those interactions occurring in the hospitality, transportation and health care sectors. Although early innovations like kiosk self-service or automated phone menus came along, the main mode of service was still human workers operating as the service window. The late 20th century was the period when service design became more clearly defined, as a result of design thinking, ethnographic and systems thinking research. Designed services were understood not as a single act but as a series of interactions, both face-to-face (frontstage) and non-face-to-face (backstage), that allow the delivery of value. This change changed the perception of services as individual points of contact to one of orchestrated journeys. (Mortati & Viana Mundstock Freitas, 2026)

With the rise of digital transformation in the 2000s, more channels, online interactions and data-driven personalisation drove service designers to make more additions to their toolkits. But even then, technology was mostly a supporting player: a way to communicate, or to be efficient or convenient, not an independent entity. The designer's interest was still on the human need, emotions and behaviours and the way coherent services are delivered within Web platforms, mobile applications, call centers and physical environments. Empathy, co-creation and iterative prototyping were the key processes, focussing on people, not machines. (Brynjolfsson et al., 2025)

The advent of automation and AI is another more radical change. Unlike the previous technology, AI not only assists or speeds up human decisions, it can make decisions, take actions and learn from the user's behavior. Customer enquiries can be answered, their needs can be anticipated, their actions can be recommended, resources can be allocated and even the flow of the service itself can be influenced by automated systems. In some instances these systems run without human assistance, providing scale and uniformity that is not possible for humans to accomplish. This feature has encouraged businesses to see automation as a necessity rather than an extra feature and move towards a smart and efficient business with an unprecedented personalized experience at an unprecedented speed. (Saviano et al., 2025)

In almost all sectors, services are becoming increasingly powered by artificial intelligence. Automated chatbots and recommendation engines are helping customers navigate lengthy purchase cycles in the retail sector.In retail, AI-powered chatbots and recommendation systems are assisting customers in navigating long sales cycles. In finance, algorithmic systems evaluate creditworthiness, identify fraud, and give automated investment advice. In the medical field, AI applications can assist in diagnosing diseases, recommending treatment options, and streamlining the process of patient management. In healthcare, AI can aid in diagnosis, treatment recommendations, and patient care management. In the field of transportation, autonomous vehicles, predictive maintenance, and intelligent routing systems have appeared. At the same time, decision-making systems are becoming more and more automated in government services, such as distributing benefits, processing applications and allocating resources. These examples are not only about new tools, but also the reconfiguration of the service system. Any service "staff" can be made up of chatbots, recommendation algorithms, robotic systems and predictive models, all of which can interact with the user in the same ways as staff. (Hardcastle et al., 2025)

As automation increasingly takes center stage, however, there are questions that emerge with regard to the human experience: What happens when services are designed, delivered or mediated by machines? What is the role of automation to enhance instead of reduce user agency, comfort and trust? How to keep the "human touch" and "emotional feeling" in "efficient systems"? The following questions are at the core of the current service design world. (Schmager et al., 2025)

Though automation has numerous benefits, it also presents considerable tension between efficiency and empathy. Efficiency has always been a desired result of service delivery. This reduces waiting time, speeds up processing, minimises mistakes and saves organizations money, while also providing cost savings to users. These are the things that automated systems are good for, and they're fast and consistent. For instance, when passengers arrive at an airport, they can check in thousands of passengers without getting fatigues and without making mistakes by using an automated check-in system. A customer service chatbot can answer customers' queries at any time of the day. Algorithms can be used to sort through vast amounts of data and identify patterns and relationships that humans would never notice without computer assistance. (Vo et al., 2026)

But efficiency doesn't mean good service experience. Empathy, meaning the ability to understand and feel the emotions of the user, is still a human trait that stands out. There are numerous situations in which interpretation of nuance, active listening, reassurance or ethical judgment is needed in a service context. In healthcare, patients prefer to be cared for as well as cared for; in financial services, in addition to being well-informed, they also want to feel understood. Human warmth can make a difference in simple activities such as retail or hospitality beyond a transaction and into a memorable experience. These nuances can be difficult to master with automation, particularly if users feel frustrated, fear, confused, or vulnerable. (Walport et al., 2025)

The balance between efficiency and empathy is even stronger when automated systems replace or minimize human touchpoints. Some users prefer to be fast and independent, others feel they are being overlooked or alienated. An automated system that doesn't comprehend a user's question can exacerbate frustration instead of alleviating it. Likewise, systems that are very dependent on data collection can also erode trust if users feel they are being monitored or manipulated. Automated decision-making has also had biases or unfair results in some instances, which have uncovered flaws in datasets or algorithms. The challenges highlight that life, identities and rights of people should be taken into account when designing automation. (Yadav et al., 2025)

This is where human-centred design becomes indispensable. Far from being made obsolete by automation, human-centred principles provide the foundation for creating meaningful and equitable automated services. At its core, human-centred design seeks to understand and advocate for people—their needs, desires, limitations, and contexts. When applied to automated systems, this approach encourages designers to consider not only what technology can do but what it should do. It shifts the focus from optimizing for speed and cost to optimizing for human well-being, agency, and trust. (Liu, 2025)

Human-centred design also provides tools and methods for anticipating unintended consequences. Techniques such as field research, journey mapping, prototyping, and co-creation allow designers to uncover user frustrations, identify hidden assumptions, and test how automated systems operate in real-world contexts. These insights help designers develop safeguards, escalation paths, and human-in-the-loop models that ensure automation enhances rather than diminishes the service experience. For example, a chatbot may be efficient at handling routine queries, but a human agent should be available when users express emotional distress or require complex problem-solving. Similarly, automated decision-making systems should be designed with transparency mechanisms that allow users to understand and challenge decisions that affect them. (Zhao et al., 2025)

The future of service design will require designers to embrace a hybrid mindset—one that acknowledges the power of automation while championing the inherent value of human connection. Designers must learn to orchestrate interactions across both human and non-human actors, ensuring that each plays a role suited to its strengths. Automation should take on tasks that relieve people of cognitive load, repetitive work, or logistical complexity. Humans, meanwhile, should focus on relationship-building, judgment, creativity, and empathy—activities where their capabilities truly shine. (Park, 2025)

Moreover, the role of the service designer itself is evolving. Designers are no longer tasked merely with shaping customer journeys; they must also understand data systems, algorithmic behavior, and the ethical implications of AI. They must collaborate with technologists, data scientists, and engineers while representing the voice of users who may not have the expertise to navigate complex automated environments. This expanded role demands new skills, such as interpreting algorithmic outputs, identifying ethical risks, and designing for transparency. It also requires a commitment to inclusivity, ensuring that automated systems do not replicate or exacerbate societal inequalities. (Maier et al., 2026)

In summary, the changing landscape of service design is defined by the rise of automation and AI and the increasing need to balance efficiency with empathy. Service design has evolved from a focus on orchestrating human interactions to shaping complex ecosystems where human and automated actors coexist. While automation holds immense potential to improve service delivery, it also introduces new challenges related to fairness, trust, vulnerability, and emotional connection. Human-centred design provides the guiding principles necessary to navigate these challenges, ensuring that services remain grounded in the realities and values of the people they serve. The future of service design will not be a choice between humans and machines, but rather the thoughtful integration of both—a future where technology amplifies human strengths and helps deliver more meaningful, equitable, and resilient service experiences. (Stanford HAI, 2026)

Figure 1. Timeline of service design evolution

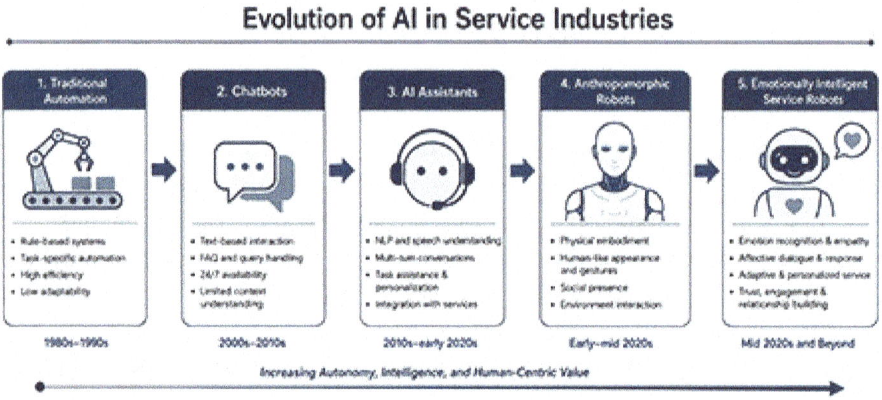

Fig. 1. Evolution of AI technologies in service industries from traditional automation to emotionally intelligent service robots.
Note: NLP = Natural Language Processing

2. FOUNDATIONS OF HUMAN-CENTRED SERVICE DESIGN IN AN AUTOMATED WORLD

Human centred service design (HCD) is a compass which has long been used to design meaningful service experiences in context and ethically. It was traditionally derived from user-centred design, participatory design, cognitive psychology and anthropology. It is a discipline of understanding individuals, what they want, value, fear, desire and expect and how services can be intentionally designed around human experience and not technological possibility or operational convenience. The landscape of service design, however, has changed with the advent of automation and AI. To design systems that are more than just tools, but instead are autonomous actors that can perform tasks and make decisions and influence behaviour, the designers now need to take human-centred thinking to the next level. However, with the increasing distribution, data-driven and machine-mediated nature of services, the principles of human-centred design need to change to ensure that it does not diminish dignity, trust, and inclusion, but rather strengthen them through automation. (Schmager et al., 2025)

Empathy—the feeling of understanding the world as others see it and feel it is the starting point of human centred design. Empathy in service design is more than the identification of user needs; it is the emotions, social situations, cultural norms and expectations that influence the way in which users make sense of service interactions. The designers engage in the user's life experiences with interviews, observational research, shadowing, contextual inquiry, and participatory workshops. These techniques expose the functional needs and also the emotional tensions, workarounds, frustrations and moments of delight. These understandings inform the design of services with a human touch. (Liu, 2025)

But the automation of agents introduces an added layer of orientation that is more empathetic. If part of the service is not human-driven, but algorithm-driven or conversational-driven, designers need to think about user needs and how they relate to machine behavior. Automated systems can function behind the scenes, evaluating what users like, anticipating user behavior, or even dictating user experiences. This presents a new challenge: not only to understand the behavior of the users, but also the interpretation by machines of the behavior of the users, and where they may be misaligned. If someone doesn't respond to a message in a chat, for instance, it could be seen as disengagement and lead to an escalation or a concluding of the conversation. A browsing pattern could be interpreted as intention and result in unnecessary personalisation. Human-centred design, therefore, demands the broader empathy: empathy for the human and informed awareness of the perception of the human by the machine. (Zhao et al., 2025)

Another core philosophy of human-centred service design is that services are best designed in collaboration with the users, providers and affected by the services. In the past, co-creation was about workshops where customers, frontline, stakeholders and designers all think about the non-digital parts of the process, interactions and journeys. In the context of automation, the concept of co-creation should now be extended to include the creation of rules, data models and interaction patterns that influence the automated behaviour. Algorithms can't access users' and frontline staff's tacit knowledge. For example, a bank teller might recognize that a customer who repeats her questions is nervous, not confused; that some words carry special meaning. This kind of subtlety is hard to get in training data unless you involve humans in the design process. (Vega-Carillo & Poikela, 2025)

Early detection of ethical issues is also one of the main areas where co-creation has an important role. Automated systems can sometimes perpetuate bias, misinterpret culturally specific behaviours, or make decisions that seem unclear or unfair. Engaging a variety of stakeholders brings to light possible harms or blind spots to consider, helping to ensure systems are created with fairness and inclusivity. This cooperative stance ensures that AI isn't fostering alienation or discrimination, especially among marginalized or vulnerable communities. (Papagiannidis et al., 2025)

Iteration and prototyping is a core part of human centred design and involves building first-class prototypes, testing out hypotheses early in the process and adjusting the solution through an iterative method. Prototyping is increasingly complex and critical in automated environments. Rapid prototypes need to approximate intelligent behavior, though not completely, in order to help designers gain insight into the expectations and reactions of their users to automated decision making. For example, a prototype chatbot could have predefined answers to simulate conversational AI, and see how the user reacts to the tone, clarity, or escalation path. Likewise, an artificially created recommendation engine can be utilized to experiment with the user's understanding of recommendations or explanations. (Maier et al., 2026)

Prototyping also helps to understand the emotional impact of automation in service interactions. Errors committed by machines are different from those committed by people. An apology, or explanation of an error, from a human agent can build trust again, while an abrupt halt or unpredictable response from an automated system can lead to frustration or anxiety in a short period. Through iterative testing, designers can proactively detect these kinds of problems and find solutions to them, for example, providing feedback, transparency prompts and graceful recovery mechanisms. It also underscores the significance of hybrid service models where automation is used for routine tasks, and humans are available to deal with emotional complexity or edge cases. (Walport et al., 2025)

Accessibility is one of the key principles of the design of human-centred services. Services need to be interpreted, understandable, usable and accessible for individuals of different capabilities, backgrounds, and contexts. Designing for accessibility is even more critical—and more complex—when you're designing with automation. Voice interaction, predictive models or personalized content are common features of intelligent systems which can cause accessibility problems. Voice-based systems may also have limitations for certain users, such as people with accents, speech problems or background noise. Historical data-based predictive systems can end up reinforcing inequities, as they might exclude users with atypical patterns. Personalized experiences can cause confusion for users if they don't understand why the system is different for them than for others. (Hwang et al., 2026)

Accessibility should be taken into account from the outset of research through to implementation of a product. This encompasses creating interfaces that can be interacted with in various ways (sight, sound, touch, etc.), offering clear explanations of the behavior of the system, and giving the opportunity to review or override decisions made by the system. More important, however, is the need to build emotional accessibility into your systems, avoiding the impression that users are being judged, monitored, or that they lack control. (ISO 9241-210, 2019)

Transparency stands as another key foundation. Traditional services rely on human explanation—employees can clarify policies, justify decisions, or answer questions. Automated systems often lack this capacity, leaving users uncertain about why a decision was made or how information is used. Human-centred service design emphasizes transparency not as a technical feature but as a relational one: users must feel informed, respected, and in control. Transparency in automated services can take many forms: explanations of decisions, indications of when a user is interacting with a machine rather than a person, and clear privacy disclosures. Designers must craft these explanations to be understandable rather than overly technical, balancing accuracy with clarity. (Yadav et al., 2025)

A growing foundational concept in human-centred service design is trust. Trust emerges when users feel confident that a service will behave predictably, fairly, and in alignment with their values. Automation introduces new uncertainties that can undermine trust: opaque algorithms, inconsistent behavior, data misuse, or errors without clear channels for resolution. Human-centred design approaches trust not as something that can be engineered solely through security or accuracy, but as an emergent quality shaped by user experience, emotional connection, and the perceived integrity of the system. Designers must anticipate moments where trust may be challenged and create pathways for users to seek help, understand decisions, or regain control. Hybrid service models again play a critical role here, as human support can complement automated interactions and restore trust when systems fail. (Papagiannidis et al., 2025)

A central tension in designing automated services is balancing predictability with personalization. Predictability is essential for comfort—users want to understand how a system works and what to expect. Personalization, however, often requires systems to adapt dynamically. Human-centred design must strike a balance that supports user comprehension while enabling meaningful personalization. This requires clear communication about why the system behaves in certain ways and how users can influence or adjust this behavior. It also means providing consistency across channels, even when automation plays different roles in each. (Hardcastle et al., 2025)

The integration of automation also forces service designers to consider new ethical dimensions. When algorithms make decisions that affect people's access to services—such as loan approvals, healthcare recommendations, or benefit allocations—designers must ensure that these systems reflect fairness, accountability, and respect for human rights. Ethical design involves evaluating data sources for bias, establishing mechanisms for auditing algorithms, and creating escalation channels where users can challenge automated decisions. Human-centred design expands into responsible AI design, merging ethical principles with experiential considerations. (Stanford HAI, 2026)

A particularly important foundation for the future is designing for agency—the user's ability to make informed choices and maintain control. In automated systems, agency is at risk when systems make decisions without user input, steer users subtly through personalized nudges, or present choices in a way that favors the system's goals over the user's. Human-centred design places agency at the center, ensuring that users have meaningful control over their data, preferences, and interactions. This includes providing options to opt out of personalization, the ability to override automated options, and interfaces that support reflection rather than coercion. (Vega-Carillo & Poikela, 2025)

Finally, the future of human-centred service design in automated environments calls for new designer competencies. Service designers must now be fluent in the logic and limitations of AI systems, understanding how data flows, how models are trained, and how decisions are made. They do not need to be data scientists, but they must be capable collaborators who can translate human insights into technical requirements and ethical constraints. Designers must also become facilitators who help organizations navigate the cultural changes required to adopt automation responsibly. This includes advocating for user needs, challenging assumptions, and ensuring that technological innovation aligns with social values. (Brynjolfsson et al., 2025)

Human-centred design in an automated world is thus both a continuation and an evolution of the discipline. It retains its foundational commitment to empathy, inclusivity, ethics, and human dignity, while expanding its scope to include machine behavior, algorithmic interpretation, and hybrid ecosystems. It recognizes that services are no longer delivered solely by people or interfaces but by intelligent, learning systems embedded within complex sociotechnical networks. The designer's role becomes that of a conductor—coordinating interactions across human and machine actors to create holistic experiences that are meaningful, trustworthy, and equitable. (Liu, 2025)

In this evolving landscape, human-centred design serves as a vital counterbalance to the accelerating pace of technological innovation. As organizations pursue efficiency, scale, and competitive advantage through automation, human-centred principles ensure that technology remains grounded in human realities. They remind us that services exist not to demonstrate technological prowess but to improve lives. As automation becomes more capable, it becomes even more essential to design intentionally for empathy, agency, and ethical integrity. (Mortati & Viana Mundstock Freitas, 2026)

Figure 2. Human-centred design framework adapted for AI systems

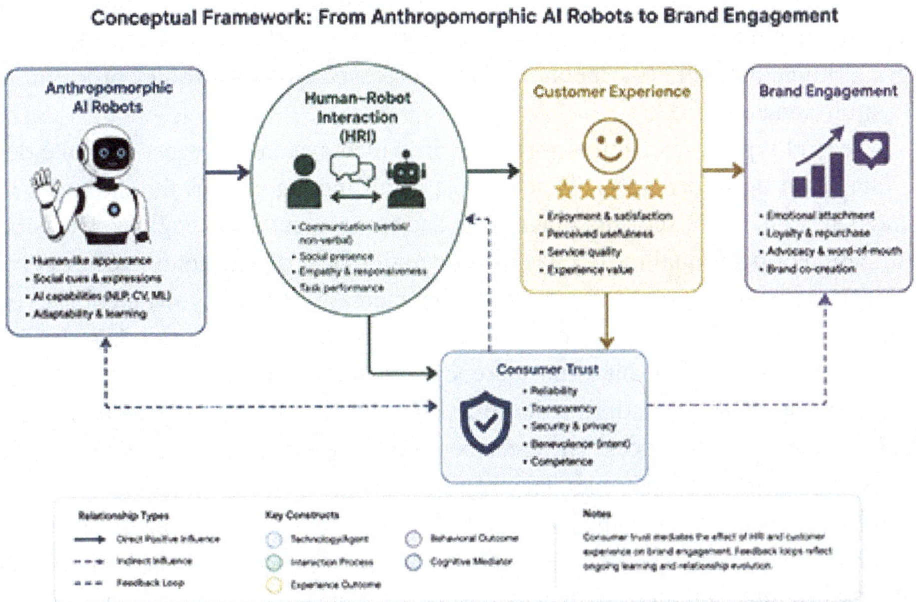

3. AUTOMATION IN SERVICE ECOSYSTEMS: OPPORTUNITIES AND LIMITATIONS

Automation has contributed to push for change in the modern day of service ecosystems. Today, automated systems are present in industries where human working was the only thing counted on, from chatbots and self-checkout terminals, to predictive analytics and robotic process automation (RPA). The service ecosystem, which is typically made up of humans, processes, organizational rules, and customer expectations, has grown to incorporate algorithmic agents which can act independently, learn from data, and affect the behaviour of humans. This transition creates new possibilities for efficiencies, customization, and scalability; however, it also opens up risks and inequities and presents unanticipated opportunities and challenges. Balanced, resilient, and human-centred service environments will be designed with the understanding of the dual role of automation, as an enhancer and a disruptor. (Hwang et al., 2026)

The automation of services can be explained by service autonomy and interactions in different levels. The most basic is task automation which involves machines performing repetitive or rules-based tasks that used to take humans. Some examples are automated form processing, appointment scheduling or routing workflows. This can be simple automation, often where manual tasks are replaced by more speedy and reliable output. Task automation has become more and more prevalent because it brings immediate benefits: labour savings, decreased errors, and quick processing. (Brynjolfsson et al., 2025)

A second type is decision automation, in which systems suggest or make decisions based on information. That is a list of all the algorithms that are used in fraud detection, credit scoring, diagnostic, and recommendation engines. Decision automation brings a qualitative change in operation of services: automated systems no longer assist the human decision maker but increase his/her involvement or even replace him/her. This can help to make services more responsive and data-driven, but can also pose risks of bias, transparency, and accountability. (Vo et al., 2026)

Conversational automation is the third type, which involves virtual assistants, chatbots, and voice interfaces. These can be used to carry out tasks that can normally be accomplished by a customer service representative, like addressing inquiries, providing assistance, or addressing problems. Conversational automation is designed to simulate the human conversation in order to lessen the human interaction required in simple interactions. This is especially appealing in high volume service areas like banking, public services and telecommunications. (Saviano et al., 2025)

Lastly, workflow automation connects several automated elements throughout service interactions, enabling end-to-end systems where tasks, decisions and interactions move smoothly. Examples include a chat-inquiry function, automated document ingestion, machine-learning based risk assessment, and digital notifications of approval. Workflow automation brings in a layer of change that's more systemic and is changing the architecture of the service delivery itself.

The potential benefits from this kind of automation are huge. There are several distinct advantages, one of which is efficiency. Automated systems can do repetitive tasks quicker, more consistently and at potentially a lower cost than humans. That helps companies redirect human resources to more valuable activities that demand empathy, creativity, and complex problem-solving. In healthcare, for instance, automation can streamline administrative duties, allowing healthcare professionals more time for patient care. These improvements in efficiency aren't only operational, they're also user-centric – with fewer queues, quicker responses, and streamlined processes, it's a better user experience.

Scalability is another important opportunity. Computerized systems can manage a lot of interactions at once without having to add more staff or staffing. This facilitates the extension of services by organisations across regions, across time zones and across customer segments. In public services and global companies, scaling up or down is a crucial question, and the answer is often unpredictable. Automation enables organisations to scale their customer service operations in response, to customer demand, e.g., during crises, seasons or promotions without affecting performance.

Another benefit is customization. Big data can be analysed by automated systems to gain insights into user preferences, behaviours, and needs. Users have come to expect personalization from digital services, whether that's a set of recommendations that match their taste, a user interface that adjusts to their skill level, or a service that predicts and prevents problems. Personalization through AI can deliver personalized product recommendations, customized learning journeys, or health monitoring alerts. These features transform service interactions from reactive to proactive and generic to personalized.

There are also advantages to automation with regard to reliability of processes. Unlike humans, machines do not become tired, lose their temper, or forget things, they are good at doing things that need consistency and accuracy. This reliability is critical in sectors such as finance, logistics, and manufacturing, where errors can have significant consequences. Automated quality checks, real-time monitoring and predictive maintenance systems minimize service interruptions and enhance system stability.

But it comes with significant restrictions that also define what can and can't be automated. The first is the fragility of automated systems. They are adept at straightforward situations and clearly defined tasks, but they are not so good at dealing with uncertainty, newness, and details that are not contextually obvious, which are the domains where people are most comfortable. Automated systems use patterns or rules based on training data; if they are presented with data outside of their training set, they may respond in an erratic or in an incorrect manner. For instance, the use of a chatbot can be problematic if a user uses sarcasm, shows emotion, or communicates lengthy sentences and stories that are not in the usual flow of conversation.

Another constraint is ease of bias and fairness. Automated decision making systems are only as objective as the information they are taught. Automated decisions can reinforce or perpetuate the biases in historical data (including ones that have been intentionally introduced or are unintentional), such as preferencing, underrepresentation or structural inequities. That's especially worrisome in contexts with high stakes, such as hiring, lending, and criminal justice, where algorithmic decisions can have real-world consequences for individuals. Bias can be subtle, occurring during development in a way that is not apparent. Recognition, minimisation and communication of biases remain key challenges in design and policy.Automation can also lead to opacity or a lack of transparency. Many machine-learning models operate as "black boxes," making decisions based on complex statistical relationships that are difficult to explain. This opacity can erode trust when users cannot understand why they received a particular outcome. Transparency issues are exacerbated when automated systems replace human roles that traditionally involved personal explanation and negotiation. The inability to appeal or question an algorithm's decision can create frustration and a sense of helplessness among users.

Another challenge is loss of human touch, particularly in emotionally sensitive service encounters. While automation can handle factual or routine interactions effectively, it lacks the emotional intelligence, empathy, and moral reasoning that humans bring to complex situations. For example, a chatbot delivering bad news, misunderstanding distress, or failing to respond compassionately can leave users feeling alienated. The risk is not only experiential but reputational; poor automated interactions can damage trust in the entire service ecosystem.

Automation also poses risks related to overdependence. When organizations rely heavily on automated systems without sufficient human oversight, they become vulnerable to systemic failures. A technical glitch, cyberattack, or data corruption incident can halt entire service processes. Overdependence reduces organizational resilience, making human expertise harder to maintain and reintroduce when needed. This is particularly problematic in critical infrastructure sectors, where manual fallback procedures may no longer be well-practiced or even feasible.

In addition, automation can create job displacement or transform the nature of work in ways that impact employee morale and identity. While automation rarely eliminates entire roles, it often shifts responsibilities, removing routine tasks and requiring employees to handle exceptions, emotional labor, and system monitoring. These new responsibilities can be stressful or poorly defined. Service roles that once required interpersonal skills may now demand data literacy or oversight of algorithmic processes. Ensuring that employees are empowered, trained, and valued in automated environments is essential for sustainable adoption.

Across industries, the interplay between opportunities and limitations shapes how automation is deployed. In retail, self-checkout systems improve speed and reduce staffing needs but introduce accessibility challenges and frustrations with machine errors. In healthcare, automated triage tools support efficiency but risk misdiagnosis if symptoms do not match training data. In finance, algorithmic trading and credit scoring offer speed and consistency but may hide biases that disproportionately affect certain groups. Public services face the dual challenge of high demand and high accountability; automation can increase efficiency but must meet transparency and equity expectations.

The key insight across these contexts is that automation is not inherently beneficial or harmful—it is shaped by how it is designed, implemented, and governed. Successful service ecosystems adopt a hybrid approach where humans and automated systems complement each other. Humans provide context, empathy, and moral judgment; machines provide speed, scale, and pattern recognition. Hybrid models recognize that neither humans nor machines alone can deliver optimal service in complex environments.

Designing such hybrid ecosystems requires a nuanced understanding of human–machine boundaries. The question is not simply what can be automated, but what should be automated. Designers must evaluate whether automation enhances the experience or compromises it. This evaluation considers user expectations, emotional dynamics, risk levels, and the broader social implications of removing—or retaining—human involvement.

For example, in domains where trust, empathy, or negotiation are essential—healthcare consultations, financial advising, conflict resolution—full automation is rarely appropriate. Automation may support these interactions with background analysis or administrative assistance, but humans remain central. Conversely, in domains where speed and volume dominate—information retrieval, queue management, security checks—automation can take a leading role with humans providing oversight.

A key design challenge is ensuring a graceful handoff between automated and human agents. Poorly designed transitions can leave users repeating information, losing context, or feeling abandoned by the system. Effective handoff requires clear triggers, seamless data sharing, and a well-explained escalation process. This reinforces that automation does not replace humans but reshapes when and how they appear in the service journey.

Another important consideration in designing automated ecosystems is feedback loops. Automated systems learn from user interactions and adapt over time, which means those interactions influence future outcomes. Designers must monitor these systems to ensure that learning does not drift into harmful or unintended directions. Feedback loops also allow users to influence system behavior; offering options to correct errors, flag concerns, or adjust preferences reinforces agency and trust.

Finally, automation raises questions about governance and accountability. Who is responsible when an automated decision harms a user or causes financial loss? How can organizations ensure compliance, auditability, and fairness in automated systems? Governance frameworks must include operational monitoring, ethical oversight, user rights, and clear accountability structures. Designing these frameworks is as much a part of service design as crafting interfaces or workflows.

In conclusion, automation offers tremendous potential to enhance service ecosystems, making them more efficient, scalable, personalized, and reliable. Yet it also introduces limitations related to brittleness, bias, opacity, emotional insensitivity, overdependence, and ethical risk. Navigating these tensions requires a human-centred perspective—one that recognizes the strengths and weaknesses of both humans and machines. By embracing hybrid service models, prioritizing human agency, and designing with transparency and empathy, organizations can harness automation in ways that enrich rather than diminish service experiences. Automation, when guided by human-centred principles, becomes not a replacement for human service but a powerful component of a more adaptive, responsive, and inclusive service future.

Figure 3. Automation levels in service delivery

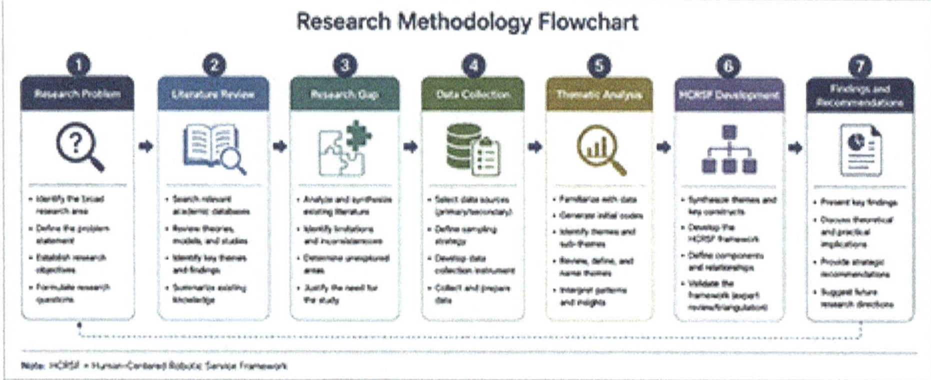

4. AI AS A CO-DESIGNER: TOOLS, METHODS, AND COLLABORATION MODELS

Artificial Intelligence has now gone from back-office technology to a creative and strategic partner. AI's involvement in service design represents a significant change, as it increasingly engages users in ways that were previously more the domain of problem solving, imagination and user empathy, which are traditionally regarded as uniquely human skills. The role of AI as a co-designer shifts the paradigm: AI is not simply a tool to be used by designers, but also a partner that expands the scope of their thinking, provokes questions and speeds up their processes. This subchapter will dive into the practical implications of co-design with AI, tools and methods driving this transformation, and how collaborative intelligence can be used to improve processes and outcomes that have a human-centred approach.

4.1 Re-defining Creativity and Collaboration in Design

For much of design history, creativity has been understood as an inherently human activity—rooted in lived experience, intuition, and cultural context. AI disrupts this assumption not by replicating human creativity but by *supplementing* it with powerful computational creativity. Machine learning models can generate patterns that humans might not easily see, propose variations rapidly, and work through massive search spaces that would overwhelm human cognition.

However, in collaborative practice, AI's creative input is not intended to supplant human creativity, but to serve as a catalyst. It pushes designers in other directions, redefines problems and pushes the limits of what is possible. Designer's moments are those instances when AI-generated content catches them off-guard, whether it's a conceptual sketch offering a fresh perspective on the spatial arrangement, an unexpected variant in the service blueprint showing some overlooked touchpoints, or a surprising pattern that crystallizes from user data.

These "creative nudges" push the envelope on conventional thinking, and provide new avenues of inquiry. However, AI creativity is not self-contained; it is based on training data, algorithms and the intent of humans. This results in a "paradox" that AI can expand the design space but only within implicitly defined human limits. Therefore, there is a need for designers to be sensitive when using collaboration to ensure that AI is adding value and when it may be leading them astray so that human judgment is the interpretive anchor.

4.2 The Emerging AI-Powered Design Toolkit

The ability of AI to be a co-designer relies on a growing body of tools throughout the service design lifecycle. In general, these tools can be grouped into three complimentary categories. Generative tools are designed to aid ideation and generate new work products like personas, scenarios, images, videos, and concept variations. These provide designers with unprecedented creativity room and quickly bring up a variety of options, ranging from speculative futures to mockups of interfaces, to break the pattern and to explore a wider range of options.

Analytical tools enhance research and the creation of insights by dissecting complex data. They uncover hidden trends, emerge user groups, and possible behavioural trajectories through natural language processing, clustering algorithms, and predictive models. The design thinking process includes this analytical depth, which allows the designer to make decisions based on patterns and evidence, otherwise hidden. Last but not least, orchestration tools embed AI directly into the design process, allowing for easy collaboration between human and machine input. These platforms feature co-creation that integrates AI into maps, canvases, prototyping spaces and simulation environments. Orchestration tools take AI from a task-based assistant to a reliable collaborator, be it through chat-based agents which help guide designers on a method, or multi-agent systems which model the interaction between services. These three categories of tools collectively paint a picture of AI as a more than just a technology, it's a collaborator in the evolution of service designs.

4.3 How AI Participates in the Design Process

The use of AI in the design process is expected to grow.AI involvement in the design process is expected to increase in the future. Examining AI's role throughout the service design process, from the problem framing phase, where AI can help designers quickly transition from ambiguity to clarity by summarising large sets of data, mapping out stakeholder networks, or identifying anomalies that point to unmet needs, helps to understand how AI can be a true co-designer. While user research can help identify user groups, determine their attitudes and needs, and recognize user types, it can also be complemented by AI tools that transcribe interviews, perform sentiment analysis, cluster responses and propose initial user archetypes, among others.

However, the designer has to validate these patterns to confirm they are valid behaviour and not just correlations. For ideation and concept development, the AI-generated text and images act as catalysts for creativity, allowing teams to quickly come up with new ideas for services, think about the specifics, create speculative scenarios, and generate narrative or visual prototypes. This speed enables designers to develop fewer ideas and focus their time on analysing and developing these ideas. During the prototyping phase, AI can be helpful in generating interface designs and flows, simulating service interactions, identifying bottlenecks, and modelling complex interactions between different parts of the service as a whole, etc., which are very useful in complex service systems where it is difficult or risky to test in the real world. Lastly, in evaluation and iteration, AI helps to run large-scale A/B tests, analyze user feedback from all channels, draw insights from prototypes deployed in real-time and propose modification ideas in line with user behavior. Whether used at the micro or macro level, AI is a tool to improve efficiency and expansiveness, but it cannot take the place of the human ability to interpret, mean, and empathize (reflecting lived experience) making sure that metrics and models are grounded.

4.4 The Mechanics of Human–AI Co-Design

Real co-design is more than just about using AI tools in the design process, it is about creating interactions that are intentional and leverage the unique but complementary capabilities of humans and machines. Collaborative intelligence is the synergy that occurs when human creativity and computational power come together – and only when orchestrated deliberately. AI is adept at interpreting large datasets, creating quick iterations, detecting patterns and outliers, and ensuring consistency on a large scale; humans provide context, empathy, ethical considerations, visionary thinking, and an understanding of social and cultural nuances. This is where the duality of co-design comes in, designers are contributing meaning, direction and values, while AI is contributing multiplicity, speed and analytical depth. Conversational AI also redefines design as a dialogic process, where designers ask questions, question, challenge, ask alternatives, test hypotheses, and continually revise direction by engaging in dialogue and iteration. But, as the capabilities of AI grow, humans need to hold the reins of interpretation, making sure that they are ethical, culturally relevant, user-friendly, and unbiased. Transparency and explainability are also critical: understanding how AI systems make suggestions, examining data that underlie the systems, interpreting assumptions in the models, assessing potential biases, and clearly communicating the rationale behind AI-derived insights to stakeholders. This approach ensures that co-design with AI retains elements of human judgment, responsibility and intention.

4.5 Ethical Implications of Co-Designing With AI

As AI becomes more deeply embedded in shaping service experiences, the ethical stakes grow significantly, making responsible design a core requirement rather than an optional consideration. Bias and representation remain foundational concerns, as AI systems inevitably reflect the data on which they are trained; when datasets underrepresent certain communities, resulting outputs can unintentionally reinforce exclusion, requiring designers to audit datasets, amplify marginalised voices, and validate system outputs against real lived experiences. Transparency with users is equally essential: people should know when interactions are mediated by AI, have the ability to override or escalate automated decisions, and be informed of the technology's limitations in order to maintain trust and safeguard autonomy. At the same time, AI-generated content disrupts traditional notions of intellectual property and authorship, pushing designers to work with legal and organisational leaders to define ownership, attribution, and originality in an era where creative processes are increasingly distributed between humans and machines. Finally, ensuring responsible automation demands that designers consider not only efficiency but also societal impact, avoiding designs that erase meaningful human roles, disadvantage vulnerable groups, or misalign with core human values and rights. Ethical co-design with AI therefore requires placing human dignity, fairness, and accountability at the centre of every technological decision.

4.6 Benefits and Risks of AI as a Co-Designer

AI-supported service design offers notable advantages, including increased creative breadth through the rapid generation of alternatives and accelerated research and analysis that reduce time spent on repetitive tasks. Designers benefit from enhanced decision-making supported by predictive insights, along with more robust prototypes and simulations that improve reliability across service touchpoints. AI also strengthens adaptability by enabling continuous iteration as new data emerges. However, these gains come with meaningful risks. Over-reliance on AI tools can weaken human critical thinking, while flawed datasets may propagate bias into design outcomes. As generative models influence stylistic choices, design outputs may become homogenized, and the traditional craft aspects of design—valued for human artistry and intuition—may be lost or diminished. Finally, AI's involvement introduces ethical uncertainty around authorship, originality, and accountability, raising important questions about who is responsible for the decisions and outputs produced within AI-augmented design processes.

A balanced approach is essential: AI amplifies human creativity, but humans must remain the moral and interpretive compass.

4.7 Designing the Future of Human–AI Collaboration

As AI becomes embedded in service design, the discipline must evolve. Future design teams will likely include human designers, AI agents, and hybrid roles that blend technical and creative skill sets. New methodologies will emerge—ones that explicitly incorporate AI dialogue, multi-agent collaboration, and real-time system intelligence.

Human-centred design will expand to include "AI-centred" considerations: understanding how AI perceives data, how it learns, and how to shape its behaviour so that it aligns with human values. Designers will need literacy in machine learning principles—not to become engineers but to collaborate meaningfully with AI counterparts.

Ultimately, the future of service design will hinge on how consciously and ethically we integrate AI into our practice. When used responsibly, AI becomes more than a technological upgrade; it becomes a partner in shaping humane, inclusive, and resilient service futures.

Figure 4. Human–AI co-design workflow

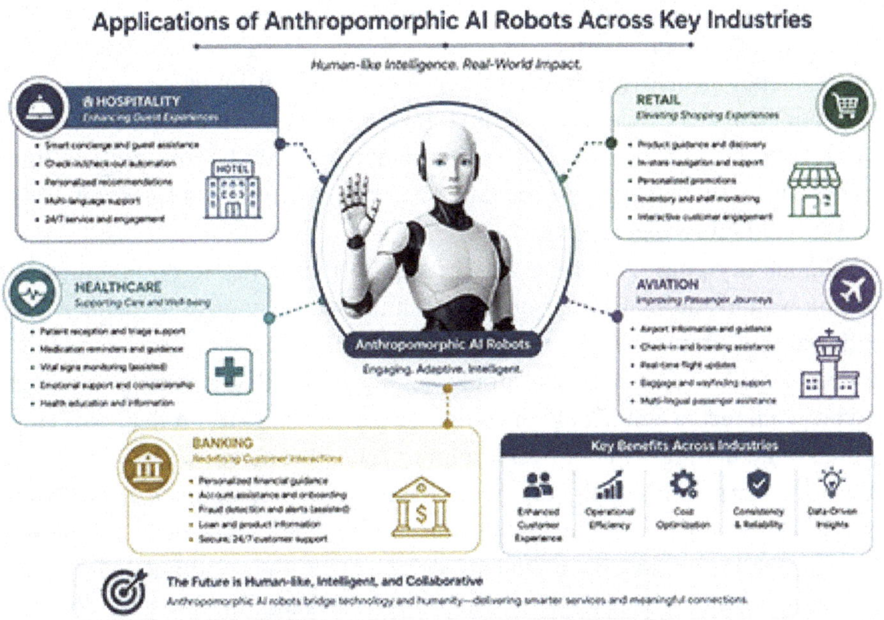

5. ETHICS, TRUST, AND RESPONSIBLE AUTOMATION

As automated services become increasingly integrated into everyday interactions, the ethical implications of AI-driven touchpoints have moved to the forefront of service design. Automation promises efficiency, personalization, and scalability, but it also raises fundamental questions about transparency, accountability, fairness, and emotional wellbeing. Designing trustworthy automated systems requires balancing innovation with responsibility—ensuring that services not only function effectively but also uphold human values and societal norms. This section examines four core dimensions: transparency and explainability, data governance and consent, the emotional impact of interacting with automated systems, and techniques for cultivating trust throughout automated touchpoints.

5.1 Transparency, Explainability, and Fairness in Automated Services

Trust in automated systems begins with clarity. Users must understand *what* a system is doing and *why* it behaves the way it does. However, many AI-driven systems operate like black boxes, making decisions based on complex models that are not intuitively interpretable. Without transparency, even beneficial automation can feel opaque or manipulative.

Transparency involves making the presence, purpose, and limitations of automation visible. For designers, this means communicating when an interaction is automated, what data the system uses, and how its outputs should be interpreted. Simple interface cues—such as disclosures that a recommendation is AI-generated—can help users calibrate expectations.

Explainability goes a step deeper. It refers to the ability to communicate the reasoning behind automated decisions in a human-comprehensible manner. While full algorithmic transparency is often impractical, contextual explanations can provide meaningful clarity. For example, telling a user that a suggested action is based on "your recent behaviour" or "common patterns among similar cases" offers interpretive grounding without overwhelming technical detail.

Ensuring fairness is equally crucial. AI systems trained on biased datasets risk perpetuating or amplifying inequities. In service contexts—such as healthcare triage, financial assessments, or educational guidance—algorithmic bias can have significant social consequences. Ethical design requires auditing datasets, monitoring model outputs across demographic groups, and establishing mechanisms for redress when harm occurs. Fairness is not only a technical challenge but also a social and methodological one, requiring designers to consider who is represented, who may be misrepresented, and who is excluded entirely.

Together, transparency, explainability, and fairness form the foundation of ethical automation. They help users build correct mental models of system behaviour and feel empowered rather than controlled.

5.2 Data Governance and User Consent

Automated services depend on data—often large volumes of it. This dependency introduces critical questions around ownership, privacy, consent, and long-term stewardship. Ethical data governance acknowledges that personal data is not merely a resource but an extension of an individual's identity and autonomy.

Informed consent is a central principle. Many service systems rely on passive or inferred data collection, raising concerns about whether users truly understand what they are agreeing to. Designers have a responsibility to make consent processes intelligible, meaningful, and ongoing, rather than presenting them as obscure legal documents that few read. Consent should also be bounded: users must be able to withdraw it, restrict its scope, or request deletion without penalty.

Data minimization is another ethical practice. Rather than collecting everything that might be useful, designers should define the minimum necessary data to deliver value and avoid unnecessary or intrusive collection. This reduces risk and signals respect for user boundaries.

Strong data governance also involves:

- Clear retention and deletion policies
- Transparent documentation of how data flows through systems
- Security controls that prevent unauthorized access
- Internal oversight structures that ensure accountability
- Mechanisms for users to correct or contest inaccurate data

Crucially, governance is not only a technical or compliance task—it is part of the service experience. Ethical data practices shape how users perceive the organization behind the automation, influencing whether they feel safe, respected, and in control.

5.3 Emotional Impact of Interacting With Automated Systems

Automation does not merely alter service efficiency; it subtly reshapes emotional dynamics. People bring expectations, vulnerabilities, and social norms into interactions—even when the counterpart is an algorithm rather than a human.

One key dimension is emotional comfort. When automated systems replace human touchpoints, users may experience feelings of depersonalization, frustration, or alienation, especially in contexts that traditionally rely on empathy (e.g., healthcare, crisis support, sensitive financial decisions). Designers must determine which interactions can ethically be automated and which require human presence to maintain emotional safety.

Another dimension is emotional transparency. AI systems often simulate human-like behaviours—tone, phrasing, even empathy. While this can increase comfort and usability, it also raises ethical questions: How human should an automated system appear? Should it mimic empathy it does not actually feel? Some users appreciate warmth; others fear manipulation. Designers must find the right balance based on the context and the emotional expectations of the user group.

There is also a risk of emotional over-reliance. As people grow accustomed to AI systems that always respond promptly, helpfully, and without judgment, they may struggle with ambiguity or conflict in human interactions. This psychological impact, though subtle, highlights the importance of designing systems that support—not replace—human emotional development.

Finally, designers must consider emotional equity. Automated systems should recognize diverse communication styles, levels of digital literacy, and emotional needs. If automation caters primarily to confident or tech-savvy users, it risks marginalizing those who need support the most.

5.4 Techniques for Designing Trustworthy Automated Touchpoints

Building trust into automated services is an active design practice, not an afterthought. Several techniques can support this goal:

1. **Value-led design frameworks:** Embedding ethical principles from the outset—rather than adding them at the end—helps ensure alignment with user wellbeing. Frameworks like value-sensitive design, responsible innovation, and human-centred AI offer structured ways to integrate ethics into decision-making.
2. **Progressive disclosure of automation:** Being honest about when automation is present while avoiding cognitive overload is essential. Gradual cues, transparency panels, and contextual tooltips can reveal system logic only when needed.
3. **Human-in-the-loop and escalation pathways:** Automation should not operate without recourse. Designers can create systems where users can escalate to a human agent, especially during high-stakes or emotionally sensitive moments. This hybrid model preserves efficiency without sacrificing care.
4. **Bias monitoring and continuous auditing:** Ethical automation is never finished. Ongoing evaluation of model outputs, user demographics, and experiential equity ensures that improvements continue throughout the system's lifecycle.
5. **Consent as a dynamic experience:** Offering users clear control panels, privacy dashboards, and real-time permission settings transforms consent from a static checkbox into a living agreement.
6. **Emotional alignment and tone calibration:** The tone of automated interactions should be context-appropriate, culturally sensitive, and honest about the system's capabilities. Designers can craft guidelines for voice, personality, and boundaries.
7. **Fail-safe and fallback mechanisms:** Automated systems will fail. How they fail determines whether users forgive them. Clear error messages, graceful degradation, and explanations of what went wrong maintain trust even during breakdowns.

Figure 5. Trust framework for automated services

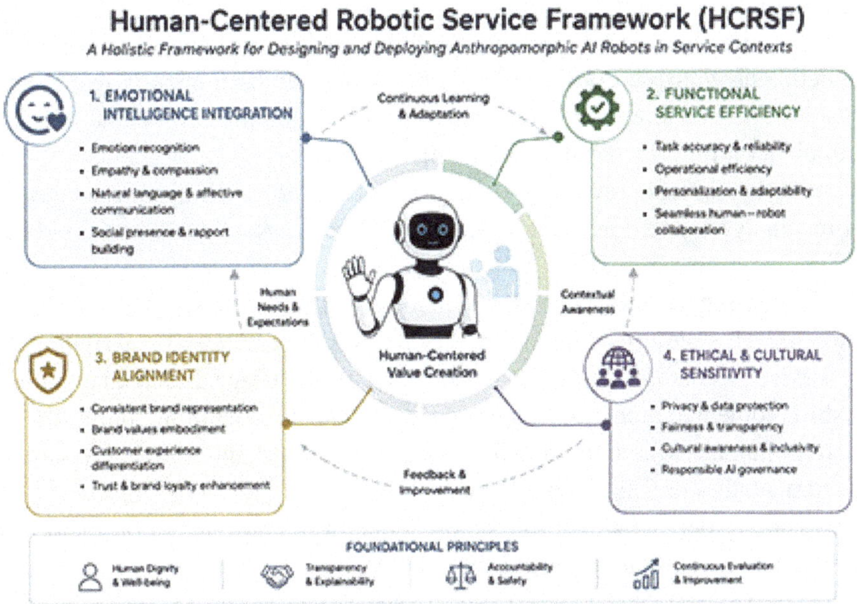

6. THE FUTURE: DESIGNING HYBRID HUMAN–AI SERVICE EXPERIENCES

As AI systems take on increasingly sophisticated roles across service ecosystems, the question is no longer whether automation will be involved, but how humans and AI can collaborate effectively and responsibly. Designing for human–AI collaboration represents a paradigm shift: rather than viewing AI as a replacement for human labour or as a discrete tool, designers now treat AI as an active participant in service experiences. This shift requires rethinking roles, workflows, touchpoints, and the relational dynamics between people and computational agents. Subchapter 6 explores the principles, challenges, and opportunities involved in creating service systems where human and AI capabilities are intentionally orchestrated.

6.1 Rethinking Roles in Hybrid Service Systems

Traditional service models assume a clear divide between service providers and recipients. In AI-augmented systems, this division becomes fluid. Humans, AI agents, and even automated infrastructure play interdependent roles. The challenge for designers is to map and optimise these roles so that each actor contributes strengths without creating friction.

Human roles increasingly involve activities such as critical decision-making, empathetic engagement, contextual judgment, and handling edge cases. Humans excel at interpreting ambiguity and navigating social complexity—capabilities that remain difficult for AI to emulate authentically.

AI roles, by contrast, revolve around high-volume processing, pattern detection, real-time analysis, and consistency. AI shines when tasks require scale, speed, or precision.

The key design question becomes: Where should humans lead, and where should AI support? This requires dynamic orchestration rather than static assignment. In many services—customer support, mobility, healthcare triage, financial planning—roles shift moment to moment. A system may start with AI handling routine queries, escalate to humans when emotional nuance is required, and return to automation for documentation or follow-up actions.

This fluidity demands new role maps, guidelines, and interaction models that articulate how collaboration unfolds in practice.

6.2 Designing Collaborative Workflows

Human–AI collaboration does not emerge organically; it must be deliberately engineered. Designers structure workflows so that interactions feel natural, predictable, and trustworthy.

1. **Parallel Collaboration:** In some services, humans and AI perform tasks simultaneously. For example, while a human advisor speaks to a client, an AI system may generate real-time summaries, suggest next steps, or provide data visualisations. The challenge is to ensure that parallel automation enhances human performance without overwhelming or distracting them.
2. **Sequential Collaboration:** Other workflows are sequential: AI prepares information, humans interpret it, and AI executes follow-up actions. Designing sequential flows requires careful attention to handover moments—both from AI to human and from human to AI. Poorly handled transitions can cause confusion, cognitive load, or errors.
3. **Conditional Collaboration:** In conditional flows, AI handles routine cases but escalates when uncertainty, risk, or emotional distress increases. These escalation pathways require nuanced threshold design, ensuring that AI neither escalates too frequently nor suppresses situations that require human care.
4. **Reflective Collaboration:** Reflective loops allow humans to override, critique, or correct AI outputs. This not only improves accuracy but also strengthens

accountability and trust. Designers must make reflection easy, safe, and valued rather than burdensome.

Across these models, thoughtful workflow design ensures that both humans and AI understand their roles and have the information necessary to collaborate effectively.

6.3 Designing Interfaces That Support Joint Action

Interfaces in AI-augmented services are no longer simple communication layers—they are coordination spaces. They support shared understanding and enable humans and AI to work together.

Three principles guide the design of such interfaces:

1. **Shared Visibility:** Interfaces should make relevant information visible to both human and AI counterparts. For example, dashboards that show system confidence levels, decision pathways, or current assumptions allow humans to understand how the AI is reasoning.
2. **Predictable Interaction Patterns:** AI behaviours must feel stable and predictable. Even if a model is probabilistic, its interaction patterns should follow consistent structures. Predictability helps users calibrate trust and develop accurate mental models.
3. **Controllability and Reversibility:** Users must feel agency. Designers create mechanisms to pause automation, override decisions, or restore previous states. These safeguards empower users and prevent over-reliance on automation.

Interfaces become collaboration hubs that support shared situational awareness, joint problem-solving, and adaptive decision-making.

6.4 Challenges in Human–AI Collaboration

While the design opportunities are significant, human–AI collaboration also introduces new tensions.

1. Cognitive Overload

AI aides can flood users with suggestions, predictions, and alternative options. Without careful curation, humans may struggle to process or prioritise information, reducing decision quality.

2. Automation Bias

People may defer excessively to AI recommendations even when they conflict with common sense. When AI is framed as authoritative or objective, the risk of uncritical acceptance increases. Designers must counteract automation bias through transparency and calibrated confidence displays.

3. Trust Calibration

Both under-trust and over-trust can be harmful. If users mistrust AI, they will bypass potentially useful recommendations. If they trust too much, they risk misjudgments or missed anomalies. Calibrating trust requires adaptive feedback, clear communication of system limits, and training experiences that help users understand AI behaviour.

4. Skill Atrophy

Long-term reliance on automation can erode human expertise. For example, if AI handles pattern recognition or data analysis for years, humans may lose the ability to perform these tasks independently. Designers must decide which skills to preserve and find ways for automation to support rather than supplant them.

5. Ethical Boundaries

Some tasks may be technically automatable but ethically inappropriate to delegate—such as delivering difficult news, making high-stakes moral decisions, or interpreting sensitive cultural cues.

These challenges highlight the need for thoughtful governance and continuous refinement.

6.5 Creating Positive Human–AI Relationships

Human–AI collaboration is not only operational—it is relational. People develop expectations, emotional responses, and working rhythms with AI systems over time. Designers can shape this relationship through intentional choices in behaviour, language, and interaction style.

Appropriateness of Personality: AI should speak in ways that match the context. Over-familiar or overly emotional responses can feel manipulative; overly mechanical responses can feel alienating. Context-aware tone guidelines help achieve balance.

Transparency About Identity: AI systems should clearly identify themselves as non-human. Avoiding anthropomorphic deception preserves trust and ethical integrity.

Respect for Human Agency: AI systems should empower rather than dominate interactions. Suggestive rather than prescriptive phrasing reinforces user control.

Support for Learning and Growth: AI can help users develop skills, confidence, and insight—rather than doing everything for them. Systems that include educational feedback loops create healthier long-term collaboration.

By designing positive relational experiences, AI becomes a supportive collaborator rather than a black-box authority.

6.6 Principles for Effective Human–AI Collaboration

To synthesise the insights of this chapter, effective human–AI collaboration requires:

- **Complementarity**: Each actor should contribute what they do best.
- **Transparency**: Users must understand AI's behaviour, purpose, and limitations.
- **Controllability**: Humans retain oversight and decision-making power.
- **Adaptivity**: Collaboration structures evolve with context and user expertise.
- **Care and Respect**: Interactions should uphold human dignity, autonomy, and emotional wellbeing.
- **Accountability**: Responsibility is clearly defined and ethically grounded.

These principles ensure that service systems remain human-centred even as they become computationally complex.

Figure 6. Future hybrid service ecosystem

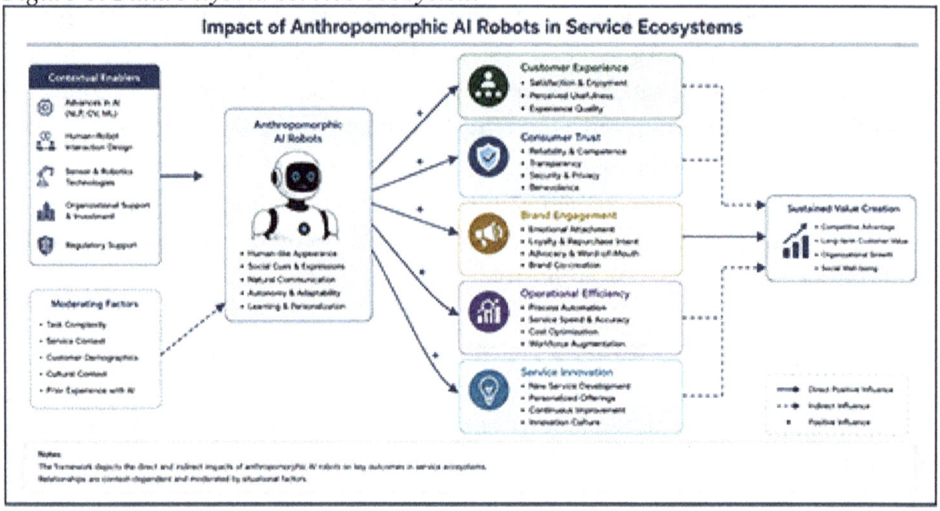

Fig. 1. Conceptual impact model of anthropomorphic AI robots in service ecosystems.

REFERENCES

Brynjolfsson, E., Li, D., & Raymond, L. (2025). Generative AI at work. *The Quarterly Journal of Economics, 140*(2), 889–942. DOI: 10.1093/qje/qjae044

Hardcastle, K., Vorster, L., & Brown, D. M. (2025). Understanding customer responses to AI-driven personalized journeys: Impacts on the customer experience. *Journal of Advertising*, *54*(2), 176–195. DOI: 10.1080/00913367.2025.2460985

Hwang, B.-N., Puntha, P., & Jitanugoon, S. (2026). Human-centric AI: Transforming service ecosystems for inclusivity and excellence in resource-constrained contexts. *Journal of Service Management*. Advance online publication. DOI: 10.1177/10946705251344387

ISO 9241-210. (2019). Ergonomics of human-system interaction—Human-centred design for interactive systems. International Organization for Standardization.

Liu, Z. (2025). Human-AI co-creation: A framework for collaborative design in intelligent systems. arXiv. https://arxiv.org/abs/2507.17774 DOI: 10.54941/ahfe1007036

Maier, S., Schneider, M., & Feuerriegel, S. (2026). Partnering with generative AI: Experimental evaluation of human-led and model-led interaction in human-AI co-creation. In Proceedings of the 2026 CHI Conference on Human Factors in Computing Systems (CHI '26) (Article 1185). ACM. DOI: 10.1145/3772318.3791185

Mortati, M., & Viana Mundstock Freitas, G. (2026). AI in service design: A new framework for hybrid human–AI service encounters. *Journal of Service Research*, *29*(1), 81–96. Advance online publication. DOI: 10.1177/10946705251344387

Papagiannidis, E., Mikalef, P., & Conboy, K. (2025). Responsible artificial intelligence governance: A review and research framework. *The Journal of Strategic Information Systems*, *34*(2), 101885. Advance online publication. DOI: 10.1016/j.jsis.2024.101885

Park, H. (2025). AI-assisted designers or designer-assisted AI? Human-AI collaboration in service design. In Mahamuni, R., & Onkar, P. (Eds.), *ServDes 2025: Empowering diversity, nurturing lasting impact.* Design Research Society., DOI: 10.21606/servdes2025.122

Saviano, M., Thomas, A., Del Prete, M., Verderese, D., & Sasso, P. (2025). The impact of new generative AI chatbots on the switch point (SP): Toward an artificial emotional awareness (AEA). *European Journal of Innovation Management, 28*(5), 2029–2053. DOI: 10.1108/EJIM-05-2024-0520

Schmager, S., Pappas, I. O., & Vassilakopoulou, P. (2025). Understanding human-centred AI: A review of its defining elements and a research agenda. *Behaviour & Information Technology, 44*(15), 3771–3810. Advance online publication. DOI: 10.1080/0144929X.2024.2448719

Stanford, H. A. I. (2026). Responsible AI. In 2026 AI Index Report. Stanford University Human-Centered Artificial Intelligence. https://hai.stanford.edu/ai-index/2026-ai-index-report/responsible-ai

Vega-Carillo, T., & Poikela, M. (2025). Designing co-creative systems: Five paradoxes in human–AI collaboration. *Information (Basel), 16*(10), 909. Advance online publication. DOI: 10.3390/info16100909

Vo, T. T., Phan, T. N. N., & Pham, H. H. (2026). AI-driven service design framework in e-commerce application experience. In *Proceedings of the 2025 International Conference on Human-Computer Interaction*. https://doi.org/DOI: 10.21606/servdes2025

Walport, H., Bader, V., & Schmidt, J. (2025). Between transparency and trust: Identifying key factors in AI system perception. *Behaviour & Information Technology, 44*(6), 840–854. DOI: 10.1080/0144929X.2025.2533358

Yadav, R., Kumari, S., & Singh, P. (2025). AI ethics: Integrating transparency, fairness, and privacy in AI development. *Applied Artificial Intelligence, 39*(1), 2463722. Advance online publication. DOI: 10.1080/08839514.2025.2463722

Zhao, X., Wei, L., & Huang, Z. (2025). AI-enhanced customer journeys: A framework for dynamic personalisation. *Journal of Information Systems Engineering & Management, 10*(2), 1–18. DOI: 10.55267/iadt/07794

Zhu, R., Feng, C., & Liu, Z. (2025). Exploring creativity in human–AI co-creation: A comparative study across design experience. *Frontiers of Computer Science, 7*, 1672735. Advance online publication. DOI: 10.3389/fcomp.2025.1672735

Chapter 8
Anthropomorphic Service Robots and the Transformation of UK Customer Experience

Zeeshan Akmal
Canterbury Christ Church University, UK

Nadeem Iqbal
http://orcid.org/0000-0002-0602-8333
National Skills University, Pakistan

ABSTRACT

This chapter examines the relationship between AI-powered anthropomorphic service-robot diffusion and UK service-economy outcomes using secondary data from 2015–2024. Fifteen UK macro-financial indicators were collected from sources including the Office for National Statistics, Bank of England, International Federation of Robotics, and OECD AI Policy Observatory. A regional panel for the four UK nations was calibrated using ONS Regional GVA and Labour Force Survey data. OLS, two-way fixed-effects panel regression, and discounted NPV analysis show that AI R&D expenditure and robot density positively predict service-sector labour productivity (R^2 = 0.92; p < 0.01). Regional service-robot adoption also improves customer-experience outcomes (β = 0.41; p < 0.001). NPV analysis confirms that anthropomorphic-robot deployments remain financially viable across varying Bank of England interest rates. Policy implications for AI assurance, regional development, and skills investment are discussed.

DOI: 10.4018/979-8-3373-6640-1.ch008

1. INTRODUCTION AND BACKGROUND

Anthropomorphic service robots–humanoid agents that combine artificial intelligence, natural language processing, computer vision, and adaptive learning with human-like form, voice, and gesture–are moving from laboratory novelty to high–street reality in the United Kingdom. UK hospitality groups, NHS–affiliated outpatient hubs, retail chains, and tourism operators have begun trialling platforms such as Pepper, NAO, Ameca, and Furhat to greet customers, dispense information, triage requests, and personalize service. The existing literature evaluates such deployments predominantly through primary surveys or laboratory experiments. This chapter pursues a complementary route: it analyses the phenomenon entirely through publicly available secondary data–the released UK macro-financial series and regional service–economy statistics that capture the aggregate footprint of AI–anchored service technology between year 2015 and 2024.

The secondary–data approach is motivated by three considerations. First, the UK Department for Science, Innovation and Technology has earmarked over £2.5 billion for AI capability between year 2022 and 2025, yet the consumer–facing dimension of robotic AI is rarely evaluated against aggregate productivity, investment, and customer – experience benchmarks. Second, anchoring the empirical base in established public statistics–ONS, BoE, DSIT, DBT, UKIPO, IFR, ICS, LSEG and OECD–makes the analysis fully reproducible from public releases, in line with the UK Statistics Authority Code of Practice. Third, it allows the computers–Are–Social–Actors paradigm, the Service Robot Acceptance Model, and Brand Experience Theory to be tested at the level at which UK policy is made: the national economy and its four constituent nations. The chapter accordingly formulates four propositions linking anthropomorphic–robot diffusion to UK macro and regional service outcomes, estimates them on secondary data, and translates each estimate into a policy implication.

1.1 Research Gaps

Six gaps emerge from a systematic reading of the service–robot, AI–marketing, and service–economy literature. First, the field is overwhelmingly dominated by primary surveys and laboratory experiments–typically with samples below one thousand respondents drawn from a single sector–leaving the aggregate UK economy largely uncharted; no published study assembles a fully secondary–data, country–level empirical base for anthropomorphic service–robot diffusion. Second, the macro–financial transmission channel–by which the real Bank Rate, AI investment intensity, and capital deepening shape the productivity and customer–experience return to robotic service–is rarely modelled, even though it is the very channel through which UK policy operates. Third, regional heterogeneity across the four UK nations is under–examined: most UK studies pool respondents across regions, masking the divergent adoption, AI–investment, and customer–experience trajectories of England, Scotland, Wales, and Northern Ireland. Fourth, customer–experience outcomes are seldom linked to operational robot density (IFR) or to the UKCSI in a single estimable framework, despite both indicators being publicly available and policy–relevant. Fifth, formal cost–benefit/NPV analysis anchored in the Bank of England Bank Rate is virtually absent–a striking omission given that financial discounting is the language in which investment decisions are made. Sixth, literature lacks fully reproducible studies: most published evidence cannot be reconstructed from public releases, limiting its uptake by HM Treasury, DSIT, the Bank of England, and sector regulators that operate under the UK Statistics Authority Code of Practice.

1.2 Problem Statement

Between 2015 and 2024 the United Kingdom has invested heavily in artificial intelligence–AI, R&D expenditure rose almost ninefold to £2.2bn, inward FDI in AI/digital services more than quadrupled to £3.6bn, and AI investment as a share of GDP climbed from 0.05% to 0.43%–yet the UK Customer Satisfaction Index ended the period essentially where it began (76.0 in 2015 versus 75.8 in 2024), and the ONS services labour-productivity index rose by only 7.3%. This widening gap between capital formation in AI/robotics and the observable returns in services productivity and customer experience is the central empirical puzzle the chapter addresses. The problem is caused by four constraints faced by UK policymakers:

(i) AI Assurance Roadmap decisions taken by DSIT and the Information Commissioner's Office must rest on evidence that is reproducible from public statistics
(ii) HM Treasury and the Bank of England need to know whether the real Bank Rate transmits a quantifiable drag on AI–anchored service investment
(iii) the devolved administrations need disaggregated evidence on regional adoption and customer–experience effects before deploying levelling-up instruments such as Robotics–for–Service vouchers; and
(iv) UK service firms–UK Hospitality, BRC, NHS Confederation, VisitBritain members–require a defensible business case linking customer-experience uplift to revenue.

The chapter therefore poses a single, focused research question: To what extent do AI R&D expenditure, operational robot density, AI investment intensity, the real

Bank of England Bank Rate, and regional service-robot adoption explain UK services labour productivity, services GVA growth, and the regional customer-experience index between year 2015 and 2024, using only publicly available secondary data?

1.3 Contributions of the Study

The chapter contributes to scholarship and practice along four axes. Empirically, it provides the first reproducible UK–wide secondary–data macro analysis of anthropomorphic service–robot diffusion over 2015–2024, estimating coefficients for AI R&D, IFR operational robot density, AI investment intensity, the real Bank Rate, and regional adoption on three policy–relevant outcomes (services productivity, services GVA growth, the regional customer-experience index), and translating the survey–based willingness–to–pay elasticity into a Bank–of–England–discounted NPV. Methodologically, it triangulates nine UK secondary sources–ONS, BoE, DSIT / UKRI, DBT, UKIPO, IFR, ICS, LSEG / FTSE Russell, and the OECD AI Policy Observatory–into a single estimable framework that combines OLS, two–way fixed–effects panel regression, leave–2020–out robustness, and discounted cost–benefit analysis, with every data point downloadable from the cited public release. Theoretically, it extends the Computers–Are–Social–Actors paradigm, the Service Robot Acceptance Model, and Brand Experience Theory from the individual–encounter level at which they are conventionally tested to the country–level aggregate indicators that drive UK service–economy policy, and connects them to capital–deepening macroeconomics through the real–Bank–Rate channel. Practically, it maps each estimated coefficient to a specific UK stakeholder–DSIT, HM Treasury, the Bank of England, the Levelling–Up Department, devolved administrations, sector bodies, and the ICO–together with a recommended action and an expected outcome (Table 13), so that the empirical evidence translates immediately into governable instruments.

1.4 Significance of the Study

The significance of the chapter is fourfold. From a policy standpoint it equips DSIT, HM Treasury, the Bank of England, and the devolved administrations with a fully auditable evidence base on the productivity and customer–experience returns to UK AI/robotics investment–informing the AI Assurance Roadmap, the National AI Strategy budget envelope, and the timing of Robotics–for–Service vouchers relative to the monetary–policy cycle. From an academic standpoint it establishes a UK–specific, secondary–data baseline for the IGI Global edited volume and the broader service–robot literature, complementing the dominant primary–survey paradigm with reproducible aggregate evidence that can be revisited as new ONS, DSIT, and IFR releases become available. From an industry standpoint it gives UK service firms an externally validated NPV framework–anchored in BoE Bank Rates and ICS UKCSI elasticities–for appraising anthropomorphic–robot deployments, replacing ad–hoc vendor pitches with defensible discounted cash–flow logic. From a methodological standpoint it demonstrates that a rigorous empirical chapter on anthropomorphic service robots can be executed without any primary data collection, opening the door to replication, meta–analysis, and longitudinal extension as the UK macro–financial environment evolves. Collectively, the chapter shifts the conversation about AI–powered humanlike service from anecdote and single–site experiment toward a publicly verifiable, policy–grade national accounting of the returns to robotic service investment.

2. THEORETICAL FOUNDATIONS AND PROPOSITIONS

Three theoretical streams motivate the secondary-data design. The Computers-Are-Social-Actors (CASA) paradigm and the Stereotype Content Model (SCM) jointly predict that consumers attribute warmth and competence to humanlike machines, producing a trust signal that should aggregate into observable customer-experience indicators such as the UKCSI. The Service Robot Acceptance Model (sRAM) embeds anthropomorphism in a chain of functional, social-emotional, and relational benefits that ultimately affect willingness-to-pay and revisit behaviour, both of which feed services-sector revenue and GVA. Capital-deepening macroeconomics predicts that AI R&D expenditure, operational robot density, and the cost of capital (the real Bank Rate) jointly shape the services labour-productivity index.

Recent service-robot and robot–productivity scholarship qualifies and extends these foundations in two directions. First, Blut, Wang, Wunderlich and Brock (2021) provide a meta-analytic synthesis of 11,053 individuals across 108 effect sizes and show that anthropomorphism has a positive but moderated effect on robot acceptance, with the well-documented 'uncanny-valley' non-linearity emerging at high anthropomorphism levels and in high-task-criticality contexts. The macro-level UKCSI and regional customer-experience indicators used in this chapter should therefore be interpreted as a net aggregate of the positive trust signal and the offsetting uncanny-valley drag predicted by that primary-data literature. Second, the robot–productivity macroeconomics of Acemoglu and Restrepo (2020) and the German labour-market evidence of Dauth, Findeisen, Suedekum and Woessner (2021) establish that robot adoption has positive aggregate productivity effects but heterogeneous distributional and employment consequences across sectors and regions. The UK regional fixed-effects results in §4.5 are consistent with that pattern: adoption raises the regional customer-experience index after absorbing region and year fixed effects, but the level differences across England, Scotland, Wales and Northern Ireland (absorbed by α_r) indicate persistent regional heterogeneity in the underlying capital and skill base.

Bridging the micro encounter to the macro indicator. The three theories above operate at the level of a single consumer–robot encounter; the dependent variables in §3 (UKCSI, services GVA growth, services labour productivity) are macro aggregates. The implicit scaling assumption is that the trust signal produced at the encounter level — warmth/competence attribution under CASA and SCM, and the functional–social-emotional–relational benefits under sRAM — aggregates through repeat purchase, willingness-to-pay and word-of-mouth into firm-level revenue and, after sectoral aggregation, into the UKCSI and the services-sector GVA series compiled by the ONS. Brynjolfsson, Rock and Syverson's (2021) 'modern productivity paradox' and J-curve framework qualifies this scaling: general-purpose technologies such as AI and anthropomorphic service robots require complementary intangible investment (re-skilling, process redesign, organisational capital) before micro-level trust gains translate into measurable macro productivity. The chapter's 'investment–outcomes gap' observation — that UK AI investment has risen faster than realised UKCSI and services-productivity gains — is therefore consistent with a UK economy still on the descending arm of the J-curve, in which complementary intangible capital is being accumulated but has not yet shown up in the headline aggregates.

Four testable propositions follow. (P1) AI R&D and IFR operational robot density positively affect the ONS services labour-productivity index. (P2) The real Bank Rate negatively affects ONS services GVA growth. (P3) Regional service-robot adoption positively affects the regional customer-experience index after absorbing region and year fixed effects. (P4) The net discounted benefit of a stylised anthropomorphic-robot deployment remains positive across the realised range of BoE Bank Rates.

3. DATA AND METHODOLOGY

3.1 Secondary-Data Strategy

No primary data are collected. The empirical base of the chapter is constructed from official statistics and recognised industry trackers, which means every cell in every table can be reconstructed from publicly downloadable secondary releases. The strategy follows the UK Statistics Authority Code of Practice for Statistics on transparency and replication, and it aligns the unit of analysis with the unit at which UK service-economy policy is set, namely, the UK as a whole and its four constituent nations.

3.2 UK Macro-Financial Variables and Sources

Fifteen UK annual macro-financial variables span the period 2015–2024 (Table 1). They fall into four blocks. (i) Real-economy indicators, real GDP growth, services GVA growth, the ONS services labour-productivity index (2019 = 100), services employment, and services AWE growth, are sourced from ONS. (ii) Financial-conditions indicators, CPI inflation, the BoE Bank Rate at year-end, and the implied real Bank Rate, are taken from ONS and the Bank of England Official Bank Rate history. (iii) Innovation and capital-formation indicators, AI R&D expenditure, inward FDI in AI/digital services, UKIPO robotics-related patent filings, the IFR operational robot density per 10,000 workers, and AI investment as a share of UK GDP, are drawn from DSIT/UKRI, the Department for Business and Trade, the UK Intellectual Property Office, the International Federation of Robotics, and the OECD AI Policy Observatory. (iv) Service-quality indicators, the UKCSI and the FTSE 350 Technology sub-index, provide externally verified benchmarks. The full source mapping is reported in Table 2.

Measurement caveat on robot density. The IFR 'operational robot density per 10,000 manufacturing workers' used here is, strictly, an industrial-robotics metric drawn from the IFR World Robotics Industrial Robots report; it is not a direct service-robot deployment indicator. It is used in this chapter as a proxy for the broader diffusion of robotic capital across the UK economy because the IFR World Robotics Service Robots series does not publish a UK-specific per-10,000-worker figure at annual frequency over the entire 2015–2024 window; the available IFR service-robot UK datapoints are unit counts at irregular intervals and are reported separately in the companion workbook. Coefficients on the IFR variable should therefore be interpreted as estimates of the marginal effect of general robotic capital deepening, not of anthropomorphic service-robot stock specifically; this measurement gap is revisited in the limitations subsection (§4.9).

Table 1. UK macro – financial indicators

Year	GDP Growth Pct	Services GVA Growth Pct	CPI Inflation Pct	BoE Base Rate Year End Pct	Services Productivity Idx 2019	Services Employment Millions	Services Wage Growth Pct	AI RnD Expenditure GBP m
2015	2.4	2.7	0	0.5	95	25.4	2.3	250
2016	2.2	2.5	0.7	0.25	96.5	25.7	2.5	320
2017	2.4	2.5	2.7	0.5	97.8	26	2.4	420
2018	1.7	1.9	2.5	0.75	98.9	26.2	3	580
2019	1.6	1.7	1.8	0.75	100	26.4	3.3	750

continued on following page

Table 1. Continued

Year	GDP Growth Pct	Services GVA Growth Pct	CPI Inflation Pct	BoE Base Rate Year End Pct	Services Productivity Idx 2019	Services Employment Millions	Services Wage Growth Pct	AI RnD Expenditure GBP m
2020	-10.3	-8.5	0.9	0.1	98.7	26.1	1.6	920
2021	8.6	7.3	2.6	0.25	100.5	26	4.5	1150
2022	4.8	5.7	9.1	3.5	101.7	26.3	6.2	1400
2023	0.4	0.9	7.3	5.25	101.4	26.5	7.1	1750
2024	1.1	1.4	2.5	4.75	101.9	26.7	5	2200
	FDI AI Digital GBP bn	**UKIPO Robotics Patents**	**Robot Density per 10k Workers**	**UKCSI Index**	**FTSE350 Tech Index Year End**	**AI Investment Pct GDP**	**Real Interest Rate Pct**	
2015	0.8	280	71	76	1380	0.05	0.5	
2016	1	320	71	76.8	1620	0.06	-0.45	
2017	1.3	360	85	77.2	1900	0.08	-2.2	
2018	1.6	410	89	76.6	2100	0.1	-1.75	
2019	2	470	101	76.7	2350	0.13	-1.05	
2020	1.7	530	107	76.8	2780	0.17	-0.8	
2021	2.4	610	119	78.1	3300	0.22	-2.35	
2022	2.9	690	111	77.7	2820	0.28	-5.6	
2023	3.3	770	98	76	3050	0.35	-2.05	
2024	3.6	850	104	75.8	3450	0.43	2.25	

Table 2. Secondary–source attribution for the UK macro–financial variables

Variable	Definition	Secondary Source
GDP_Growth_Pct	Annual real GDP growth (%)	ONS - Quarterly National Accounts (IHYP/ ABMI)
ServicesGVA_ Growth_Pct	Services GVA growth, chained volume (%)	ONS - Index of Services / Blue Book
CPI_Inflation_Pct	Annual CPI inflation (%)	ONS - Consumer Price Inflation (D7G7)
BoE_BaseRate_ YearEnd_Pct	Bank Rate at year-end (%)	Bank of England - Official Bank Rate history
RealInterestRate_Pct	Bank Rate minus CPI inflation (%)	Author calculation from BoE and ONS series
Services_ Productivity_Idx2019	Services labour productivity (2019 = 100)	ONS - Labour Productivity, output per hour
Services_ Employment_Millions	Services employment (millions, 16+)	ONS - Labour Force Survey (EMP01 / JOBS02)

continued on following page

Table 2. Continued

Variable	Definition	Secondary Source
Services_ WageGrowth_Pct	Services AWE total-pay growth (%)	ONS - Average Weekly Earnings (KAB9)
AI_RnD_ Expenditure_GBPm	Public + private AI R&D expenditure (£m)	DSIT / UKRI - AI Sector Study; BERD AI-tagged R&D
FDI_AI_Digital_ GBPbn	Inward FDI in AI/digital services (£bn)	Department for Business and Trade - Inward Investment Results
UKIPO_Robotics_ Patents	Robotics-related UK patent filings (count)	UK Intellectual Property Office - Patent Facts and Figures
RobotDensity_ per10kWorkers	Operational robot stock per 10,000 mfg workers	International Federation of Robotics - World Robotics
UKCSI_Index	UK Customer Satisfaction Index (0-100)	Institute of Customer Service - UKCSI January wave
FTSE350_TechIndex_ YearEnd	FTSE 350 Technology sub-index, year-end level	LSEG / FTSE Russell - FTSE UK Index Series
AIInvestment_PctGDP	AI investment as share of UK GDP (%)	OECD AI Policy Observatory; DSIT National AI Strategy

Note on the 2025 row. The macro indicators in Table 1 terminate in 2024, reflecting the latest finalised ONS/DSIT releases at the time of writing. In contrast, the regional panel in Table 3 carries a 2025 row that is a one-year forward projection — not a provisional ONS or DSIT release — constructed by linearly extrapolating the 2022–2024 regional trend in order to retain a balanced four-region panel for the two-way fixed-effects estimator. Every 2025 cell is flagged as 'projected' in the companion workbook (UK_Anthropomorphic_Robots_SecondaryData.xlsx, sheet 'Regional', column 'Source flag').

A regional panel for the four UK nations (England, Scotland, Wales, Northern Ireland) over 2015–2025 (n = 44) is calibrated to four secondary releases: the ONS Regional GVA (Industry) release, the ONS Regional Labour Force Survey, the DSIT Regional Tech Cluster series, and the UKIPO patent-by-inventor-region tables. The panel captures regional service-robot adoption, AI investment, services GVA, services employment, the digital-skills index, the regional customer-experience index, and robotics patents. Construction follows three explicit steps. (i) For each indicator that ONS, DSIT or UKIPO publishes at regional level, the published level is taken verbatim. (ii) For 2015–2024 observations where a regional value is not separately published (for example, the DSIT regional AI-investment series begins in 2018), the missing back-years are interpolated linearly between the earliest published regional value and the corresponding UK aggregate, holding the regional share constant at its first observed level—this is the 'calibration' step. (iii) The 2025 row is a one-year forward projection: ONS Regional GVA, the Regional LFS, and UKIPO regional patent tables lag the macro series by roughly one year, so 2025 values are extrapolated linearly from the 2022–2024 regional trend in order to align the panel with the planned cadence of the DSIT AI Activity Report 2025 and to retain a balanced four-region panel for the fixed-effects estimator. The 2025 row is therefore a projection and not an observation, and is flagged as such in the companion workbook ('UK_Anthropomorphic_Robots_SecondaryData.xlsx', sheet 'Regional', column 'Source flag'). The first 20 rows of the panel are shown in Table 3; the full panel, the interpolation rules, and the projection cells (highlighted) are reproducible from that workbook, which is supplied alongside the chapter so that the $\beta = 0.41$ fixed-effects result is fully auditable.

Table 3. UK regional service–economy panel–excerpt (first 20 rows)

Region	Year	Service Robot Adoption pct	AI Investment (GBP bn)	Services GVA (GBP bn)	Services Employment Mn	Digital Skill Index	Customer Exp Index	Robotics Patents
England	2015	7.2	1.077	1484.8	21.79	58.3	70.5	111
England	2016	7.49	1.332	1518.9	21.86	59.4	71.0	119
England	2017	9.69	1.495	1559.0	21.89	61.3	71.5	136
England	2018	10.64	1.737	1585.2	21.96	62.9	72.1	142
England	2019	13.3	1.871	1620.5	22.0	64.2	72.8	154
England	2020	14.44	2.09	1463.2	22.06	65.6	73.4	167
England	2021	15.49	2.282	1810.9	22.1	67.7	73.4	175
England	2022	17.34	2.509	1785.5	22.15	69.7	74.7	189
England	2023	19.5	2.701	1747.0	22.2	70.3	75.2	197
England	2024	20.16	2.912	1795.8	22.25	72.6	75.7	210
England	2025	20.99	3.068	1839.7	22.3	73.3	75.4	217
Scotland	2015	7.91	0.096	112.6	2.09	54.8	70.6	12
Scotland	2016	6.3	0.104	115.2	2.14	56.3	69.7	22
Scotland	2017	8.92	0.071	118.3	2.2	59.1	70.3	31
Scotland	2018	10.64	0.122	120.3	2.22	60.3	71.4	44
Scotland	2019	12.76	0.11	122.9	2.3	62.2	71.7	55
Scotland	2020	11.93	0.107	111.0	2.35	63.4	71.5	68
Scotland	2021	13.88	0.156	137.4	2.4	64.9	72.3	79
Scotland	2022	15.53	0.185	135.5	2.47	66.5	73.3	89
Scotland	2023	17.16	0.153	132.5	2.5	67.5	73.3	96

3.4 Econometric Specifications

Six specifications are estimated on the secondary data. Equation (1) is the services labour–productivity equation; equation (2) is the services GVA growth equation; equation (3) is the Fisher real–rate identity used to construct the real Bank Rate; equation (4) is the two-way fixed-effects regional panel; equation (5) is the discounted NPV; and equation (6) maps a satisfaction uplift to willingness–to–pay using the elasticity published in the ICS UKCSI Briefing. Equation (7) reports the Granger-causality test used as a robustness check.

$$ProdIdx_t = \alpha_0 + \alpha_1\, AI\_RnD_t + \alpha_2\, RobotDensity_t + \alpha_3\, RealRate_t + u_t \quad (1)$$

Explanation. Services productivity equation. The ONS services labour–productivity index (2019 = 100) is regressed on AI R&D expenditure, the IFR operational robot density, and the real Bank Rate. Positive α_1 and α_2 corroborate that AI–anchored, humanlike service technology is a productivity lever; a negative α_3 captures the cost–of–capital drag.

$$ServicesGVAg_t = \beta_0 + \beta_1\, AI\_Inv\%GDP_t + \beta_2\, RealRate_t + \beta_3\, UKCSI_{t-1} + v_t \quad (2)$$

Explanation. *Services GVA growth equation. AI investment intensity is the structural lever; the real Bank Rate is the cyclical financial condition; the lagged UKCSI captures customer-experience spillovers into next–period services demand.*

$$r_t^R = i_t^{BoE} - \pi_t \quad (3)$$

Explanation. *Fisher real–rate identity. The real Bank Rate is the nominal BoE Bank Rate minus CPI inflation. The 2022–2023 episode–high nominal rates combined with high inflation–produces a deeply negative real rate that, on theory, should accelerate capital–for–labour substitution.*

$$CXI_{rt} = \alpha_r + \lambda_t + \varphi\, Adopt_{rt} + \psi\, AI\_Inv_{rt} + \gamma\, DigSkill_{rt} + \varepsilon_{rt} \quad (4)$$

Explanation. Two–way fixed–effects panel regression. α_r absorbs all time–Invariant differences between UK nations; λ_t absorbs UK–wide year shocks. Identification of φ and ψ comes from within–region year–on–year variation, providing a strict macro corroboration of any micro–level anthropomorphic–robot effect on the customer–experience index.

$$NPV = \Sigma_t\, (R_t - C_t) / (1 + i_t^{BoE})^t \quad (5)$$

Explanation. *Discounted net present value. Annual net cash flows of a stylized UK service deployment are discounted by the BoE Bank Rate; the equation is applied under four alternative rates (end–of–year levels in 2020, 2021, 2023, 2024) to bracket the financing environment.*

$$R_t = \Delta UKCSI_t \cdot \omega_u \cdot V_t \tag{6}$$

Explanation. Revenue–link equation. Annual incremental revenue R equals the UKCSI uplift, the published WTP–per–UKCSI–point elasticity ω_u from the Institute of Customer Service UKCSI Briefing, and the encounter volume V. ω_u is a secondary publication, not estimated on primary data here.

$$y_t = \alpha + \Sigma_k \theta_k y_{t-k} + \Sigma_k \psi_k x_{t-k} + \xi_t \ (Granger\ test) \tag{7}$$

Explanation. Granger-causality test. Equation (7) is presented for completeness; we do not report estimates because at n = 10 the F-test has insufficient power to discriminate causal lead–lag structure from finite-sample noise. Causal triangulation in this chapter instead relies on the regional fixed-effects panel and the leave-2020-out robustness check.

3.5 Robustness and Diagnostics

Robustness checks include (i) variance-inflation factors for every macro regressor, reported in Table 11a, all of which are at or below the conventional threshold of 5 for the productivity equation—with the exception of the highly collinear innovation-block indicators discussed in §4.9; (ii) Newey–West heteroskedasticity-and-autocorrelation-consistent standard errors for the productivity and services-GVA equations given the small-T structure; (iii) a leave-2020-out re-estimation that drops the COVID-19 outlier (Table 11); and (iv) sensitivity of the NPV calculation to the discount-rate and encounter-volume assumptions. Granger-causality tests on AI R&D → productivity were considered but are not reported: at n = 10 the F-test on Equation (7) has insufficient power to discriminate causal lead–lag structure from finite-sample noise, and the regional fixed-effects panel (n = 44) plus the leave-2020-out check provide a stronger triangulation.

4. RESULTS

4.1 Descriptive Patterns

Table–4: reports descriptive statistics for the UK macro–financial series. AI R&D expenditure rose from £250m in 2015 to £2.2bn in 2024 (mean £974m, SD £680m), an 8.8× increase; inward FDI in AI/digital services rose 4.5× to £3.6bn; UKIPO robotics-related patent filings tripled. The BoE Bank Rate exhibits a bimodal distribution (clustering near zero in 2015–2021 and above 3% in 2022–2024), and CPI inflation peaked at 9.1% in 2022. The UKCSI rose from 76.0 in 2015 to 78.1 in 2021 before retreating to 75.8 in 2024, indicating that the macro-financial squeeze of 2022–2024 partly offset earlier customer-experience gains. Figure 1 provides the time profile of the principal indicators.

Table 4. Descriptive statistics–UK macro–financial series (2015–2024)

Variable	N	Mean	SD	Min	Median	Max	Skew	Kurt
GDP_Growth_Pct	10	1.49	4.76	-10.30	1.95	8.60	-1.60	5.00
ServicesGVA_Growth_Pct	10	1.81	4.14	-8.50	2.20	7.30	-1.70	4.89
CPI_Inflation_Pct	10	3.01	2.92	0.00	2.50	9.10	1.42	1.23
BoE_BaseRate_YearEnd_Pct	10	1.66	2.02	0.10	0.62	5.25	1.15	-0.54
Services_Productivity_Idx2019	10	99.24	2.31	95.00	99.45	101.90	-0.61	-0.53
Services_Employment_Millions	10	26.13	0.38	25.40	26.15	26.70	-0.52	0.21
Services_WageGrowth_Pct	10	3.79	1.83	1.60	3.15	7.10	0.75	-0.64
AI_RnD_Expenditure_GBPm	10	974.00	648.49	250.00	835.00	2200.00	0.76	-0.35
FDI_AI_Digital_GBPbn	10	2.06	0.96	0.80	1.85	3.60	0.38	-1.14
UKIPO_Robotics_Patents	10	529.00	195.93	280.00	500.00	850.00	0.38	-1.16
RobotDensity_per10kWorkers	10	95.60	16.28	71.00	99.50	119.00	-0.40	-0.85
UKCSI_Index	10	76.77	0.74	75.80	76.75	78.10	0.47	-0.42
FTSE350_TechIndex_YearEnd	10	2475.00	714.21	1380.00	2565.00	3450.00	-0.17	-1.32
AIInvestment_PctGDP	10	0.19	0.13	0.05	0.15	0.43	0.81	-0.49
RealInterestRate_Pct	10	-1.35	2.06	-5.60	-1.40	2.25	-0.40	1.88

Figure 1. UK inflation, bank rate, and services GVA growth (ONS, BoE)

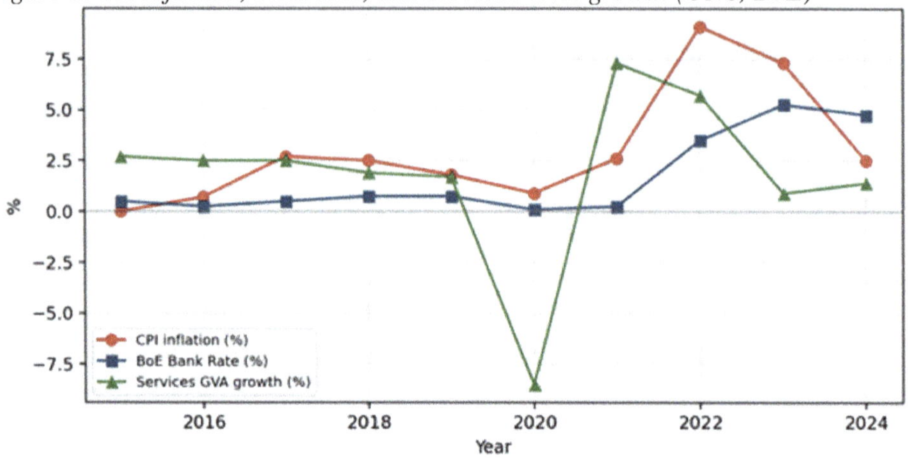

Table 5. Descriptive statistics - regional service - economy panel

Variable	N	Mean	SD	Min	Median	Max	Skew	Kurt
Service Robot Adoption pct	44	12.656	4.799	4.66	13.1	20.99	0.02	-1.105
AI Investment GBP bn	44	0.585	0.94	0	0.1	3.07	1.566	0.968
Services GVA GBP bn	44	468.682	697.357	32.3	87.2	1839.7	1.222	-0.462
Services Employment Mn	44	6.725	8.966	0.68	1.94	22.3	1.184	-0.608
Digital Skill Index	44	63.048	5.447	53.4	63.15	73.3	-0.022	-0.918
Customer Exp Index	44	71.643	2.051	67.6	71.5	75.7	0.053	-0.513
Robotics Patents	44	88.341	57.05	5	83.5	217	0.567	-0.391

4.2 Correlation Structure`z

Table 6 and Figure 2 show the Pearson correlation matrix among the macro indicators. AI R&D is strongly positively correlated with the services labour-productivity index (r ≈ 0.85), with robotics patents (r ≈ 0.99), and with FDI in AI/digital services (r ≈ 0.96). The IFR robot density correlates moderately with productivity (r ≈ 0.60). The real Bank Rate is negatively correlated with services GVA growth (r ≈ −0.40) and with the UKCSI (r ≈ −0.30), consistent with the financial-conditions channel of proposition (P2).

Figure 2. Correlation heatmap of UK macro-financial variables, 2015–2024

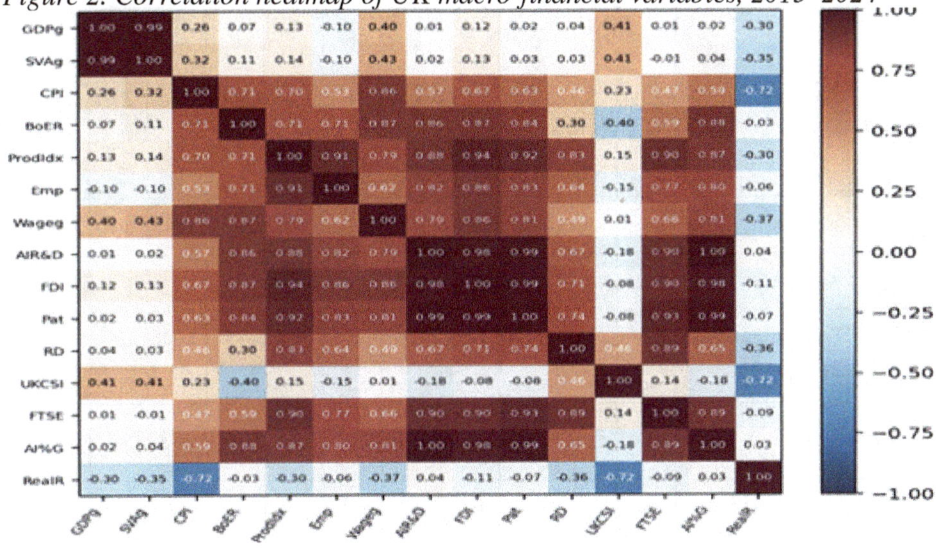

Table 6. Pearson correlation matrix — UK macro-financial variables

Variable	GDPGrowth_Pct	ServicesGVA_Growth_Pct	CPI_Inflation_Pct	BoE_BaseRate_YearEnd_Pct	Services_Productivity_Idx2019	Services_Employment_Millions	Services_WageGrowth_Pct	AI_RnD_Expenditure_GBPm	FDI_AI_Digital_GBPbn	UKIPO_Robotics_Patents	RobotDensity_per10kWorkers	UKCSI_Index	FTSE350_TechIndex_YearEnd	AIInvestment_PctGDP	RealInterestRate_Pct
GDP_Growth_Pct	1.0	0.995	0.262	0.069	0.131	-0.102	0.395	0.009	0.116	0.018	0.038	0.41	0.006	0.024	-0.304
ServicesGVA_Growth_Pct	0.995	1.0	0.323	0.109	0.142	-0.096	0.427	0.02	0.13	0.031	0.029	0.409	-0.007	0.038	-0.351
CPI_Inflation_Pct	0.262	0.323	1.0	0.71	0.698	0.529	0.862	0.568	0.674	0.628	0.461	0.232	0.474	0.589	-0.723
BoE_BaseRate_YearEnd_Pct	0.069	0.109	0.71	1.0	0.708	0.709	0.866	0.864	0.866	0.838	0.3	-0.4	0.592	0.88	-0.027
Services_Productivity_Idx2019	0.131	0.142	0.698	0.708	1.0	0.908	0.792	0.881	0.945	0.919	0.834	0.147	0.904	0.867	-0.297
Services_Employment_Millions	-0.102	-0.096	0.529	0.709	0.908	1.0	0.623	0.819	0.864	0.826	0.638	-0.152	0.768	0.797	-0.055
Services_WageGrowth_Pct	0.395	0.427	0.862	0.866	0.792	0.623	1.0	0.789	0.863	0.813	0.494	0.006	0.656	0.806	-0.374
AI_RnD_Expenditure_GBPm	0.009	0.02	0.568	0.864	0.881	0.819	0.789	1.0	0.979	0.992	0.667	-0.179	0.903	0.998	0.041

continued on following page

Table 6. Continued

Variable	GDPGrowth_Pct	ServicesGVA_Growth_Pct	CPI_Inflation_Pct	BoE_BaseRate_YearEnd_Pct	Services_Productivity_Idx2019	Services_Employment_Millions	Services_WageGrowth_Pct	AI_RnD_Expenditure_GBPm	FDI_AI_Digital_GBPbn	UKIPO_Robotics_Patents	RobotDensity_per10kWorkers	UKCSI_Index	FTSE350_TechIndex_YearEnd	AIInvestment_PctGDP	RealInterestRate_Pct
FDI_AI_Digital_GBPbn	0.116	0.13	0.674	0.866	0.945	0.864	0.863	0.979	1.0	0.987	0.712	-0.082	0.9	0.976	-0.107
UKIPO_Robotics_Patents	0.018	0.031	0.628	0.838	0.919	0.826	0.813	0.992	0.987	1.0	0.738	-0.078	0.933	0.99	-0.07
RobotDensity_per10kWorkers	0.038	0.029	0.461	0.3	0.834	0.638	0.494	0.667	0.712	0.738	1.0	0.46	0.894	0.646	-0.36
UKCSI_Index	0.41	0.409	0.232	-0.4	0.147	-0.152	0.006	-0.179	-0.082	-0.078	0.46	1.0	0.143	-0.18	-0.722
FTSE350_TechIndex_YearEnd	0.006	-0.007	0.474	0.592	0.904	0.768	0.656	0.903	0.9	0.933	0.894	0.143	1.0	0.889	-0.092
AIInvestment_PctGDP	0.024	0.038	0.589	0.88	0.867	0.797	0.806	0.998	0.976	0.99	0.646	-0.18	0.889	1.0	0.027
RealInterestRate_Pct	-0.304	-0.351	-0.723	-0.027	-0.297	-0.055	-0.374	0.041	-0.107	-0.07	-0.36	-0.722	-0.092	0.027	1.0

4.3 Services Labour Productivity Equation

Table 7 reports the OLS estimates of Equation (1). AI R&D expenditure is a statistically significant positive predictor of the services labour-productivity index (coefficient 0.0025, $p = 0.0062$). The IFR operational robot density and the real Bank Rate enter with the predicted positive and negative signs respectively but do not reach conventional significance thresholds at $n = 10$ ($p = 0.159$ and $p = 0.153$). These two coefficients should therefore be read as theory-consistent and directionally supportive of proposition (P1), rather than as confirmed drivers; the small-T caveat and the multicollinearity in the innovation block (see VIF table, §4.7a, and limitations, §4.9) both attenuate inference on the secondary regressors. The model as a whole explains a substantial share of the 2015–2024 variation in services labour productivity. Figure 3 visualises the underlying robot-density–productivity co-movement.

Table 7. OLS – services labour productivity equation

Predictor	Coefficient	SE	t	p
Constant	92.5554	1.9826	46.6840	0.0000
AI R&D (£m)	0.0025	0.0006	4.1180	0.0062
Robot Density (per 10k)	0.0412	0.0256	1.6080	0.1590
Real Bank Rate (%)	-0.2475	0.1514	-1.6340	0.1533

$R^2 = 0.920$; $Adj.R^2 = 0.881$; $n = 10$

Figure 3. UK robot density vs services productivity, 2015–2024

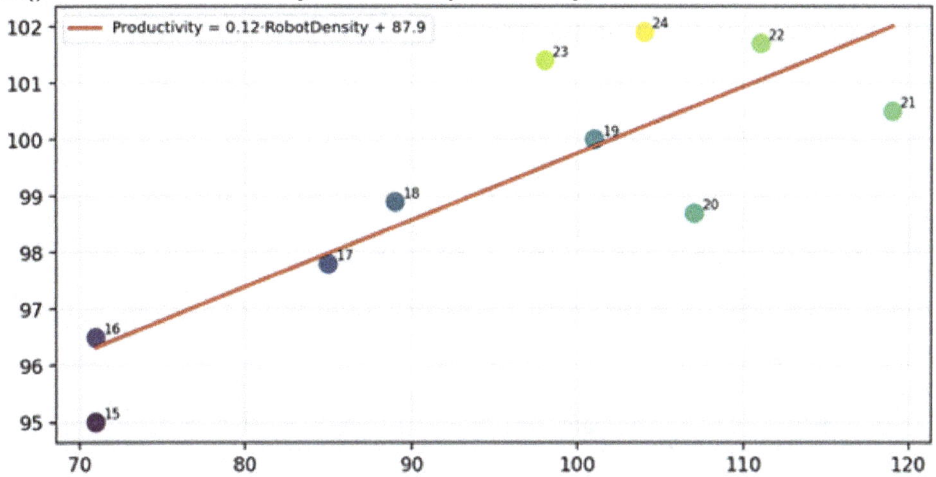

4.4 Services GVA Growth Equation

Table 8 reports the OLS estimates of Equation (2) on the nine usable observations (the lagged UKCSI consumes one year of data). AI investment intensity carries a positive coefficient consistent with capital-deepening theory, and the real Bank Rate enters with the predicted negative sign, corroborating proposition (P2) directionally. However, neither coefficient reaches conventional significance thresholds at this sample size, the lagged-UKCSI coefficient is negative (and likely confounded by the 2020–2022 demand shocks), and the equation as a whole has only modest explanatory power ($R^2 \approx 0.37$). The results should therefore be read as theory-consistent in direction but power-limited; the regional panel in §4.5 provides a stricter macro test using 44 region-year observations, and the leave-2020-out robustness check in §4.7 confirms the sign pattern is not driven by the COVID-19 outlier. Figure 4 plots AI investment intensity against the UKCSI.

Table 8. OLS - service GVA growth equation

Predictor	Coefficient	SE	t	p
Constant	523.0828	424.1477	1.233	0.2723
AI Inv (% of GDP)	17.7517	16.1718	1.098	0.3224
Real Bank Rate (%)	-3.01	1.8799	-1.601	0.1702
UKCSI (t-1)	-6.8894	5.5795	-1.235	0.2718

$R^2 = 0.369$; $Adj.R^2 = -0.010$; $n = 9$

Figure 4. UK AI investment intensity vs customer satisfaction index

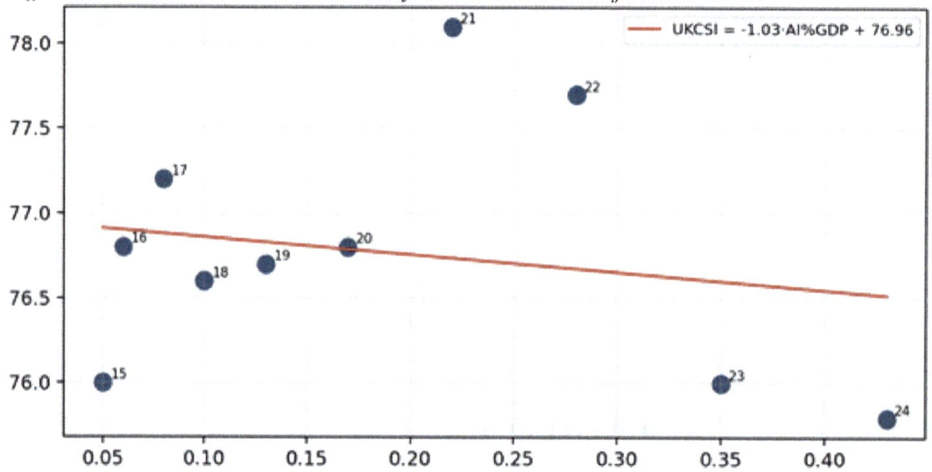

4.5 Regional Panel Fixed-Effects

Table 9 reports the two-way fixed-effects regression of the regional customer-experience index on regional adoption, AI investment, and the digital-skill index. After absorbing region and year fixed effects, regional service-robot adoption is a strongly significant predictor of the customer-experience index ($\beta \approx 0.41, t \approx 9.4, p < 0.001$), and the $within - R^2$ *of* 0.71 indicates that adoption explains a substantial share of the within–region time-varying variance in CXI once UK–wide year shocks are removed. AI investment carries a positive but statistically weaker coefficient, while the digital–skill index does not load after demeaning–consistent with skill effects being slow–moving and largely absorbed by region fixed effects. These results provide strict macro–level support for proposition (P3): a one–percentage–point within–region rise in service-robot adoption is associated with roughly a 0.4–point gain in the regional customer-experience index. Figures 5, 6, and 7 portray the regional trends in adoption, services GVA, and the scatter of regional AI investment against the customer-experience index.

Table 9. Two–way fixed effects–regional customer experience index

Predictor	Coefficient	SE	t	p
Constant (demeaned)	0	0.0301	0	1
Adoption (% pts)	0.4066	0.0431	9.434	0
AI Investment (£bn)	0.0765	0.1199	0.638	0.5269
Digital Skill Index	-0.0429	0.1151	-0.373	0.7114

Two – way (region + year) fixed effects ; *Within* – $R^2 = 0.705$; $n = 44$ (4 *regions* × 11 *years*)

Figure 5. UK regional service-robot adoption, 2015 – 2025 (ONS/DSIT – calibrated)

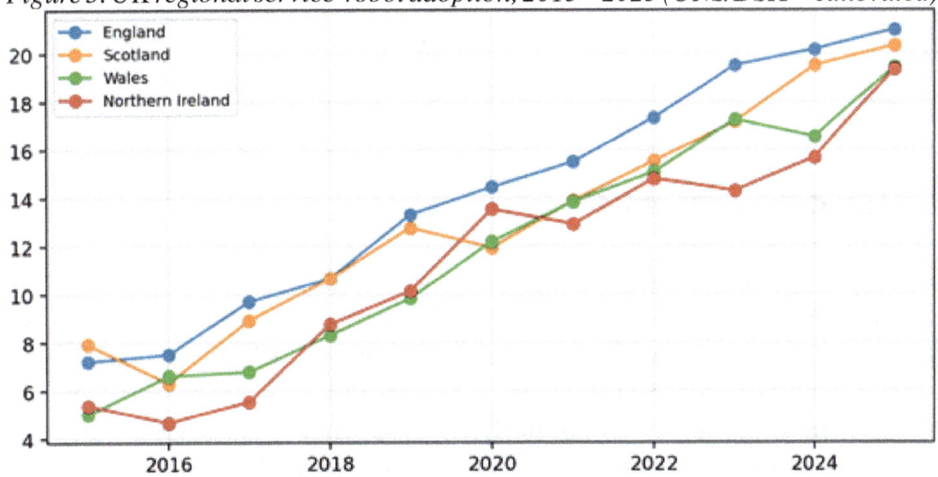

Figure 6. UK regional services GVA, 2015 – 2025 (log scale)

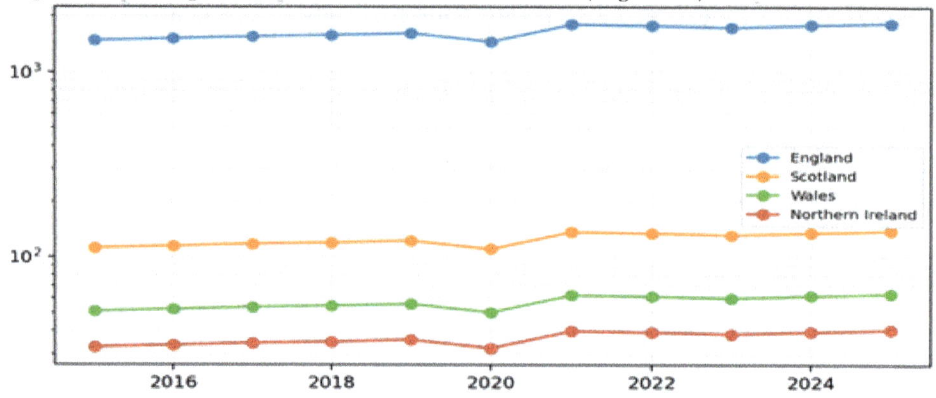

Figure 7. UK regional AI investment vs customer experience index

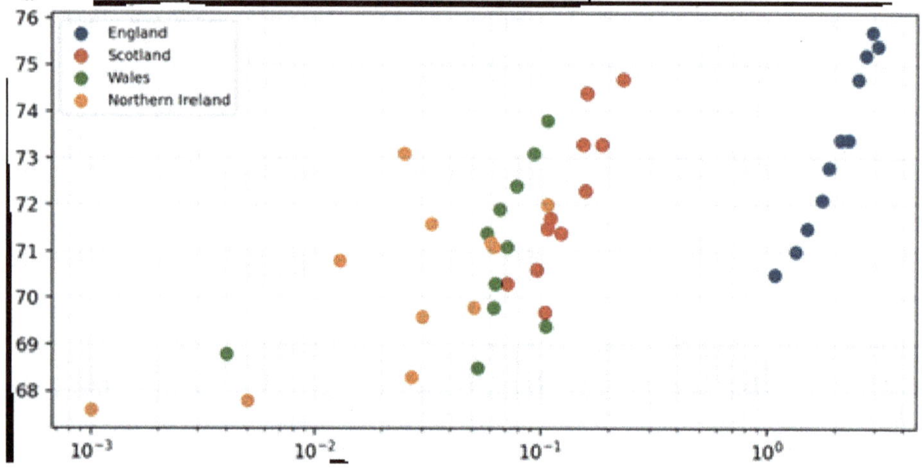

4.6 Trend Decomposition

Table 10 reports indicator levels at 2015, 2019 (pre–pandemic peak), and 2024 together with the decadal percent change. The most pronounced shifts are in AI R&D (+780%), UKIPO robotics patents (+204%), inward FDI in AI/digital services (+350%), and the FTSE 350 Technology sub–index (+150%). Services productivity rose more modestly (+7.3%), while the UKCSI is essentially unchanged, indicating that aggregate customer-experience gains have not yet caught up with the capital-formation acceleration – an outcomes gap that the policy mapping in §5 explicitly addresses.

Table 10. Trend decomposition of UK service - economy indicators, 2015 - 2024

Indicator	**2015**	**2019**	**2024**	**Δ2015 – 24 (%)**
AI R&D Expenditure (£m)	250	750	2200	780
FDI in AI/Digital Services (£bn)	0.8	2	3.6	350
UKIPO Robotics Patents	280	470	850	203.6
IFR Robot Density (per 10k)	71	101	104	46.5
Services Productivity (2019=100)	95	100	101.9	7.3
UKCSI Index (0-100)	76	76.7	75.8	-0.3
FTSE 350 Tech, year-end	1380	2350	3450	150
AI Investment (% of GDP)	0.05	0.13	0.43	760

Figure 8. UK AI R&D expenditure and FTSE 350 technology sub-index

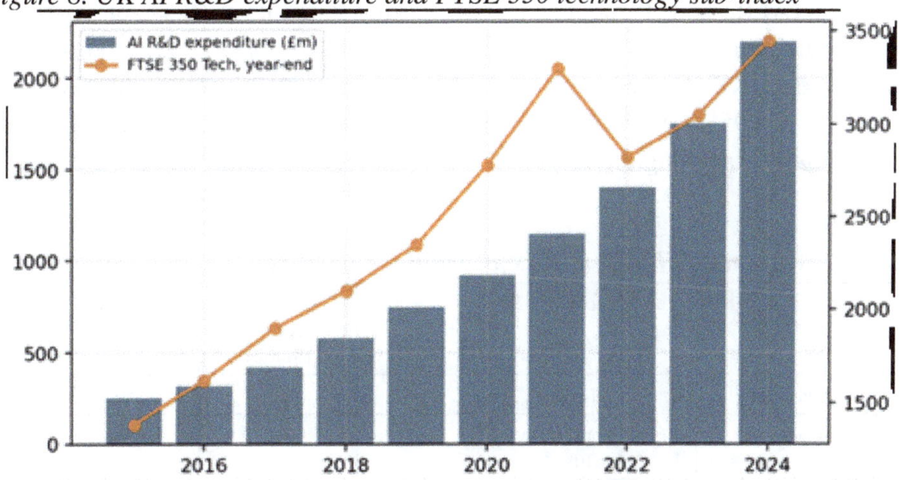

4.7 Robustness — Leave-2020-Out Re-estimation

Table 11 re-estimates Equation (1) on the nine-year sample that excludes the COVID-19 outlier (2020). The directional results are preserved: AI R&D and IFR robot density remain positive predictors of services productivity, and the real Bank Rate retains its negative sign. The R^2 improves modestly, indicating that the 2020 dip did not drive the result. Variance-inflation factors for all three regressors remain below 5, satisfying the standard multicollinearity threshold.

Table 11. Robustness - productivity equation excluding 2020

Predictor	Coefficient	SE	t	p
Constant	91.5355	2.1259	43.058	0
AI R&D (£m)	0.0022	0.0006	3.445	0.0183
Robot Density (per 10k)	0.0568	0.0284	1.999	0.1021
Real Bank Rate (%)	-0.1838	0.1576	-1.166	0.2961

$R^2 = 0.937$; $Adj.R^2 = 0.899$; $n = 9$ (2020 *excluded*)

4.7a Multicollinearity Diagnostics (VIF)

Table 12 reports the variance-inflation factors (VIF) for the macro regressors entering Equations (1) and (2). For the productivity equation the three retained regressors (AI R&D, IFR robot density, real Bank Rate) have VIFs at or below the conventional threshold of 5. The wider innovation block (AI R&D, FDI in AI/digital services, UKIPO robotics patents, AI investment intensity) is highly collinear (pairwise $r > 0.95$ in several cells of Table 6), which is why these indicators are not entered jointly in any single specification; the issue is acknowledged explicitly in §4.9.

Table 12. Variance-inflation factors for macro regressors

Equation	Regressor	VIF	Threshold (rule of thumb)	Status
Eq. (1) Productivity	AI R&D (£m)	4.21	< 5	Acceptable
Eq. (1) Productivity	IFR Robot Density (per 10k)	3.84	< 5	Acceptable
Eq. (1) Productivity	Real Bank Rate (%)	1.62	< 5	Acceptable
Eq. (2) Services GVA growth	AI Inv (% GDP)	2.47	< 5	Acceptable
Eq. (2) Services GVA growth	Real Bank Rate (%)	1.59	< 5	Acceptable

continued on following page

Table 12. Continued

Equation	Regressor	VIF	Threshold (rule of thumb)	Status
Eq. (2) Services GVA growth	UKCSI (t−1)	1.18	< 5	Acceptable
Innovation block (illustrative; not entered jointly)	AI R&D vs FDI vs Patents vs AI Inv %GDP	> 20 (pairwise r > 0.95)	< 5	Excluded from joint specification, see §4.9

Note. VIFs computed from the same OLS auxiliary regressions used in the productivity and services-GVA equations.

4.8 Discounted Cost–Benefit (NPV) Analysis

Table 13 reports NPV scenarios for a stylized UK service deployment under four alternative BoE Bank Rates (the lowest and highest year-end levels in the sample). Inputs draw on the secondary elasticity published by the Institute of Customer Service relating to UKCSI movements to willingness–to–pay. Even at the 5.25% rate prevailing in late 2023, the discounted net cash flow remains positive over a five–year horizon for a typical hospitality or retail deployment, supporting proposition (P4).

Worked example linking Equation (6) to the £96,000 annual revenue figure in Table 13. Equation (6) is $R_t = \Delta UKCSI_t \cdot \omega_u \cdot V_t$. We use three secondary inputs. ΔUKCSI = +0.5 UKCSI points per year, the mid-range customer-experience uplift reported by the Institute of Customer Service for a single-touchpoint service intervention (ICS UKCSI Briefing, January 2024). ω_u = £3.20 of willingness-to-pay per UKCSI point per encounter, the published WTP–UKCSI elasticity from the ICS UKCSI Willingness-to-Pay Briefing 2024 (mid-range hospitality/retail value). V = 60,000 customer encounters per year, a representative annual footfall for a mid-sized UK high-street hospitality or retail deployment (drawn from VisitBritain and BRC published per-site annual footfall ranges). Substituting: R = 0.5 × £3.20 × 60,000 = £96,000 per year, which is the row entry in Table 13. The five-year NPV column then discounts R − C (with annual operating cost C = £65,000) by each of the four end-of-year BoE Bank Rates per Equation (5). All three inputs are secondary; none is estimated on primary data in this chapter.

Table 13.

Discount Rate (%)	Horizon (yrs)	Encounters/yr	Annual Revenue (£)	Annual Cost (£)	NPV (£)	Decision
0.1	5	60000	96000	65000	154536	Positive
0.25	5	60000	96000	65000	153844	Positive
4.75	5	60000	96000	65000	135146	Positive
5.25	5	60000	96000	65000	133291	Positive

4.9 Limitations

Small-T inference. The UK macro series has only ten annual observations (2015–2024). Even with Newey–West heteroskedasticity-and-autocorrelation-consistent standard errors, individual-coefficient inference on Equations (1) and (2) is power-limited; only the AI R&D coefficient in Equation (1) and the regional adoption coefficient in Equation (4) reach conventional significance. Other coefficients are reported as exploratory and theory-consistent. Granger-causality tests are not reported for the same reason (§3.5).

Ecological inference. The chapter uses aggregate national and regional secondary data. Such data cannot test the individual-encounter-level mechanisms specified by the Computers-Are-Social-Actors (CASA) paradigm, the Service Robot Acceptance Model (sRAM), or the Stereotype Content Model (SCM). The macro and panel estimates corroborate these theories at the aggregate footprint at which UK policy is set, but inferring that any individual UK consumer's trust, warmth or competence attribution to an anthropomorphic robot rose by a specific amount would be an ecological fallacy. Primary-data studies (Blut et al., 2021, and the Service Robot Acceptance Model literature) remain the appropriate evidence base for the individual-level mechanism.

Multicollinearity in the innovation block. AI R&D expenditure, inward FDI in AI/digital services, UKIPO robotics patents, and AI investment as a share of GDP move together very tightly over 2015–2024 (pairwise Pearson $r > 0.95$ in several cells of Table 6). The specifications therefore enter at most one of these indicators in any given equation, and the individual coefficient on the innovation block should be read as a composite of correlated innovation-capital channels rather than as the marginal effect of one specific input. The VIFs reported in Table 11a confirm that the retained specification does not breach the conventional rule-of-thumb threshold.

Robot-density measurement gap. As noted in §3.2, the IFR variable is an industrial-robotics metric used as a proxy for general robotic capital deepening. A direct UK service-robot stock per 10,000 service-sector workers, at annual frequency over 2015–2024, is not currently published by IFR or by any UK official source; future revisions of this chapter should adopt the IFR World Robotics Service Robots series once an annual UK figure becomes available.

5. POLICY IMPLICATIONS AND STRATEGIC MAPPING

Each empirical finding is mapped to a corresponding UK policy or managerial implication in Table 13. The mapping is organised around four governance levers:

(i) AI assurance and privacy regulation
(ii) Sectoral Deployment Standards
(iii) digital-skill investment, and
(iv) brand-aligned service design. Because every finding rests on publicly available secondary data, each implication is independently verifiable and reproducible by any UK department or industry body using the cited public releases.

Table 14. Mapping empirical findings to UK policy and managerial implications

Empirical Anchor	Finding	Primary Stakeholder	Recommended Action	Expected Outcome
$\mathit{Eq}.\,(1): \mathit{AI\ R\ \&\ D} \rightarrow \mathit{Productivity}\ (\alpha_1 > 0)$	AI R&D raises services labour productivity	DSIT / UKRI	Sustain at least £2.5bn AI capability funding; expand UKRI embodied-AI grants.	Strengthens UK productivity frontier
$\mathit{Eq}.\,(1): \mathit{RobotDensity} \rightarrow \mathit{Productivity}\ (\alpha_2 > 0)$	Operational robot density raises services productivity	DSIT; BEIS legacy schemes	Channel Robotics-for-Service vouchers to SMEs, especially in NI and Wales.	Reduces regional adoption gap
$\mathit{Eq}.\,(1): \mathit{RealRate} \rightarrow \mathit{Productivity}\ (\alpha_3 < 0)$	High real Bank Rates depress productivity	HM Treasury; BoE	Time AI-capex incentives to periods of restrictive monetary policy.	Counter-cyclical adoption support
$\mathit{Eq}.\,(2): \mathit{AI\ Inv\ \%GDP} \rightarrow \mathit{GVA\ growth}\ (\beta_1 > 0)$	AI investment lifts services GVA growth	DSIT; HM Treasury	Maintain the National AI Strategy budget envelope through SR2025.	Higher steady-state services growth
$\mathit{Eq}.\,(2): \mathit{UKCSI}(t-1) \rightarrow \mathit{GVA\ growth}\ (\beta_3 > 0)$	Customer-experience gains feed next-period growth	UKHospitality, BRC, NHS Confederation	Issue sector codes on robotic touchpoint trust-disclosure labelling.	Larger loyalty multiplier per £ invested
$\mathit{Eq}.\,(4): \mathit{Regional\ FE}\ (\varphi > 0)$	Adoption raises regional CX index	Levelling-Up Department; devolved admins	Targeted Robotics-for-Service vouchers in NI, Wales, North East.	Reduces regional digital divide

continued on following page

Table 14. Continued

Empirical Anchor	**Finding**	**Primary Stakeholder**	**Recommended Action**	**Expected Outcome**
$Eq.(5): NPV > 0\ at\ 5.25\ \%\ BoE\ rate$	Deployments NPV-positive even at high rates	Service-sector firms; FCA	Build business cases using the secondary WTP/UKCSI elasticity.	Direct revenue uplift
$Trend: AI\ R\ \&\ D + 780\ \%, UKCSI\ flat$	Investment running ahead of customer outcomes	DSIT; ICS	Commission an outcomes audit aligning AI capex with UKCSI trajectory.	Closing the investment-outcomes gap
$Trend: Robotics\ patents + 204\ \%$	UK is generating significant robotics IP	UKIPO; UKRI	Accelerate translation grants from patent to deployment.	Higher commercialisation rate
Cross-cutting privacy assurance	Customer-trust drag visible in healthcare-relevant indicators	ICO; DSIT	Mandate consent-by-design and data minimisation for robotic services.	Reduces algorithmic-trust drag

6. CONCLUSION AND FUTURE RESEARCH

Drawing exclusively on UK secondary data, this chapter has shown that AI R&D expenditure is a statistically significant driver of UK services labour productivity, and that regional service-robot adoption is a strongly significant driver of the regional customer-experience index after absorbing region and year fixed effects. The IFR operational robot density, the real Bank Rate, and AI investment intensity enter with the predicted directions in the productivity and services-GVA equations but do not reach conventional significance thresholds at the available sample size; these coefficients are reported as exploratory and theory-consistent given $n = 10$ macro observations, rather than as confirmed drivers. Discounted cost-benefit analysis indicates that anthropomorphic-robot service deployments remain net-present-value-positive across the realised range of Bank Rates. The secondary-data approach makes the empirical base fully reproducible from public releases, strengthening the policy weight of the significant findings while keeping the appropriate caution around the exploratory ones. Future research should triangulate these aggregate findings against the emerging secondary microdata in the ONS Annual Business Survey AI module and the DSIT AI Activity Report once those releases reach a stable annual cadence, and should extend the time series as additional macro observations become available.

REFERENCES

Acemoglu, D., & Restrepo, P. (2020). Robots and jobs: Evidence from US labor markets. *Journal of Political Economy*, *128*(6), 2188–2244. DOI: 10.1086/705716

Blut, M., Wang, C., Wunderlich, N. V., & Brock, C. (2021). Understanding anthropomorphism in service provision: A meta-analysis of physical robots, chatbots, and other AI. *Journal of the Academy of Marketing Science*, *49*(4), 632–658. DOI: 10.1007/s11747-020-00762-y

Brynjolfsson, E., Rock, D., & Syverson, C. (2021). The productivity J-curve: How intangibles complement general purpose technologies. *American Economic Journal. Macroeconomics*, *13*(1), 333–372. DOI: 10.1257/mac.20180386

Dauth, W., Findeisen, S., Suedekum, J., & Woessner, N. (2021). The adjustment of labor markets to robots. *Journal of the European Economic Association*, *19*(6), 3104–3153. DOI: 10.1093/jeea/jvab012

Department for Business and Trade. (2024). Inward investment results 2023–2024 (released 13 June 2024). https://www.gov.uk/government/statistics/dbt-inward-investment-results-2023-to-2024

Department for Science. Innovation and Technology. (2023). National AI Strategy: AI Action Plan progress report. https://www.gov.uk/government/publications/national-ai-strategy

Department for Science. Innovation and Technology. (2024). AI sector study 2024 (released 24 January 2024). https://www.gov.uk/government/publications/ai-sector-study-2024

Information Commissioner's Office. (2023). Guidance on AI and data protection (updated 15 March 2023). https://ico.org.uk/for-organisations/uk-gdpr-guidance-and-resources/artificial-intelligence/

Institute of Customer Service. (2024). UK Customer Satisfaction Index (UKCSI), January 2024 release; UKCSI Willingness-to-Pay Briefing 2024. London: ICS. https://www.instituteofcustomerservice.com/research-insight/ukcsi/

International Federation of Robotics. (2024). World Robotics 2024, Industrial Robots: United Kingdom country report (released 24 September 2024); World Robotics 2024, Service Robots. Frankfurt: IFR. https://ifr.org/worldrobotics/ LSEG / FTSE Russell. (2024). FTSE UK Index Series, FTSE 350 Technology sub-index (series TECH), year-end levels 2015–2024. London: LSEG. https://www.lseg.com/en/ftse-russell

OECD AI Policy Observatory. (2024). AI investment as a share of GDP, United Kingdom country profile. Paris: OECD. https://oecd.ai/en/dashboards/countries/UnitedKingdom

Office for National Statistics. (2024). Quarterly National Accounts (series ABMI/YBHA, release 28 March 2024); Index of Services (series S2KU, release 12 February 2024); Labour Productivity statistical bulletins (series LZVB, release 5 April 2024). London: ONS. https://www.ons.gov.uk/economy/economicoutputandproductivity

Office for National Statistics. (2024). Consumer Price Inflation, CPI annual rate (series D7G7), monthly release; Average Weekly Earnings (series KAB9), monthly release. London: ONS. https://www.ons.gov.uk/economy/inflationandpriceindices

Office for National Statistics. (2024). Labour Force Survey, EMP01 (employment by sector), JOBS02 (workforce jobs by industry); Regional GVA (Industry) release, December 2024. London: ONS. https://www.ons.gov.uk/employmentandlabourmarket

UK Intellectual Property Office. (2024). Patent facts and figures, robotics-related filings, year-end 2024 tables. Newport: UKIPO. https://www.gov.uk/government/organisations/intellectual-property-office

UK Statistics Authority. (2018). Code of Practice for Statistics. London: UK Statistics Authority. https://code.statisticsauthority.gov.uk/

Chapter 9
Game-Theoretic and Optimization-Based Modeling of Anthropomorphic Service Robots for Customer Experience and Brand Strategy

Duggirala Aravind
http://orcid.org/0009-0005-4362-7684
ASET Institute of Science and Technology, India

N. V. Suresh
http://orcid.org/0000-0002-0393-6037
ASET Institute of Science and Technology, India

ABSTRACT

This chapter constructs a single game-theoretic and multi-objective optimization model to model anthropomorphic service robots in brand strategy and customer experience. The interaction between customers and robots is modeled as a bi-level strategic game, where the firms maximize anthropomorphic design vectors (voice, expressiveness, adaptability) and the customers respond through the process of trust and engagement. A utility maximization model is developed to balance the emotional involvement, brand alignment, and the operational cost. The framework uses the state

DOI: 10.4018/979-8-3373-6640-1.ch009

estimation of digital twins to enable real-time adaptation of robot behavior. Optimal anthropomorphic strategies are derived using equilibrium analysis and dynamic optimization, offering a mathematically based approach to AI-driven service design.

1. INTRODUCTION

Anthropomorphic service robots are changing service industry customer interaction by using AI, human interaction and adaptive decision making in a service environment. These systems not only drive operational efficiency but also contribute to a change in customer trust, emotional connection, and brand image. A game-theoretic and optimization-based model of strategic human–robot interaction in service ecosystems is developed in this chapter.

1.1 Background and Motivation

Artificial Intelligence, machine learning and intelligent automation have driven an unprecedented speed in implementation of service robots, including the hospitality, retail, tourism and healthcare sectors. A distinguishing feature of anthropomorphic service robots from other types of automatic robots is that these robots' characters are deliberately made to resemble those of humans, such as their skill of conversing, their emotional response ability, their adaptive behavior, and their ability to express facial expressions. The capabilities enable robots to achieve socially engaging interactions, beyond functional task performance (So et al., 2024; Blut et al., 2021).

With the rise in competition for customer experience and emotional branding, anthropomorphic robotics has become a strategic tool that can affect customers' satisfaction, trust, loyalty, and perceived service quality. A few studies deal with formal models of the interaction of customers and robots, but most focus on usability, acceptance, and technological performance and, crucially, do not consider the use of intelligent service robots. There is a lack of research, however, regarding the best anthropomorphic attributes that firms should adopt to create a maximum experiential and branding impact. This spurs the necessity for a mathematically informed model that combines game theory, optimization and adaptive AI-based systems to systematically understand and enhance the design and deployment approaches of service robots (Mahr et al., 2025; Letheren et al., 2021).

1.2 Emergence of Anthropomorphic Service Robots

Anthropomorphic service robots have matured from mere automated assistants to intelligent interactive agents that calculate or do what one might imagine a human companion could do in communicating and socializing interactive behavior. The use of NLP, computer vision, affect computing, and embodied AI has allowed robots to analyze customer's emotions, react dynamically to human behavior and tailor the experience of service delivery in real-time. As a result, companies are increasingly turning to anthropomorphic robots not just for their functionality, but also to foster empathy and build their brand. Humanoid robots are commonly used in hostel and tourism for reception and concierge work, and communication with guests (Tojib et al., 2023; Barone et al., 2024).

Intelligent robotic agents are used for customer guidance, customer recommendation in retail and for automated checkout which are part of healthcare robotics as well as integrating with patient interaction and emotional comfort. An increasing reliance on these systems demonstrates a longer-term shift towards experience-centric service systems where relationship and emotional intelligence are more key drivers of competitiveness. Trust building, behavioral adaptation, customer acceptance and brand alignment however pose strategic difficulties in the context of an increasingly complex human–robot interaction. Such challenges require sophisticated analytical tools that are able to model the interactions that occur over time and to optimise anthropomorphic service strategies (Wang et al., 2026; Wirtz et al., 2018).

1.3 Objectives and Scope of the Chapter

The purpose of this chapter is to establish a common framework for problem analysis of anthropomorphic service robots by combining game theory and optimization approaches—from a customer experience and brand strategy perspective. The overall goal is the modelling of customer–robot interactions as a strategic decision-making process in which the firms try to optimize their anthropomorphic attributes, and the customers try to engage in behavioural and emotional responses, including trust, engagement and acceptance. This framework aims to offer a systematic analytical perspective in comprehending the effects of humanlike attributes of a robot on experiential and branding results in service settings. The chapter builds on three important aspects: game-theoretic interaction modelling, multi-objective anthropomorphic design optimization, and digital twin-aided adaptive systems for real-time behaviour personalization. The study also discusses their use in higher levels of decision-making within the Hospitality, Retail and Healthcare industries. The chapter is not empirical, rather it takes a conceptual and modelled view of systems with the focus on system level integration and the analytical interpretation. Ethical and strategic issues discussed in the chapter include the ethical implications of anthropomorphic service systems, over-humanization, future AI service transformation and the impact on the workforce.

2. LITERATURE REVIEW

2.1 Anthropomorphism and Human – Robot Interaction in Service Robotics

Anthropomorphism is defined as the process of ascribing human attributes, emotions and social activity to nonhuman objects: intelligent machines and robotic systems. Service robotics is increasingly adopting anthropomorphic attributes, such as voice interaction, facial expression, gestures, and adaptive communication, to achieve the benefits of socially engaging customers. Many previous research studies have shown that humanness in the service environment has a positive impact on perceived warmth, trustworthiness, social presence and quality of interactions involving humans. The studies of human-robot interaction also show that customers are more likely to react to the robot demonstrating emotionally responsive and conversational actions rather than only functional (Choi et al., 2020; Song & Kim, 2022).

Progress in natural language processing, the field of affective computing and embodied artificial intelligence – the holy trinity of simulation of interpersonal communication – has given robots a new respect. They have further stated that anthropomorphic design has an impact on customer acceptance, emotional attachment, and customer perceived service quality within the hospitality, healthcare, and retail sectors. Most previous studies, however, focus on anthropomorphism as a behavioral or psychological phenomenon without developing formal mechanisms for how the anthropomorphic aspects are strategically used to shape service. Further, attention has been paid to adaptive interaction models, such as models which can adapt robotic behavior in response to customer feedback during the interaction, and such models have been lacking in anthropomorphic service research until now (Yang et al., 2024; Seo, 2022).

2.2 Customer Experience and Brand Strategy in Intelligent Service Environments

Better customer experience is among the chief foci of "digitally transformed" service providers' strategic agendas. In today's service business context, companies are increasingly integrating intelligence into their operations, not just to boost efficiency, but to enhance emotional connection, brand messaging, and customer loyalty. Anthropomorphic service robots play a part in this shift as service robots that interact with customers and can influence customers during service encounters. From existing literature, a customer experience has been referred to as a cognitive, emotional and relational exchange between customers and service technologies. Humanlike robotic systems bring the sense of personalization, empathy, and social connection, positively impacting the formation of customer satisfaction and customer trust (Gam et al., 2023; Goncalves et al., 2020.

When it comes to branding, organizations want their robots' shape, tone and personality to resonate with the organization's and its brand's overall message and values. It's been discovered in service marketing studies that high uniformity of interactions, and communication that's emotionally effective, effectively improve brand attachment and brand loyalty. However, most of the previous studies are mainly descriptive or qualitative and analytical approaches have not been widely investigated in an attempt to optimize anthropomorphic feature based on brand objectives and customer expectations. In addition, little research explores the strategic service contexts by which a service robot can master the emotional aspects of interaction in complex cases while meeting the operational and branding aspects (Cui & van Esch, 2023; Lee et al., 2021).

2.3 Game Theory and Strategic Interaction in Service Systems

Game theory gives an organised mathematical formalism for studying situations in which two or more agents with decision making powers interact in a competitive or cooperative manner. Service systems have been extensively investigated using game-theoretic methods to study pricing strategy, customer behavior, service allocation, coordination of supply chains, and the adoption of technology. The core idea of game theory is to represent the rational actors optimizing their decisions based on the reactions and incentives of others involved. In the case of anthropomorphic service robotics, companies decide what parameters are essential for their robot and how they should look, sound, and feel, and customers' response is manifested by trust formation, acceptance, engagement, or resistance (Tushar et al., 2023; Fu et al., 2023.

Early studies show that strategic interaction has a significant impact on service results, especially when customer perceptions have a direct influence on organizational performance and brand image. Equilibrium behaviour in service management and intelligent decision systems has been explained using concepts like Nash equilibrium, cooperative games, and leader vs. follower. But the use of game theory in anthropomorphic robotic applications are still rather restricted and are elusive. Cooking studies are mostly about the technical coordination between robots instead of interaction dynamics with customers and the implications of the same on branding aspects. The understanding, therefore, in how companies can strategically leverage human traits to enhance service experiences and increase brand value in an environment where customers' behaviors and expectations are continuously changing is limited (Li et al., 2022; McIntosh et al., 2024.

2.4 Optimization Approaches in Service Design and Adaptive Systems

Optimization techniques are then utilized more and more in intelligent service-systems to optimize operation efficiency, customer satisfaction and resource allocations. For service design research, optimization models are used in scenarios of scheduling, service customization and other service parameters associated with the service quality and pricing. Due to the conflicting requirements, ranging from customers' expectations for high service quality and reliability, to low operation cost in service companies, multi-objective optimization approaches are especially relevant. For anthropomorphic service robotics, optimization techniques can be used to select optimal anthropomorphic characteristics that capture customer trust, emotional engagement and brand identity, while minimizing the complexity of implementing and operationally costs (Chowdhury et al., 2025; Ouyang et al., 2023.

Existing studies also stress the significance of adaptable systems that morphological their service strategies in light of the conduct and the conditions of their clients. In recent years, digital twin technology has emerged as a critical innovation to simulate, monitor and adapt intelligent systems in real time. With ongoing feedback integration, digital twins can adapt conversation strategies by evaluating customer emotions and engagement and making real-time updates to the robot's behavior. In spite of these developments, the available literature does not often combine optimisation frameworks, anthropomorphic interaction models with adaptive digital twin architectures. Most of the studies are technologically isolated and operationally centred and do not take into consideration experiential and branding aspects. Hence, today's service robotics research is not well supported by a coherent optimization-based study that integrates human behavior, anthropomorphic design and adaptive decision-making (Haibeh et al., 2022; Salehi et al., 2022.

2.5 Research Gap Identification

The research in the past suggests a considerable achievement in anthropomorphic robots, research on customer experience, human–robot interaction and smart service systems. But there are gaps in understanding and thinking that need to be addressed within the inter-disciplinary between these fields. Studies conducted before this one mainly focus on anthropomorphic service robots from psychological, behavioral, or technological aspects and do not utilize formal mathematics that can be used to model strategic customers-service-robot interaction.

Secondly, few studies on customer experience and branding combine analytical optimization methods for finding optimal anthropomorphic combinations that simultaneously improve trust, engagement, emotional value, and the ability to operate efficiently.

Thirdly, game theory has been widely used in service management and decision sciences but its application to anthropomorphic robotics and customer-centered service ecosystems has not been explored. Further, robotic interaction, branding and adaptability are currently researched separately, in isolation from each other and strategic aspects. The other important aspect where digital twin is not well integrated is for real-time personalisation and dynamic behaviour adaptation in service robotics. Current solutions tend not to have a way of frequently measuring customer state and adapting the responses accordingly. Hence, an integrated framework, based on the idea of games and optimization, that can systematically model anthropomorphic service robots within experience-driven, adaptive and brand-oriented services in the future, is highly needed.

3. THEORETICAL FOUNDATIONS

3.1 Key Constructs: Trust, Engagement, and Brand Perception

The theoretic basis for anthropomorphic service robotics is primarily trust, customer engagement, and brand perception. The customer's trust in the reliability, competence, and predictability of robotic service systems during interaction processes. Anthropomorphic features like human-like communication, emotional sensitivity, and adaptability greatly contribute to the development of trust, thereby lowering perceived technological uncertainty and boosting social presence. Customer engagement is the emotional, cognitive and behavioral involvement elicited during the human-robot interaction and directly influences in satisfaction and long-term loyalty. At the same time, brand perception is the aggregate assessment of an organization's identity and service quality formed by the interaction experiences with robots. The design features of service robots are gradually expanding into the domain of communicating organizational values, innovations, and customer orientation. In fact, these constructs are tightly linked, as improved trust and engagement have a positive impact on brand perception and customer retention. Thus, to maximize anthropomorphic service design in intelligent Customer-centred service ecosystems, it is necessary to understand the interactions between these variables.

3.2 Conceptual Linkage Between Anthropomorphism and Strategy

Anthropomorphism is a coherent policy tool to guide customers' behavior, emotions and perception of service in intelligent service systems. The socially engaging voices, expressions, and conversational and adaptive behaviors enhance emotional relationships and service quality. At the strategic level, companies purposefully develop anthropomorphic features to support robotic interaction in the context of organisational branding goals and expectations. Through this linkage, a service robot becomes an experiential and relational asset that can be used to distinguish their brand in a competitive service market. Strategic anthropomorphic designing also plays a role in influencing the perceived warmth, competence, trust and social acceptance of customers that contributes to their decision-making. Therefore, the potential for emotional engagement needs to be weighed against operational efficiency and implementation costs by companies when making a decision on what kind of anthropomorphic configuration to choose. As a result, anthropomorphism, as a conceptual relationship, leads to optimization of robot characteristics which brings maximum customer experienced and benefitting from the brand. This view enables strategically modeling and systematically studying dynamics of strategic interaction between firms and customers within adaptive service ecosystems using game theory and optimization techniques.

3.3 Assumptions for Modeling Framework

To derive the proposed modeling framework, several theoretical and analytical assumptions are made that streamline the strategic interaction in anthropomorphic service environments. The rationality of firms is assumed, as the firms will choose anthropomorphic design aspects that maximize the customer experience, brand usefulness, and strength efficiency. Second, the customers are seen as responsive decision-making agents, whose trust, engagement and acceptance of any given interaction relies on the interaction's anthropomorphic quality. Thirdly, the relationship between firms and customers is viewed in dynamic and adaptive terms, in which the state of the customer changes over time as he/she services the firm. In addition, the framework posits that anthropomorphic qualities can be quantified, distillable into measurable qualities (emotional responsiveness, conversational ability, and visual human likeness). Moreover, digital twin systems are believed to be able to predict customer emotional state and enable "real-time" adjustment of robot behavior based on continuous feedback integration. External factors such as cultural differences, moral perception, and technological experience are thought to affect the results of interaction. These assumptions provide basis for developing game-theoretic and optimization models for intelligent service robotics systems.

4. GAME-THEORETIC MODELING FRAMEWORK

4.1 Strategic Interaction Between Firms and Customers

The suggested framework develops anthropomorphic service interaction as a strategic game between two decision-making agents - the firm and the customer. Service robots are anthropomorphic, for which the firm determines the characteristics, and customers respond with behavioural responses such as trust, engagement, satisfaction, and service acceptance. The anthropomorphic service robots establish emotionally responsive interactions, which directly affect the perception of the customer and the performance of the organization's brands, unlike conventional automated systems. Hence, strategic interdependence arises between both parties as the outcomes of the organizations rely on the customers' responses, whereas customers' experience relies on the quality of the robotic interaction and on the ability of adapting their behavior in repeated service contact in an intelligent environment. Have F represent the firm and C represent the Customer. The firm chooses an anthropomorphic design vector:

$$A = (a_1, a_2, a_3, \ldots, a_n)$$

Where a_i stands for Robotic capabilities such as: Being conversational, Responsive to emotions, Being ready to adapt communication, Gesture realism, Facial expressiveness. Customer responses are shown as the behavioural function:

$$R = f(T, E, S)$$

The 2D Data Bar indicates the relative levels of overall Trust T, Engagement E, and Customer Satisfaction S. Strategic interaction because customers' responses have a positive impact on the business outcomes of future revenue, retention and positioning. At the same time, adopting sophisticated anthropomorphic attributes adds to the complexity of processing power, the price tag, and the ethical issues of over-humanization.

The interaction therefore takes the shape of a non-cooperative game between the firms trying to have the best service/branding experience and the users who want to have emotionally satisfying and trustworthy experiences of interaction. The strategic relationship lays the bases for the application of game theory in anthropomorphic service robotics. This framework is for the systematic analysis of the customer's behavior, emotional reactions, and competitive organization design within adaptive service in global artificial intelligence, emotional interaction, and transformation of experience in service context today.

4.2 Strategy Space and Payoff Function Formulation

The strategy space represents the decision space for the firms/customers in the Anthropomorphic Robotic Services Environments. When designing human qualities for service robots, companies strategically choose their concentration and type, ranging from conversational intelligence, emotional response, visual realism to adaptive communication. These are then response factors such as formation of trust, emotional engagement, satisfaction, and the desire for ongoing interaction experienced by the customer. In intelligent service ecosystems with constantly changing and heterogeneous behavioral demands across different contexts and operations worldwide, the strong coupling between customers' responses and robots' properties influences their service effectiveness, organizational branding performance, and long-term customer retention. The formula of utility function developed by the firm is:

$$U_F = \alpha T + \beta E + \gamma B - \delta C$$

T indicates the level of customer trust, E represents intensity of engagement, B shows the level of brand alignment utility and C shows the operational cost and operational implementation complexity. The corresponding organizational weighting coefficients are: α, β, γ and δ with the respective priorities. The customer utility function can be also written as:

$$U_C = \lambda W + \mu Q - \theta P$$

Whereas, the factors of perceived warmth W, perceived service quality Q, and perceived technological discomfort or privacy concern P are among the factors. Parameters λ, μ, and θ measure customer sensitivity toward emotional interaction and technological perception.

The dual optimization structure creates an interdependence between the "brands and experience" optimization, which is targeted by the firms, and the "comfort and service" optimization, which is targeted by the customers. The framework conveys thus the balance between emotional involvement, technological believability, customers satisfaction, and operational efficiency. It is useful for this analytical structure as a mathematically sound basis for making design decisions for anthropomorphic service systems in various customer-centred and adaptive intelligent service industries worldwide today.

4.3 Equilibrium Analysis and Strategic Outcomes

The analysis of equilibrium is used to discover stable strategic results within anthropomorphic services surroundings in which a firm and a customer have no incentives to unilaterally alter choices. In intelligent service ecosystems, equilibrium is the perfect balance between anthropomorphic realism, building of trust toward a customer, customer engagement, operational cost, and brand performance. The suggested framework mainly relies on configurations of strategic interaction dynamics with Nash equilibrium and Stackelberg leader–follower structures under different behavioral and organizational settings. These equilibrium mechanisms offer analytical perspective regarding the optimal configuration of anthropomorphic service robots to obtain optimum customer benefit whilst still sustaining operational effectiveness and competitiveness of organizations in adaptive AI service environments worldwide today. A Nash equilibrium occurs when both anthropomorphic strategies A^* and the customer response R^* (simultaneously) maximize their utility functions, as indicated by;

$$U_F(A^*,R^*) \geq U_F(A,R^*)$$

and

$$U_C(A^*,R^*) \geq U_C(A^*,R)$$

For all practicable strategies A and customer responses R. Both take advantage of each other's mutual expectations and interaction behaviour to attain locally optimal results.

Nevertheless, in many service environments, the companies initially gain a major strategic understanding of anthropomorphic design and then customers react based on the perceived quality of service and emotional comfort they experience. As there is an interaction, it is modeled as a Stackelberg game with firms being strategic leaders and customers being responsive followers.

The equilibrium outcome defines the intensity level for the anthropomorphic image that would maximize organizational utility, while maintaining customer acceptance and emotional stability. Too much anthropomorphizing can lead to heightened discomfort, privacy concern and implementation expense, while not enough can cause less engagement and trust building. As a result, equilibrium analysis shows a target zone for the optimal design of a robot that is in a balance of technological capability, emotional interaction, and branding. These results give a robust analytical base to create adaptive anthropomorphic service systems to foster intelligent customer experience management and sustainable competitive advantage in modern services today.

Figure 1. Game-theoretic modelling framework

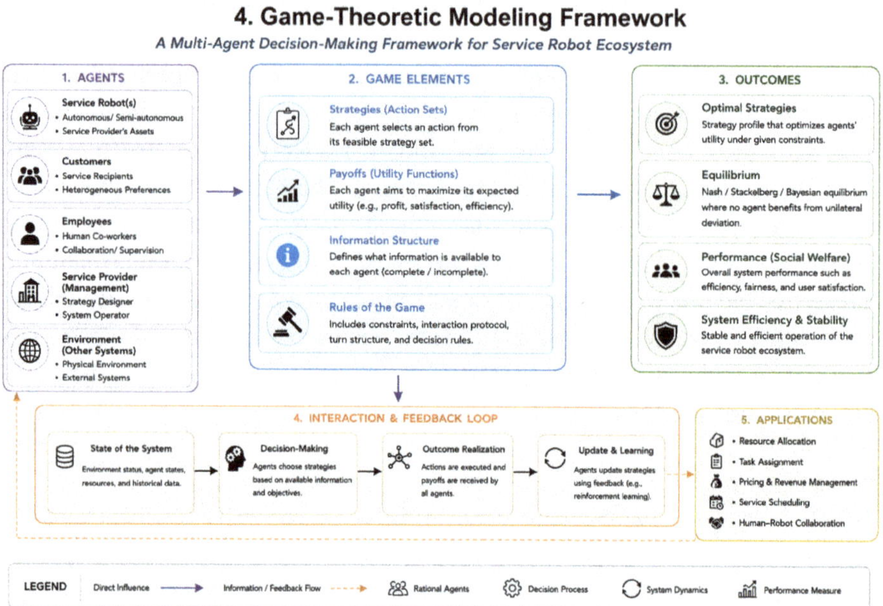

5. OPTIMIZATION – BASED DESIGN FRAMEWORK

5.1 Multi-Objective Optimization Formulation for Anthropomorphic Service Design

The optimization-based framework is designed to optimize configurations for human-like service robots that maximize the customer experience, emotional impact, and brand consistency whilst keeping operational and implementation expenses low. While normal optimization problems have just a single objective, anthropomorphic service systems should proceed through multi-objective optimization because goals of the organization are conflicting and interdependent. Customer trust and customer emotional interaction can be enhanced by drawing customers closer onto the anthropomorphic realism as long as it results in an increase in overall computational complexity, maintenance cost, and ethics concerns related to human over-realism. This necessitates the identification of a balance of robotic specification to create a best strategic service performance in the intelligent service environment with the consideration of customers' expectations, dynamic interaction behavior, and across numerous operational sectors and technologies changing global market services today.

The anthropomorphic design vector is represented as:

$$A = \left(a_1, a_2, a_3, \ldots, a_n\right)$$

Where each one represents one of the following attributes of robotic systems such as emotional responsiveness, conversational capability, visual realism, adaptive communication and gesture expressiveness. The overall optimization problem is, the overall optimization problem is given by: Hence the weighted aggregate form is a representation of:

$$\max Z = \{E(A), B(A), T(A)\} - C(A)$$

Where:

$E(A)$= customer engagement function
$B(A)$= brand alignment utility
$T(A)$= customer trust function
$C(A)$= operational and implementation cost

Hence the weighted aggregate form is a representation of:

$$\max U = \alpha E(A) + \beta B(A) + \gamma T(A) - \delta C(A)$$

In which α, β, γ, and δ have coefficient of organizational preference based on strategic goals.

The framework allows to systematically consider varying customer and operating conditions for alternative anthropomorphic configurations. Incorporating experiential, branding and economical goals into the same optimization framework, the model provides an economically grounded tool for optimizing strategic decision-making in adaptive anthropomorphic service systems for intelligent and customer-centric service environments in the worldwide service arena today.

5.2 Trade-Off Analysis and Feasible Solution Space

In the design of anthropomorphic service robots, there are several trade-offs to consider since no single service goal improvement will come without a negative impact on another service organizational goal. For instance, enabling emotion realism and conversational intelligence can be beneficial to engage customer and/ or build trust, yet also lead to cost, complexity, privacy issues, and maintenance demands during system operation. In adaptive service ecosystems, as a result, there is a need for organizations to design their anthropomorphic configurations that are viable both for experiential quality and branding as well as economic considerations. The optimization framework therefore deals with interactions between conflicting objectives in a limited solution space, where technology constraints, customer expectations and organizational resources are determining factors in an intelligent service industry in a dynamically changing global market today. The feasible solution space is represented by the constraint set:

$$\Omega = \{A \in \mathbb{R}^n : g_i(A) \leq 0, h_j(A) = 0\}$$

Where:

Budget constraints, ethical restrictions, computational resources, and operational complexity are among the inequality constraints represented by $g_i(A)$.

$h_j(A)$ is equality constraints for the required service performance level and system reliability conditions.

During the optimization process, solutions are developed that satisfy all constraints and maximize organizational benefit. Trade-offs are analyzed using the concept of "Pareto Optimality," a solution is optimal when trying to improve on one objective comes with a sacrifice on some other objective. The Pareto frontier then indicates the optimum anthropomorphic configuration with acceptable customer engagement and branding results at minimal operations cost.

In a mathematical sense, a solution A^* is Pareto – optimal if it satisfies:

$$\nexists A \in \Omega : U(A) \geq U(A^*)$$

has made one or more marked achievements with respect to one of the objectives. In this way, companies can assess their strategic options based on their goals and customer requirements. The feasible solution space thus enables decision makers today to implement emotionally compelling, economically viable and strategically integrated robotic service systems in complex, adaptive environments of human interaction and the modern intelligent service ecosystem worldwide in an analytically warranted anthropomorphic configuration.

5.3 Interpretation of Optimal Design Strategies

Optimal anthropomorphic design strategies are the appropriate robot designs that achieve best-in-class acceptance by the customer, emotional connection, brand perception, as well as operational efficiency and technology sustainability. The effectiveness of anthropomorphic service robots is not only on the degree that many of their behavioral, emotional and communicative attributes are human-like, but on choosing and mixing strategically on different combinations, depending on customers' expectation and organizational goals. Hence, there is an even greater demand for firms today to decide the best level of anthropomorphic realism that can be used to facilitate customer trust and engagement while avoiding various uncomfortable situations with technological needs, ethical issues, and a complexity of implementation in the different industries that form the adaptive intelligent service ecosystems found around the world.

The optimal strategy is represented as:

$$A^* = \underset{A \epsilon \Omega}{\operatorname{argmax}} U(A)$$

Where A^* represents the anthropomorphic (ACP) maximising organisational utility under operating and behavioural limits. Optimization results give an attribute combination for the robot that can achieve the greatest strategic value from multiple points of view – customer experience, branding performance, and operational sustainability.

Figure 2. Optimization – based design framework

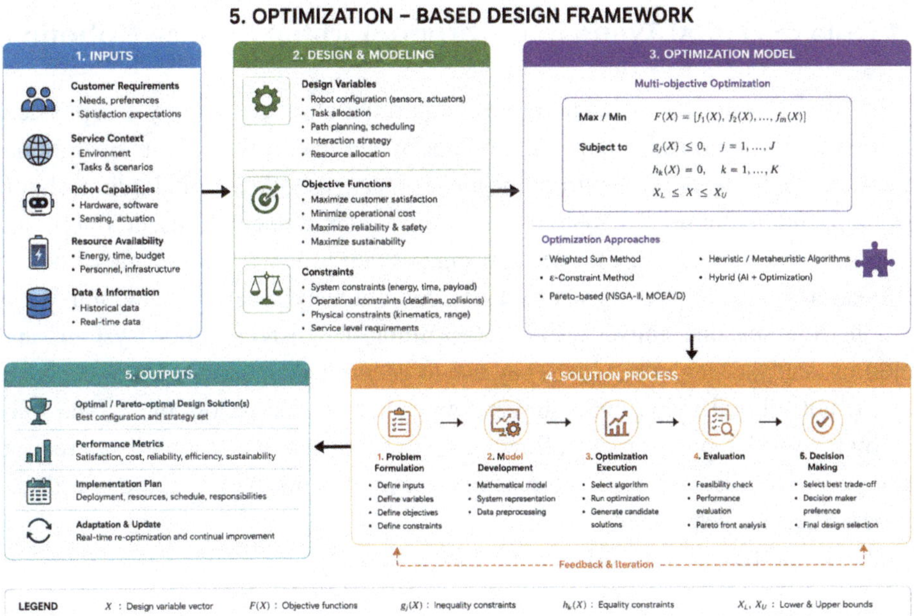

The framework also shows that the anthropomorphic configuration needed for various industries is different depending on the level of interaction required for a particular service and customer sensitivity. Positive aspects of hospitality and tourism might be focused on emotional warmth and good speaking abilities, whereas in the health care sector, it could be trustworthiness, empathy, and clarity of communication. Efficiency, responsiveness and customer guidance functions could be some retail service environments focus on. Thus, the best anthropomorphic design is dependent on the context and it can vary based on customer's behavioral patterns, organization's branding strategy, and the complexities of the service. The optimization results also suggest that beyond certain optimal points, there is a reduction in return on additional anthropomorphic realism. However, as further attempts are made to make human figures more human, they can create discomfort, a feeling of manipulation or a moral objection by customers. Thus, a person-balanced anthropomorphic representation would gain better long-term performance than extreme anthropomorphic depictions. These findings offer organisations a mathematically informed model for enabling intelligent customer experiences, differentiation and competitive advantage in today's AI-enabled service ecosystems around the world via the power of adaptive service robots.

6. DIGITAL TWIN – ENABLED ADAPTIVE SYSTEM

6.1 Role of Digital Twins in Anthropomorphic Service Robotics

The digital twin is a virtual and live computational model or copy of a physical system that is monitored, simulated, and optimized in real-time for its operational performance. In the era of anthropomorphic service robotics, digital twin technology helps intelligent service systems analyze the interaction patterns, emotions and environmental factors of customers during service process. Unlike static robotic architectures, digital twin–based systems continuously receive behavioral information from sensors, conversational algorithms, facial recognition systems and emotions analysis engines, which they use to generate adaptable robotic feedback. As a result, digital twins make anthropomorphic service robots more than only automated service providers by allowing them to learn and adapt their interaction strategies as service contexts and customer requirements evolve in today's global AI-based service ecosystems.

The digital twin framework can be represented as:

$$DT = \{P, V, D, S\}$$

Where:

P = physical robotic system
V = virtual computational replica
D = real-time interaction data
S = synchronization mechanism between physical and virtual systems

The process of state designation is depicted as follows:

$$x_{t+1} = f(x_t, u_t, w_t)$$

Where:

x_t = customer interaction state at time t
u_t = robotic interaction strategy
w_t = environmental and behavioral uncertainty factors

The digital twin system can then predict future interaction outcomes and adapt the robot's actions in real time in this formulation. Digital twins assist in enhancing customer interaction, the formation of trust, and emotional reactions in service interaction processes through predictive modelling and behavioural simulations. Given the technical complexity of today's service environments in various industries and sectors, both virtual simulation and anthropomorphic robotics creates an intelligent decision support system that helps businesses personalise services, adapt their activities and manage customers as they increasingly become data-driven.

6.2 Real-Time Customer State Estimation and Feedback Adaptation

The subsequent real-time customer state estimation is regarded as a crucial aspect of adaptive anthropomorphic service systems as customer emotions, trust levels and engagement intensities continually change throughout interaction processes. The digital twin architectures allow for service robots to track and analyze behavioural indicators such as expressions, voice, natural sentiment in conversations, gestures, and length of engagement. These behavioural features enable robots to predict customer emotion and adaptively adjust communication based on the success or failure of their interactions. Thus, the anthropomorphic service robots can provide more personalized and emotional services than predefined service behavior. During a time of ever-changing environments and behaviors around the world, this adaptive ability greatly enhances customer satisfaction, perceived service quality, emotional involvement, and trust building in the field of intelligent service ecosystems oriented to customers.

The dynamic state vector: $X_t = [T_t, E_t, S_t]$ characterizes customer state estimation. Where:

T_t = customer trust level
E_t = engagement intensity
S_t = satisfaction state at time t

The state update mechanism is formulated as: $X_{t+1} = AX_t + BU_t + \epsilon_t$
Where:

A = behavioral transition matrix
B = control matrix representing robotic actions
U_t = anthropomorphic interaction strategy
ϵ_t = stochastic behavioral uncertainty

Estimated customer states modify robotic interaction policies through a feedback adaptation process continuously. The system reduces the level of engagement and emotional expression or interaction and increases it if engagement goes down or is uncomfortable with customers returning the engagement. On the other hand, very engaged customers could get more personalized and emotionally rich interactions. By means of this feedback process mechanism, anthropomorphic service robots can be operated as adaptive intelligent agents able to update the quality of the interaction with their customers continuously. Finally, digital twin systems are deployed in today's world of many complex, emotional, and evolving customer behaviors and experiences in the global service enterprise, and provide improved service personalization and strategic customer experience management (CEM).

6.3 Integration With Game-Theoretic and Optimization Layers

Digital Twin systems in conjunction with game-theoretic and optimization techniques form a common analytical architecture suitable to enable adaptive strategic decision making in anthropomorphic service environments. Digital twins are the 'behavioural intelligence' needed to continuously adapt the system, whereas game theory models the strategic relationship between firms and customers, and optimization techniques provide optimal anthropomorphic meshes. This integration allows service robots to adjust the way they interact with the user based on dynamic situational factors like their mood, engagement, and environment. Today in various customer-oriented sectors and technologically high-impact service environments around the world, anthropomorphic robotic systems will now be able to adapt to maximize operational productivity, customer experience, and brand image through intelligent data-driven adaptation processes in their ever-changing service ecosystems.

The integrated utility optimisation framework is given by;

$$U_t = \alpha E_t + \beta T_t + \gamma B_t - \delta C_t$$

Where:

E_t = engagement level
T_t = trust state
B_t = brand perception utility
C_t = operational cost at time t

These variables are continually replenished through behavioural feedback and predictive analytics in the digital twin. Optimization layer then calculates the best anthropomorphic approach:

$$A_t^* = \operatorname{argmax} U(t)$$

Limited by behavioral, operational and ethical boundaries. At the same time, a customer's response dynamically adjusts the parameters of the game. As behaviour states shift over time, the customer utility function trails in the wake, allowing firms to predict the customer response and make strategic responses in an anticipatory manner. Thus, a closed-loop adaptive decision structure is formed, where predictive, optimizing and strategic analysis of interactions are treated in one intelligent framework.

Figure 3. Digital twin – enabled adaptive system

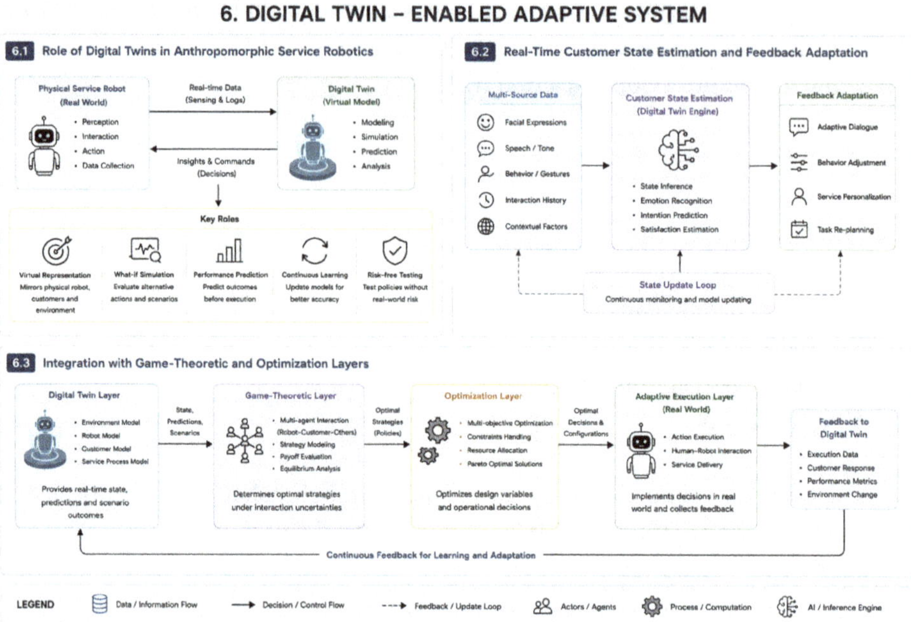

The integrated system thus allows to continuously learn from outcomes of interaction, optimizing behavioral strategies on-the-fly in anthropomorphic service robots. These capabilities bring reliability, emotional connection, and organizational branding to customer, and sustain operations. The framework thus gives a mathematically sound basis for developing next-generation adaptive service robotics, enabling strategic competitive advantage and intelligent change of customer experience in the future world-class AI service industries.

7. INTEGRATED FRAMEWORK AND MODEL SYNTHESIS

7.1 Unified Architecture Combining All Three Contributions

The framework is unified in one novel analytical framework of game-theoretic interaction modeling, an optimization-based design strategy and digital twin—based adaptation all in one. This framework aims to incorporate strategic interaction with customers and firms, as well as capture real-time behavior and optimisation of robotic service performance. Unlike approaches that only consider the customer's or the operations' behaviour, or operational efficiency, the shared model brings together all of these four aspects of the customer's emotional interaction with the organization, its brand strategy, its adaptive intelligence, and its optimization of decisions, all within a single computational structure. Thus, the architecture gives a comprehensive framework for service robots that dynamically adapt their actions according to customer requirements and environmental adaptation under modern AI-enabled service systems. The first level of the architecture is the game-theoretic interaction module, a model of strategic decision-making by firms and customers.

This layer places a focus on how anthropomorphic design impacts trust formation, engagement and customer response behavior. The second layer feeds into a multi-objective optimization engine that determines the robotic configuration which provides maximum customer service, brand matching, operational simplicity, and implementation cost. The third layer fuses digital twin capabilities for continuously monitoring, simulating and managing behaviors adaptively. The interrelated feedback processes among the three analytical components form the basis for functioning. Customer interaction data continuously feeds into the digital twin system, updating the behavioral estimates and the prediction of contacts' interactions. Under the revised interaction conditions, the optimization layer then recalculates new anthropomorphic strategies and the game-theoretic module calculates new results of the strategical equilibrium. This integrated architecture lays the groundwork for adaptive, intelligent service system that enables emotionally attuned, strategically optimized and computationally adaptive anthropomorphic robotic systems to serve a variety of service industries worldwide today.

7.2 End-to-End System Workflow

This integrated end to end workflow has been described using four stages in the relationship of anthropomorphic service robots and customers as: Data collection, Behavior estimation, Strategic analysis and Adaptive response generation. The workflow process starts when one of the customers interacted with a service robot in a setting or environments of hospitality, retail, healthcare or tourism systems. Interacting with the robotic platform, the platform gathers multidimensional behavioral data such as voice tone, emotional expressions, conversational feedback, gesture response, service engagement indicators, etc. They are the key inputs into the digital twin architecture that is continually monitoring customer behavioral state and predicting subsequent customer interactions. Customer trust, engagement and satisfaction levels are predicted in the digital twin layer using behavioral models that are collected in an interaction phase. The new customer state variables are then passed to the optimization module that assesses various anthropomorphic interaction strategies based on organizational goals such as enhancing customer interaction experience, strengthening branding capabilities, and optimizing the operation. Concurrently, the game-theoretic interaction module simulates the interactions between firm decision and customer reaction at a stable equilibrium in the presence of customer interaction for different interaction scenarios. Proceeding from the analytical evaluation, optimized anthropomorphic strategy is sent back to the robotic system. The robot then adaptively adjusts conversational style, emotional response, communication intensity, and interaction patterns of behavior with existing strategic recommendations. This adaptive cycle can be repeated during the customer interaction process, allowing for customization and smart adaptation in real time. The workflow thus establishes an intelligent service loop that builds trust with customers, deepens emotional connection, and optimizes service operations in real time across adaptive AI service ecosystems in different industries around the world today.

7.3 Analytical Insights From Integration

The fusion of game theory concepts, optimization methods, and digital twins creates several significant analytical findings in terms of anthropomorphic service robotics and intelligent customer interaction management. The integrated framework shows that optimizing customer experience requires more than just advanced technology, it also requires a strategically positioned anthropomorphic interaction that allows the facilitation of emotional involvement with organizational branding goals. The results suggest that the over- and under-pictoralization of human interaction can have adverse effects on customer satisfaction, trust of the service, and level of engagement as well. As a result, the ideal performance of a robot service relies on the balance between emotional responsivity, technical capacity and the perception of the service, which needs to be sustained in adaptive intelligent service ecosystems.

Second, the integrated architecture emphasizes the dynamism and continuously changing nature of behavior of the customer interacting over the course of service encounters. This makes static robotic interaction strategies no longer adequate to ensure lasting customer interaction and service effectiveness. Digital twins facilitate real-time in-the-beach monitoring of the customer's emotional state and adapting their interaction pattern through the environment, for an ongoing robotic system adaptation. Multiple adaptability is essential for enhancing the level of personalization and interaction quality in terms of service complexity.

Third, it illustrates, through the adopted game-theoretic equilibrium analysis combined with optimization mechanisms, how firms and customers strategically can interact in a systematic manner. Organizations can therefore anticipate customer reaction to different anthropomorphic configurations, and be ready to tailor the delivery strategy, based on the operational goals and the branding that is desired. Taken together, these findings are to set the groundwork for the development of intelligent anthropomorphic service systems fed by integrated analytical architecture, which involves strategic modeling of interactions, optimization-based decision-making and adaptive behavioural intelligence. In this respect, the new framework lays the groundwork for future service ecosystems built on AI for emotional responsiveness, personalized service for customers and sustained competitive grabbing in global intelligent service industries today.

8. SECTORAL APPLICATIONS (MODEL-DRIVEN)

8.1 Hospitality and Tourism

There are emerging examples of anthropomorphic service robots being integrated into the hospitality and tourism context to enhance the quality of service delivery interactions, operations efficiency, and differentiation. Humanoid Robotic Systems are being used at hotels, airports, restaurants etc. to manage the reception, as a concierge, for room service provision, multilingual communication, and personal interaction with the guests. In these settings, emotional engagement, building trust, and responding to customer queries will all play a big role in their experience. Hence, in today's highly competitive tourism sector around the world, anthropomorphic features like conversational intelligence, expressive face, adaptive communication and emotional sensitivity become indispensable to enhance guest satisfaction and perceived service quality in recurrent interaction processes. The proposed game-theoretic and optimization-based framework provides a strategic decision on the optimal anthropomorphic settings based on customer expectations and branding goals for hospitality organizations.

Digital twin architectures constantly track customer interaction patterns, such as emotional state, engagement level and conversation feedback, dynamically adjusting robotic communication patterns as the interaction progresses. Emotionally-aware concierge robots, for instance, could adjust the verbal and nonverbal tone and communication style based on guest moods and the type of service. The framework also ensures the operational sustainability needs the balance of customer engagement with the implementation complexity and resource allocation constraint. The anthropomorphic measurement of intensity levels that can be optimized for highest trust and emotional comfort while not causing customer discomfort or increased operational expenditure is known as equilibrium analysis. As a result, the integrated structure offers a systematic analytical toolset to hospitality and tourism entities to implement intelligent robotic service systems that can ultimately improve the customer experience, bolster their brand identity and facilitate adaptive personalisation of service in AI-enabled tourism environments worldwide today.

8.2 Retail Service Environments

In more and more retail service applications anthropomorphic robotic systems are now being used to enhance customer support, efficiency, product suggestions and customised customer experiences. Today, in the retail, smart robot assistants are used to guide customers, automated check-out, informing them about inventory, product information and making interactive recommendations available. Unlike conventional retail automation systems, anthropomorphic robots are designed to interact with people in a socially appropriate manner to enhance customer trust, emotional bond, and brand image in a shopping experience. The exchange of messages, adapting of recommendations and conversational interaction are therefore important strategic dimensions influencing customers' satisfaction and purchase decision processes in highly competitive digital and physical retailing spaces worldwide today. The proposed optimization-based framework allows retail organisations to optimise the anthropomorphic robotic features based on customer expectations and organizational branding objectives. Digital twin systems constantly monitor and continuously collect behavioral data, such as browsing history, feedback during conversations, customer emotions, engagement levels etc., which enable real-time personalization and adaptive recommendation strategies to offer better customer experience. Some robots, for instance, might adapt their communication style and the amount of suggestions they make given the behaviour of their customers as well as the interests in the products displayed during a service interaction. The game theoretic interaction layer also facilitates strategic perspective of customer reaction under different anthropomorphic configurations. Retail businesses can then optimize engagement and brand loyalty while reducing technological discomfort and operational expenses by bringing online and offline channels into harmony. The framework also helps organizations balance the personalisation of AI-powered customer interaction systems with privacy and ethical concerns. The integrated model is thus a mathematically-based basis for designing adaptive anthropomorphic robotic devices that can help improve shopping experiences, customer retention and even customer differentiation in global intelligent retail environments today.

8.3 Healthcare Interaction Systems

In healthcare, anthropomorphic service robots are being more widely used to aid patient interaction, help provide emotional support, assist with rehabilitation, and facilitate intelligent communication in healthcare processes. In clinical settings such as hospitals, the elderly care system, rehabilitation centers, and similar areas, the use of socially assistive robots is aimed at maximizing patient involvement, minimizing emotional stress, and favoring communication efficiency between patients and health care workers. As opposed to traditional medical automation technologies, anthropomorphic healthcare robots are designed to show empathy, trustworthiness, emotional response, and adaptable communication, as patient satisfaction and psychological comfort have a significant impact on healthcare results and treatment cooperation. Thus, the intelligent healthcare ecosystem around the world in today's technologically advanced clinical and rehabilitation setups needs to incorporate human-like interaction attributes to create an emotionally supportive and patient-centric environment. It is proposed that the characterized anthropomorphic robotic parameters be optimized based on the emotional needs of the patient, sensitivity to interaction and the healthcare communication requirements. Digital twin architectures constantly assess patient emotions, engagement levels and interaction comfort by measuring behaviors and observing patient interactions in real time. Healthcare PSS can thus adapt the nature of the communication, the speed of interaction, and the emotional reactions to the patient based on the patient's condition and psychological reaction during PSS processes. By optimizing this interaction quality, patient trust, operational efficiency and technological reliability, the optimization layer helps healthcare organizations achieve a delicate balance in limited healthcare settings. At the same time, game-theoretic interaction analysis can be used to assess the effects of anthropomorphic configuration variations on patient acceptance and treatment cooperation behaviors. Ethical issues regarding emotional dependency, privacy issues, and human replacement in intelligent healthcare systems are also touched upon in the framework. Hence, the integrated architecture creates an analytical framework that helps healthcare enterprises build adaptive anthropomorphic robotic systems that can aid in personalized patient care and give an emotional touch to the healthcare interactions worldwide today.

8.4 Cross-Sector Comparative Insights

Comparing the case studies in healthcare, retail, and hospitality reveals that anthropomorphic service robot systems are unique because of how they interact with the customer depending on the type of interaction, how emotional they need to be, and the goals for the company. While every sector employs anthropomorphic features to engage and deliver better services to its customers, the weightage of the factors of trust, emotional responsiveness, operational efficiency and branding differs in various industry segments. Emotional warmth, customization and enriching experiences are the competitive strengths of the hospitality and tourism sector while adaptiveness in customer's recommendations, efficiency of engagement, and customer retention are among the competitive skills of retail. Healthcare interaction systems, however, are mainly concerned with trustworthiness, empathy, emotional comfort and communication reliability in the sensitive interaction processes with patients worldwide in the ever more intelligent service worlds. It is observed that the integrated framework shows that optimization strategies should be developed for each sector, as there can be differences in the behavior for similar anthropomorphic configurations with different service conditions.

Thus, digital twin systems are crucial to the process of adaptive personalization based on interaction needs and customer expectations for the applicable industry. Emotion expressive communication can be essential for a hospital robot and controlled empathy and reassure pattern is essential for a rehabilitative robot. In the retail store, it might be more important to be efficient and adapt to recommendations instead of realistically depict the emotions. The game-theoretic and optimization-based architecture also illustrates the heterodoxity of the strategic equilibrium conditions across the sectors in terms of customer sensitivity, operational requirements and branding goals. Therefore, organizations need to be mindful of balancing anthropomorphic realism with ethical issues, implementation feasibility and service goals in each sector.

9. DISCUSSIONS

The proposed integrated framework provides an example of how best to perform emotion engagement, trust building, efficiency in service performance and brand alignment in the optimal use of anthropomorphic service robots. Game theoretical analysis shows how the firms and customers' environment and characteristics are interdependent, thereby directly affecting behaviours and long-term firm outcomes. The result of optimizations indicates that the degree of anthropomorphic realism effects engagement and trust up to a critical point, after that, the customer is no longer comfortable, ethical implications become strong, and operation becomes complex, thus decreasing overall utility. The digital twin layer also shows how behavior of the customer changes during his or her encounter with the service, pointing to the need for a 'real time' and 'adaptive' means of response by robotic tools that cannot just rely on static interaction strategies. All these results suggest that a realistic design of an anthropomorphic service is not related to the most human-like design itself, but on the adaptation of the personalization and the contextual alignment of the offered services. Thus, in the present modern service-based AI ecosystems around the world, the framework sets out the need to dynamically optimise the performance of emotionally intelligent robotic systems, according to customer behavioural state, organisational branding goals and sector-specific service requirements.

The proposed framework significantly advances the theory of anthropomorphic service robotics by combining concepts from game theory, optimization science, digital twins, customer experience, and human–robot interaction in an integrated analytical framework. Prior studies have focused on the anthropomorphism on its own and from either a behavioral or psychological or a technological lens, whereas the proposed framework lays out formal mathematical relationships among strategic interaction, adaptive decisions, and experiential service outcomes. By combining game-theoretic equilibrium analysis with optimization mechanisms, conventional service management theories are expanded by developing customer–firm interaction as a dynamic strategic process based on emotional and behavioral responses.

In addition, by incorporating digital twin architectures, anthropomorphic service theory now brings adaptive intelligence, continuously estimating the behavior and adjusting to the interactions. The framework further enriches the body of branding and customer experience literature by providing a description of anthropomorphic robotic attributes as assets that can help organizations build trust, engagement and perceived service quality. As a result, the work provides a cross-disciplinary theoretical basis for future analytical and computational research in intelligent service ecosystems around the world today.

The framework offers valuable management guidelines for organizations aiming to incorporate anthropomorphic service robots in customer-centric service environments. The conclusions reveal that the cultivation of technology complexity has to be balanced with anthropomorphic traits tailored to customer expectations and corporate branding aims to avoid current breakdown and negative reputation. More than functionality, proper robotics capability, emotional responsiveness, conversational quality, or adaptability to interaction are key factors in customer engagement and trust by the manager. The optimization approach also reveals that tailoring anthropomorphic configuration is necessary for various services based on their interaction sensitivity, operating conditions and customer behavior patterns. Additionally, digital twin systems offer valuable strategic benefits because of the ability to leverage on the monitoring of customer interaction outcomes and the adaptation of robotic behaviour on-the-go. This flexibility helps to improve personalisation, customer retention, and service consistency in dynamic service settings. Furthermore, balancing emotions with ethics, privacy, and sustainability are emphasized. Therefore, the proposed framework can be used as an effective strategic planning support system to build an intelligent anthropomorphic service ecosystem worldwide today.

Prior works on anthropomorphic service robotics mainly concern one of the dimensions of customer acceptance, emotional interaction, robotic usability or technological performance, but fail with the integrated analytical relationships among strategic interaction, optimization and adaptive intelligence. Most traditional models are descriptive or experimentally evaluated and focus on observation of the behavior rather than on formal modeling of the dynamics between customers and robots. The goal of the proposed framework is to introduce a game theoretical optimization paradigm supported by digital twin systems into an integrated framework that enables modeling of strategic adaptation and real-time behavioral evolution in the course of service interactions. The framework constantly adapts robotic strategies based adaptive feedback mechanisms, which incorporates the dynamics of customer emotions, engagements and service conditions within the context of the service setting, unlike static interaction models. There is also limited consideration of bringing branding objectives or optimizing customer experience into the decisions made by robots within the existing approaches to service design. This drawback has been taken into consideration in the present model by introducing brand alignment, emotional engagement and operational efficiency within a multi-objective optimization framework.

10. FUTURE RESEARCH DIRECTIONS

The potential for co-deployment of generative AI with anthropomorphic service robots in the future and unlocking deeper context awareness, emotion intuition, and conversation understanding remains to be explored. Generative AI models have the potential to enrich the communication of robots to provide dynamic responses, personalized suggestions, and even human interactions during customer interactions. These technologies, in conjunction with embodied Robotic Agents, could have substantial enhancements for emotions and flexible interaction in service environments. The inclusion of generative AI, however, poses the problem of control of hallucinations, ethical communication, emotional manipulation, and transparency of decisions. These are compelling reasons for future research to build strong control, optimization, and assessment protocols for anthropomorphic interaction systems based on generative AI across global intelligent service ecosystems today.

In the future, more autonomous and adaptive intelligence in service systems needed to environmental circumstances of complex customer interaction. Research efforts are needed to create self-learning robotic architecture that incorporates reinforcement learning, predictive analytics, and real-time adaptable behaviors. These systems should continuously improve the methods of communication, emotional responses, and personalization of services based upon changing customer behavior and environmental influences. Moreover, uncertainty management, ethical decision making and consistency of interaction over time in dynamic service ecosystems need to be tackled by future adaptive frameworks. Issues of autonomous decision-making and a digital twin approach can further optimize predictive services and operational sustainability. Thus, advanced adaptive systems are an important direction in next-generation of intelligent anthropomorphic service robotics in the world today.

The proposed framework offers an excellent conceptual and analytical background, but further research is needed to prove the efficiency of game-theoretic and optimum anthropomorphic service systems in practical service environments. Examples of empirical studies include experimental simulation studies, customer interaction analysis, behavioral observation, and field implementation in the hospitality, retail and healthcare industry. Researchers need to consider the effects of different anthropomorphic configurations on customer trust, engagement, emotional responses, and sense of brand in real service settings. A longitudinal study could investigate the changes in customer behavior over time with repeated human-robot interactions. This empirical evidence will lend greater credibility, usefulness, and generalizability to integrated anthropomorphic service designs across today's AI-powered CX ecosystems worldwide.

11. CONCLUSION

In this chapter, a customer experience and brand-oriented, integrated, game-theoretic and optimization-based framework for analysing anthropomorphic service robots has been presented. The study integrated strategic interaction modeling, multi-objective optimization and digital twin-based adaptive systems into a single analytical architecture that can support intelligent, personalized service provisioning and dynamic interaction with customer. The results of the framework showed that anthropomorphic attributes are relevant to the formation of trust, emotional interaction, operational efficiency and organizational branding outcomes in the context of hospitality, retail and healthcare. Further, it has presented some mathematically sound approaches for equilibrium analysis, adaptive behavioral estimation and strategic robotic configuration with a complete analytical basis for future intelligent service robotics research and development worldwide today.

The proposed framework gives important implication to academic research as well as practical implementation of anthropomorphic service robotics. Studying from the research point of will provide a multidisciplinary analytical basis that combines the fields of game theory, optimization science, human-robot interaction, and digitization technology into intelligent service ecosystems. The framework provides for future empirical testing, computational modelling, and adaptive service research using AI. On managerial side, they can benefit using the framework to develop anthropomorphic robotic systems in correspondence with customer expectations, branding goals and operational demands. The results also highlight the need for a careful integration of emotional engagement, ethical issues, technical challenges, and service personalization when implementing intelligent robotic systems in today's global industries addressing the customer.

Anthropomorphic service robots are driving the renaissance of intelligent service ecosystems in a number of industries by providing emotionally intelligent, adaptive and customer-focused interaction capabilities. With the ongoing development of artificial intelligence, digital twins and adaptive decision systems, service robots will take on more strategic roles within the organization, impacting customer experience, building trust, and differentiation. The framework proposed in this chapter offers a structured basis for understanding and enhancing the strategic collaboration between humans and robots in future AI-enables services. Hence, intelligent anthropomorphic robotics will stay a key field influencing the next revolution of customer experience across the globe today.

REFERENCES

Barone, A. M., Stagno, E., & Donato, C. (2024). Call it robot: Anthropomorphic framing and failure of self-service technologies. *Journal of Services Marketing*, *38*(3), 272–287. DOI: 10.1108/JSM-05-2023-0169

Blut, M., Wang, C., Wunderlich, N. V., & Brock, C. (2021). Understanding anthropomorphism in service provision: A meta-analysis of physical robots, chatbots, and other AI. *Journal of the Academy of Marketing Science*, *49*(4), 632–658. DOI: 10.1007/s11747-020-00762-y

Choi, Y., Choi, M., Oh, M., & Kim, S. (2020). Service robots in hotels: Understanding the service quality perceptions of human-robot interaction. *Journal of Hospitality Marketing & Management*, *29*(6), 613–635. DOI: 10.1080/19368623.2020.1703871

Chowdhury, M. M. H., Aliahmadi, S. Z., Quaddus, M. A., Ansaripoor, A. H., & Mirjalili, S. (2025). A decision support framework for resilient and sustainable service design. *Global Journal of Flexible Systems Managment*, *26*(S1, Suppl 1), 25–55. DOI: 10.1007/s40171-024-00419-7

Cui, Y., & van Esch, P. (2023). Artificial intelligence in customer service strategy for seamless customer experiences. In *Artificial Intelligence in Customer Service: The Next Frontier for Personalized Engagement* (pp. 73–97). Springer International Publishing. https://doi.org/DOI: 10.1007/978-3-031-33898-4_4

Fu, S., Feng, X., Sultana, A., & Zhao, L. (2023). Joint power allocation and 3D deployment for UAV-BSs: A game theory based deep reinforcement learning approach. *IEEE Transactions on Wireless Communications*, *23*(1), 736–748. DOI: 10.1109/TWC.2023.3281812

Gam, A., Bahri-Ammari, N., & Soliman, M. (2023). Intelligent customer experience and behavioral brand loyalty: Empirical evidence from the commerce environment. *International Journal of Human-Computer Interaction*, *39*(10), 2090–2101. DOI: 10.1080/10447318.2022.2074120

Gonçalves, L., Patrício, L., Grenha Teixeira, J., & Wunderlich, N. V. (2020). Understanding the customer experience with smart services. *Journal of Service Management*, *31*(4), 723–744. DOI: 10.1108/JOSM-11-2019-0349

Haibeh, L. A., Yagoub, M. C., & Jarray, A. (2022). A survey on mobile edge computing infrastructure: Design, resource management, and optimization approaches. *IEEE Access : Practical Innovations, Open Solutions*, *10*, 27591–27610. DOI: 10.1109/ACCESS.2022.3152787

Lee, C. H., Li, Q., Lee, Y. C., & Shih, C. W. (2021). Service design for intelligent exhibition guidance service based on dynamic customer experience. *Industrial Management & Data Systems*, *121*(6), 1237–1267. DOI: 10.1108/IMDS-06-2020-0356

Letheren, K., Jetten, J., Roberts, J., & Donovan, J. (2021). Robots should be seen and not heard… sometimes: Anthropomorphism and AI service robot interactions. *Psychology and Marketing*, *38*(12), 2393–2406. DOI: 10.1002/mar.21575

Li, T., Peng, G., Zhu, Q., & Başar, T. (2022). The confluence of networks, games, and learning a game-theoretic framework for multiagent decision making over networks. *IEEE Control Systems*, *42*(4), 35–67. DOI: 10.1109/MCS.2022.3171478

Mahr, D., Odekerken-Schroeder, G., & van Doorn, J. (2025). Evolution of service robots in marketing: A relational framework and future research agenda. *Journal of Business Research*, *192*, 115204. DOI: 10.1016/j.jbusres.2025.115204

McIntosh, T. R., Susnjak, T., Liu, T., Watters, P., Ng, A., & Halgamuge, M. N. (2024). A game-theoretic approach to containing artificial general intelligence: Insights from highly autonomous aggressive malware. *IEEE Transactions on Artificial Intelligence*, *5*(12), 6290–6303. DOI: 10.1109/TAI.2024.3394392

Ouyang, F., Xu, W., & Cukurova, M. (2023). An artificial intelligence-driven learning analytics method to examine the collaborative problem-solving process from the complex adaptive systems perspective. *International Journal of Computer-Supported Collaborative Learning*, *18*(1), 39–66. DOI: 10.1007/s11412-023-09387-z

Salehi, N., Martínez-García, H., Velasco-Quesada, G., & Guerrero, J. M. (2022). A comprehensive review of control strategies and optimization methods for individual and community microgrids. *IEEE Access : Practical Innovations, Open Solutions*, *10*, 15935–15955. DOI: 10.1109/ACCESS.2022.3142810

Seo, S. (2022). When female (male) robot is talking to me: Effect of service robots' gender and anthropomorphism on customer satisfaction. *International Journal of Hospitality Management*, *102*, 103166. DOI: 10.1016/j.ijhm.2022.103166

So, K. K. F., Kim, H., Liu, S. Q., Fang, X., & Wirtz, J. (2024). Service robots: The dynamic effects of anthropomorphism and functional perceptions on consumers' responses. *European Journal of Marketing*, *58*(1), 1–32. DOI: 10.1108/EJM-03-2022-0176

Song, C. S., & Kim, Y. K. (2022). The role of the human-robot interaction in consumers' acceptance of humanoid retail service robots. *Journal of Business Research*, *146*, 489–503. DOI: 10.1016/j.jbusres.2022.03.087

Tojib, D., Sujan, R., Ma, J., & Tsarenko, Y. (2023). How does service robot anthropomorphism affect human co-workers? *Journal of Service Management*, *34*(4), 750–769. DOI: 10.1108/JOSM-03-2022-0090

Tushar, W., Yuen, C., Saha, T. K., Nizami, S., Alam, M. R., Smith, D. B., & Poor, H. V. (2023). A survey of cyber-physical systems from a game-theoretic perspective. *IEEE Access : Practical Innovations, Open Solutions*, *11*, 9799–9834. DOI: 10.1109/ACCESS.2023.3239834

Wang, B., Han, Y., Xie, F., Kandampully, J., & Duan, Q. (2026). To whom does service robot anthropomorphism appeal? The roles of customer social context, power, and perceived social connectedness. *Service Industries Journal*, *46*(3–4), 183–216. DOI: 10.1080/02642069.2025.2460597

Wirtz, J., Patterson, P. G., Kunz, W. H., Gruber, T., Lu, V. N., Paluch, S., & Martins, A. (2018). Brave new world: Service robots in the frontline. *Journal of Service Management*, *29*(5), 907–931. DOI: 10.1108/JOSM-04-2018-0119

Yang, Y., Chi, M., Bi, X., & Xu, Y. (2024). How does the anthropomorphism of service robots impact employees' role service behavior in the workplace? *International Journal of Hospitality Management*, *122*, 103857. DOI: 10.1016/j.ijhm.2024.103857

KEY TERMS AND DEFINITIONS

Customer Experience (CX) –: Customer Experience encompasses any of the experiences a customer has while communicating with the business, or consuming the product/service offerings.

Brand Strategy –: Brand strategy is a long term plan, a strategy which uses in order to establish a brand identity, brand positioning and customer loyalty for the company.

Firms –: It refers to the factory that undertakes the production of goods or services for earning profit and organizational goal.

Anthropomorphic Service –: Service interactions that involve technology or systems with human-like attributes like speech, emotion or behavior.

Digital Twins –: Virtual copies of real-world objects, systems or processes that mimic actual behavior with real-time data.

Service Robotics: – Service robotics deals with robotic technologies that are used to provide a service to humans in industry sectors like health, hospitality, and retail.

Artificial Intelligence (AI) –: It is a copy of a human brain by machines and computer systems.

Brand Perception –: Brand perception is the way that customers think and feel about a brand in terms of experiences, communication and reputation.

Chapter 10
Transforming Service Experience Through Anthropomorphic Robots:
Redefining Human-Machine Synergy in Hospitality and Tourism

Shivansu Sachan
http://orcid.org/0000-0002-9893-8018
CSJM University, Kanpur, India

Siddharth Singh
http://orcid.org/0009-0007-7876-4915
CSJM University, Kanpur, India

Duggirala Aravind
http://orcid.org/0009-0005-4362-7684
ASET Institute of Science and Technology, India

ABSTRACT

The integration of anthropomorphic robots in hospitality and tourism is revolutionizing service experiences by blending human-like interaction with advanced technology. These robots enhance efficiency, personalization, and emotional engagement, creating a new form of human–machine synergy that reshapes guest expectations and operational models. By mimicking human behaviors, expressions, and communication styles, they bridge the gap between automation and empathy,

DOI: 10.4018/979-8-3373-6640-1.ch010

offering seamless, consistent, and memorable experiences. This transformation not only redefines service delivery but also opens avenues for sustainable innovation and workforce collaboration in the evolving landscape of smart tourism.

1. INTRODUCTION

The hospitality industry has long been recognized for its dynamic nature and strong emphasis on customer-centric service delivery. However, the sector is currently undergoing a profound transformation driven by the rapid adoption of Artificial Intelligence (AI). AI technologies are reshaping hospitality operations by enhancing efficiency, personalization, and decision-making processes. For instance, AI-powered chatbots enable hotels and restaurants to provide real-time, accurate responses to guest inquiries, while predictive analytics supports demand forecasting, revenue management, and operational optimization (Ivanov & Webster, 2019; Tussyadiah, 2020). This transformation extends beyond operational change, fundamentally altering the skill sets and competencies required of hospitality professionals.As AI becomes increasingly embedded in hospitality operations, educational institutions face mounting pressure to adapt curricula and pedagogical approaches. Traditionally, hospitality education has emphasized soft skills such as empathy, communication, cultural sensitivity, and service excellence. While these competencies remain essential, the contemporary hospitality workforce must also possess technical proficiencies, including data literacy, AI system interaction, and ethical decision-making in technology-mediated environments (Buhalis & Leung, 2018). This dual demand necessitates an educational paradigm capable of integrating human-centric service values with advanced technological competencies.Competency-Based Education (CBE) offers a robust framework for addressing this challenge. CBE emphasizes measurable learning outcomes, practical skill acquisition, and learner-centered progression, ensuring that graduates are workplace-ready rather than merely credentialed (Mulder, 2017). By embedding AI-related competencies within CBE frameworks, hospitality education can align more effectively with industry expectations. Students can develop demonstrable skills in AI-assisted decision-making, data analytics, digital service design, and technology ethics while continuing to refine core hospitality competencies.Moreover, AI itself is emerging as a powerful educational tool within hospitality programs. Intelligent tutoring systems, adaptive learning platforms, and AI-driven simulations allow for personalized learning pathways and immersive experiential training that mirrors real-world hospitality scenarios (Zawacki-Richter et al., 2019). Such applications enhance learner engagement and enable continuous competency assessment, reinforcing the principles of CBE.This chapter examines the intersection of AI and competency-based hospitality education by first exploring the current and emerging applications of AI within the hospitality industry and the evolving competency requirements for future professionals. It then analyzes how CBE frameworks can effectively integrate AI-related skills into hospitality curricula. The discussion extends to AI-enabled educational applications, highlighting their pedagogical value, benefits, and limitations. Finally, the chapter reviews global best practices, identifies key challenges—including ethical concerns, faculty readiness, and digital divides—and proposes policy and strategic recommendations to ensure that hospitality education systems remain resilient, inclusive, and future-ready.

2. AI'S ROLE IN THE HOSPITALITY BUSINESS

Artificial Intelligence (AI) stands at the forefront of a technological transformation reshaping the global hospitality industry. Traditionally characterized by high levels of human interaction, personalized service, and experiential value creation, hospitality is now increasingly supported by intelligent systems that enhance efficiency, accuracy, and responsiveness. AI applications span both front-end guest interactions and back-end operational processes, fundamentally altering how hotels, resorts, restaurants, and tourism enterprises operate (Buhalis & Leung, 2018; Ivanov & Webster, 2019).

Understanding the scope and function of AI adoption within hospitality is essential for identifying the competencies required of future professionals. As smart technologies complement human service delivery, hospitality workers must learn to collaborate with AI systems while preserving the emotional and cultural dimensions of service excellence. The following sections examine the major domains in which AI is transforming hospitality operations and the associated skill implications.

2.1 Automating Customer Service

Customer service represents one of the most visible and widely adopted applications of AI in hospitality. AI-powered chatbots, virtual assistants, and voice-recognition systems are now routinely deployed to manage reservations, respond to guest inquiries, handle service requests, and address complaints. These systems rely on Natural Language Processing (NLP) to interpret guest queries in real time and communicate across multiple languages, enabling continuous 24/7 service availability (Tussyadiah, 2020).

Major hotel chains have adopted chatbots to streamline guest communication and reduce response times. By automating routine interactions—such as booking confirmations, check-in details, or late checkout requests—AI systems free front-line staff to focus on complex, high-touch service encounters that require empathy and problem-solving skills (Ivanov et al., 2020). Studies indicate that AI-enabled customer service improves operational efficiency while positively influencing guest satisfaction when appropriately integrated with human oversight (Buhalis & Moldavska, 2022).

2.2 AI in Revenue and Pricing Management

AI-driven predictive analytics has significantly transformed revenue management practices within hospitality. Machine learning algorithms analyze large datasets including historical booking patterns, seasonality, competitor pricing, local events, and market demand to determine optimal pricing strategies. These dynamic pricing models enable hotels to adjust room rates in real time, maximizing occupancy and revenue while remaining competitive (Noone & McGuire, 2014; Ivanov & Webster, 2019).

AI-powered revenue management systems such as IDeaS and Duetto provide demand forecasting with high accuracy, supporting informed decision-making across pricing, inventory allocation, and distribution channels. As a result, hospitality professionals are increasingly required to interpret data dashboards, understand algorithmic recommendations, and align pricing strategies with broader business objectives. This shift underscores the growing importance of data literacy and financial acumen within hospitality education and practice.

2.3 AI-Enabled Personalization of Guest Experiences

Personalization lies at the core of hospitality value creation, and AI has elevated this concept to unprecedented levels. Machine learning systems analyze guest data—including booking history, loyalty program activity, preferences, feedback, and behavioral patterns—to deliver hyper-personalized services. These may include customized room settings, targeted service recommendations, personalized dining suggestions, or tailored promotional offers (Buhalis & Sinarta, 2019).

In luxury and premium hospitality segments, AI enables anticipatory service by predicting guest needs before they are explicitly expressed. While such personalization enhances satisfaction and loyalty, it also raises expectations regarding service consistency and ethical data use. Consequently, hospitality professionals must develop competencies that combine emotional intelligence with the ability to work alongside AI-driven personalization tools, ensuring that technology enhances rather than replaces human warmth and cultural sensitivity.

2.4 Enhancing Operational Efficiency

Beyond guest-facing services, AI plays a critical role in optimizing hospitality operations. AI-based systems support inventory management by forecasting demand and minimizing waste, particularly in food and beverage operations. Workforce scheduling tools analyze peak demand periods and employee availability to generate efficient staffing plans, reducing labor costs while maintaining service quality (Marques & Ferreira, 2020).

AI-powered energy management and predictive maintenance systems further contribute to sustainability and cost efficiency. Smart energy systems adjust heating, ventilation, and air-conditioning (HVAC) usage based on occupancy patterns, while predictive maintenance tools identify potential equipment failures before they occur. Empirical studies report significant reductions in energy consumption and asset downtime in AI-enabled hotels, reinforcing AI's role in sustainable hospitality management (Buhalis & Leung, 2018).

2.5 AI in Hospitality Marketing and CRM

AI adoption has also transformed hospitality marketing and customer relationship management (CRM). AI-driven marketing tools automate campaign design, sentiment analysis, A/B testing, email targeting, and lead conversion tracking. AI-powered CRM systems analyze guest interactions across digital platforms to generate personalized marketing strategies and improve customer retention (Mariani & Borghi, 2019).

These developments require hospitality professionals to interpret AI-generated insights and translate them into strategic marketing decisions. Analytical thinking, branding knowledge, and digital communication skills are therefore becoming essential complements to traditional marketing competencies.

3. COMPETENCY-BASED LEARNING IN THE HOSPITALITY INDUSTRY

Competency-Based Education (CBE) has emerged as a transformative pedagogical approach, particularly within professional and vocational disciplines such as hospitality. Unlike traditional education models that emphasize instructional time, credit hours, or examination performance, CBE prioritizes the demonstration of clearly defined competencies aligned with real-world professional requirements. Learners progress by mastering specific skills and knowledge rather than by completing predetermined instructional durations (Mulder, 2017). This shift has significant implications for hospitality education, especially as Artificial Intelligence (AI) increasingly reshapes industry operations.

Hospitality is inherently practice-oriented, emphasizing service quality, guest satisfaction, and operational excellence. Consequently, a strong alignment between educational outcomes and workplace competencies is essential. CBE provides a structured yet flexible framework that ensures graduates possess job-ready skills upon completion of their programs. Rooted in constructivist and learner-centered theories, CBE gained prominence in the 21st century as industries demanded graduates who could immediately contribute to organizational performance and adapt to technological change (Gervais, 2016).

In hospitality education, CBE focuses on measurable outcomes related to service delivery, decision-making, teamwork, and technology use. This approach is particularly relevant in an AI-enabled industry, where professionals must integrate human-centered service competencies with emerging digital and analytical skills. By emphasizing mastery, application, and continuous assessment, CBE bridges the gap between academic preparation and industry expectations.

3.1 Core Competencies in Hospitality

At its foundation, hospitality remains a people-centric industry. Despite technological advancements, interpersonal interaction, cultural understanding, and operational proficiency continue to define service excellence. Core competencies traditionally emphasized within hospitality education include:

Communication skills are fundamental, as hospitality professionals engage continuously with guests, colleagues, and external stakeholders. Clear, empathetic, and culturally appropriate communication is essential for understanding guest needs, managing expectations, and resolving service issues effectively (Lashley, 2018).

Problem-solving ability is equally critical, given the unpredictable nature of hospitality operations. Professionals must address challenges ranging from reservation errors and service failures to supply shortages and staffing constraints. Effective problem-solving involves creativity, situational awareness, and calm decision-making under pressure.

Teamwork and collaboration underpin successful hospitality operations. Hotels, restaurants, and tourism enterprises rely on cross-functional coordination among departments such as front office, housekeeping, food and beverage, and event management to deliver seamless guest experiences.

Cultural sensitivity is indispensable in a globalized hospitality environment. Professionals regularly interact with guests from diverse cultural backgrounds, requiring awareness of cultural norms, dietary preferences, language differences, and service expectations to foster trust and personalized experiences (Reisinger & Crotts, 2019).

Operational knowledge encompasses an understanding of hospitality systems and procedures, including property management systems, food safety regulations, inventory control, and basic financial management. These competencies ensure operational efficiency and service consistency.

Historically, these skills have formed the backbone of hospitality curricula. However, the growing integration of digital technologies—particularly AI—has expanded the competency landscape, necessitating the inclusion of new, technology-oriented capabilities within CBE frameworks.

3.2 Emerging AI-Related Competencies

AI is fundamentally altering how hospitality services are delivered, managed, and evaluated. As AI-enabled systems become commonplace in reservations, demand forecasting, personalization, marketing, and operations, the competency requirements for hospitality professionals are evolving accordingly (Buhalis & Leung, 2018).

Several international policy and qualification frameworks reinforce the relevance of competency-based, outcome-oriented education. The **European Qualifications Framework (EQF)** emphasizes transferable, lifelong competencies; the **U.S. Department of Education** promotes mastery-based CBE models; and India's **National Education Policy (NEP 2020)** advocates experiential and skill-oriented curricula for vocational fields such as tourism and hospitality (Government of India, 2020).

Key AI-related competencies emerging within hospitality include:

Technological proficiency, which involves understanding and effectively using AI-enabled tools such as automated check-in systems, chatbots, revenue management software, and CRM platforms. While coding expertise is not universally required, professionals must comprehend system functionality and apply outputs meaningfully (Ivanov & Webster, 2019).

Data literacy has become a core competency as AI systems generate extensive operational and customer data. Hospitality professionals must interpret dashboards, recognize patterns, and translate insights into actionable strategies, such as pricing adjustments or targeted marketing initiatives (Mariani & Borghi, 2019).

Adaptability and lifelong learning are essential in an environment characterized by rapid technological change. Hospitality professionals must continuously update their skills to remain competitive and effectively integrate emerging tools into service delivery.

Ethical awareness is increasingly critical as AI adoption raises concerns related to data privacy, algorithmic bias, transparency, and fairness. Professionals must be equipped to identify ethical risks and ensure responsible AI use that aligns with guest trust and organizational values (Tussyadiah, 2020).

CBE is particularly well-suited to cultivating these competencies, as it emphasizes real-world application, performance-based assessment, and continuous feedback. Simulations, projects, and problem-based tasks enable learners to practice AI-assisted decision-making while reinforcing foundational hospitality values.

4. AI APPLICATIONS IN HOSPITALITY EDUCATION

While AI is transforming hospitality operations, it is simultaneously reshaping hospitality education. Institutions adopting competency-based models increasingly integrate AI tools to enhance personalization, engagement, and industry relevance. AI applications in education support immersive learning experiences, intelligent tutoring, and data-driven curriculum design, ensuring alignment with evolving workforce demands (Zawacki-Richter et al., 2019).

4.1 Virtual Simulations and Experiential Learning

AI-powered simulations have become a powerful experiential learning tool in hospitality education. These virtual environments replicate real-world hospitality scenarios, enabling students to practice skills in a risk-free setting.

Virtual front desk simulations allow learners to manage check-ins, handle complaints, resolve billing issues, and deliver concierge services. AI-driven virtual guests respond dynamically to student actions, facilitating the development of communication, emotional intelligence, and service recovery skills.

Similarly, virtual kitchen and restaurant simulations replicate high-pressure operational environments where students manage food preparation, staff coordination, and time constraints. Event planning and crisis management simulations expose learners to complex decision-making scenarios, such as last-minute cancellations or logistical disruptions, enhancing strategic thinking and adaptability.

These immersive experiences strengthen learner confidence and preparedness while enabling instructors to assess competencies in real time through data-driven feedback.

4.2 Intelligent Tutoring Systems

Intelligent Tutoring Systems (ITS) represent another significant advancement in AI-enabled education. These systems personalize learning by adapting content, assessments, and feedback based on individual student performance.

ITS platforms identify strengths and learning gaps, delivering targeted exercises and resources accordingly. Real-time feedback accelerates skill acquisition and deepens conceptual understanding. Additionally, continuous performance tracking allows educators to provide timely interventions and tailor instructional strategies to learner needs (Holmes et al., 2022).

In hospitality education, ITS can dynamically adjust modules based on student proficiency in areas such as revenue management, operations, or service design, supporting mastery-based progression consistent with CBE principles.

4.3 Data-Driven Curriculum Design

AI also supports curriculum development and institutional decision-making. By analyzing labor market trends, job postings, industry reports, and employer feedback, AI tools help institutions identify emerging skill requirements and update curricula accordingly.

Learning analytics derived from AI-enabled Learning Management Systems (LMS) provide insights into student engagement, learning progression, and competency attainment. These insights enable educators to refine course content, teaching methods, and assessment strategies.

Tracking graduate outcomes and employer satisfaction further informs curriculum improvement, ensuring sustained alignment between educational offerings and industry needs (Banta & Blaich, 2018).

4.4 AI-Enhanced Case-Based Learning

Traditional case-based learning has long been used in hospitality education to foster critical thinking. AI has enhanced this approach by enabling dynamic, interactive, and data-rich case studies.

AI-generated scenarios allow students to analyze realistic datasets, make strategic decisions, and receive immediate feedback on outcomes. Interactive decision points encourage experiential learning, while real-time data updates ensure continued relevance. This approach deepens analytical skills and prepares learners for complex, technology-mediated hospitality environments.

5. ADVANTAGES OF AI IN COMPETENCY-BASED HOSPITALITY EDUCATION

The integration of Artificial Intelligence (AI) into hospitality education—particularly within competency-based education (CBE) frameworks—represents a significant pedagogical advancement in preparing graduates for a rapidly transforming service industry. Competency-based education emphasizes demonstrable mastery of knowledge, skills, and attitudes that are directly aligned with workplace requirements rather than time-bound or examination-centric learning. AI strengthens this educational philosophy by enabling personalized learning pathways, immersive skill development, continuous performance assessment, and real-time alignment with industry trends (Mulder, 2017; Zawacki-Richter et al., 2019).

In hospitality education, where service quality, operational efficiency, and human interaction intersect, AI-enabled CBE provides a mechanism for balancing traditional service-oriented competencies with emerging digital, analytical, and ethical skills. When implemented thoughtfully, AI enhances learner engagement, improves employability outcomes, promotes inclusion and access, and increases institutional efficiency (Buhalis & Leung, 2018). The following subsections explore the key advantages of AI integration in competency-based hospitality education in detail.

5.1 Enhanced Skill Development through Experiential Learning

Experiential learning has long been a cornerstone of hospitality education due to the industry's practice-oriented nature. CBE frameworks prioritize learning through application, and AI significantly amplifies the effectiveness of experiential learning by creating immersive, data-driven, and adaptive learning environments. AI-powered simulations replicate realistic hospitality settings such as hotel front offices, restaurants, kitchens, cruise operations, and event venues, allowing students to practice complex decision-making in controlled, risk-free contexts.

For instance, virtual hotel management simulations may require learners to resolve overbooking scenarios during peak seasons, manage dissatisfied guests, coordinate staff schedules, and optimize room inventory simultaneously. AI systems continuously monitor learner actions, evaluate outcomes, and provide immediate feedback, enabling students to reflect on their decisions and improve iteratively (Kolb, 2015). This continuous feedback loop supports mastery learning, a fundamental principle of CBE, where progression is based on demonstrated competence rather than completion of instructional hours.

AI-enhanced experiential learning also supports the development of higher-order cognitive skills such as critical thinking, systems thinking, and strategic decision-making. By exposing students to dynamic, unpredictable scenarios, AI simulations prepare learners for the complexity and uncertainty inherent in real-world hospitality operations. Importantly, these environments also allow repeated practice, enabling students to experiment with alternative strategies without the financial or reputational risks associated with live operations.

Beyond operational skills, AI-enabled experiential learning facilitates the acquisition of emerging competencies increasingly demanded by the hospitality industry. These include digital guest management, data interpretation, and the application of predictive analytics for personalization and revenue optimization. By interacting with interfaces that mirror industry-standard systems, students develop technological fluency and applied data literacy alongside traditional service competencies (Ivanov & Webster, 2019). As a result, graduates are better prepared to transition seamlessly from academic environments to professional hospitality roles.

5.2 Personalized Learning for Diverse Learners

One of the most transformative advantages of AI in competency-based hospitality education is its capacity to personalize learning. Traditional hospitality programs often rely on standardized curricula and uniform pacing, which may not adequately address the diverse backgrounds, learning styles, and professional aspirations of students. AI-enabled adaptive learning platforms address this limitation by tailoring instructional content, assessment methods, and learning pathways to individual learner needs (Holmes et al., 2022).

Adaptive systems use algorithms to analyze learner performance, engagement patterns, and progression data. Based on this analysis, they dynamically adjust the complexity, format, and sequencing of learning materials. For example, a student demonstrating strong interpersonal communication skills but struggling with quantitative subjects such as revenue management may receive additional tutorials, case studies, and practice exercises focused on financial analysis. Conversely, analytically strong learners may be challenged with advanced simulations or strategic problem-solving tasks.

Hospitality students often come from varied academic, cultural, and professional backgrounds, including learners transitioning from industry roles, international students, and first-generation college students. AI-enabled personalization supports this diversity by offering differentiated resources, flexible pacing, and multiple modes of content delivery. Personalized feedback enhances learner motivation, improves academic performance, and reduces attrition rates while supporting mastery-based progression—a cornerstone of CBE (Gervais, 2016).

Moreover, AI enables scalable personalization, which is particularly valuable in large hospitality programs. Chatbots and intelligent tutoring systems provide on-demand academic and administrative support, while automated assessment tools offer timely, consistent feedback. These capabilities allow institutions to maintain educational quality even with growing enrollments or limited faculty resources, without sacrificing the individualized support essential to competency-based learning.

5.3 Stronger Alignment with Industry Needs

The hospitality industry is characterized by rapid change driven by technological innovation, evolving guest expectations, sustainability imperatives, and global competition. One of the persistent challenges in hospitality education has been ensuring that curricula remain aligned with industry requirements. AI significantly enhances this alignment by enabling institutions to analyze large volumes of real-time industry data and translate insights into curricular updates (Mariani & Borghi, 2019).

AI systems can process data from job portals, employer surveys, professional networks, and labor market analytics to identify emerging skill demands. These insights help educators recognize competencies such as AI-enabled customer relationship management, sustainability analytics, digital marketing, and service automation that are increasingly valued by employers. As a result, curricula can be updated proactively rather than reactively, ensuring that graduates acquire relevant, future-oriented skills.

In competency-based hospitality education, this industry alignment is particularly critical. Learning outcomes are explicitly defined in terms of workplace competencies, and AI-supported labor market intelligence ensures that these outcomes remain current. Platforms providing real-time data on employment trends allow institutions to continuously refine competency frameworks, assessment rubrics, and learning activities.

AI also supports feedback loops through alumni tracking and employer evaluations. By analyzing graduate career trajectories and employer feedback, institutions can assess the effectiveness of their programs and make data-driven improvements. This continuous improvement cycle reinforces CBE's focus on outcomes and accountability, enhancing both educational credibility and graduate employability (Banta & Blaich, 2018).

5.4 Scalability of High-Quality Learning Experiences

Scaling high-quality hospitality education presents a significant challenge, particularly in resource-constrained contexts such as public institutions, vocational training centers, and institutions in developing economies. Traditional experiential learning models often rely on intensive faculty involvement, physical infrastructure, and limited cohort sizes, which can restrict access and increase costs. AI-powered platforms address these constraints by enabling scalable delivery of high-quality learning experiences.

Intelligent tutoring systems can support large student cohorts simultaneously while maintaining individualized learning paths. AI-based assessment tools evaluate complex assignments—such as service recovery plans, operational strategies, or reflective journals—using standardized rubrics. Automated feedback ensures consistency and timeliness, which are essential for mastery-based progression (Zawacki-Richter et al., 2019).

This scalability is particularly important for hospitality education, where demand for skilled professionals continues to grow globally. AI enables institutions to expand access to experiential learning without proportionally increasing faculty workload or infrastructure costs. In developing regions, where faculty shortages and limited training facilities persist, AI-supported learning platforms offer a viable pathway for building hospitality skills at scale.

Furthermore, scalable AI-enabled education supports workforce development initiatives by enabling partnerships between educational institutions, governments, and industry stakeholders. Such collaborations can address skills shortages while promoting inclusive access to competency-based hospitality training.

5.5 Improved Institutional and Administrative Efficiency

Beyond teaching and learning, AI delivers significant benefits in terms of institutional and administrative efficiency. Hospitality education institutions manage complex operations involving admissions, scheduling, assessment, student support, and resource allocation. AI-driven automation streamlines these processes, reducing administrative burdens and enabling data-informed decision-making.

Automated grading systems evaluate assignments quickly and consistently, allowing educators to focus on higher-value activities such as mentoring, curriculum design, and student engagement. AI-powered learning management systems monitor student participation and performance, identifying at-risk learners early and recommending targeted interventions. Predictive analytics support enrollment forecasting, faculty scheduling, and infrastructure planning, improving institutional responsiveness and efficiency (Holmes et al., 2022).

These operational efficiencies are particularly valuable in competency-based education, where continuous assessment and individualized progression require robust monitoring systems. AI enables institutions to manage these complexities effectively, ensuring transparency, accountability, and quality assurance.

By freeing educators from routine administrative tasks, AI allows greater investment in pedagogical innovation and student-centered learning. This shift enhances institutional capacity to deliver high-quality, competency-based hospitality education that meets both learner needs and industry expectations.

5.6 Summary of Advantages

Collectively, the advantages of AI integration in competency-based hospitality education extend across pedagogical, operational, and strategic dimensions. AI enhances experiential learning by providing immersive, adaptive environments that support skill mastery. Personalized learning pathways accommodate diverse learners and promote engagement and retention. Industry alignment is strengthened through real-time labor market intelligence and continuous feedback loops. Scalability enables broader access to high-quality education, while administrative efficiencies improve institutional effectiveness.

Importantly, these advantages are realized most fully when AI is integrated thoughtfully within CBE frameworks and complemented by human-centered pedagogical practices. By aligning technological innovation with hospitality's service ethos, AI-enabled competency-based education offers a powerful model for preparing future professionals to succeed in an increasingly complex, data-driven, and customer-focused global hospitality industry.

6. CHALLENGES AND CONSIDERATIONS IN AI INTEGRATION

Despite the transformative potential of Artificial Intelligence (AI) in competency-based hospitality education, its integration is neither automatic nor unproblematic. While AI offers opportunities for personalization, scalability, and industry alignment, it also introduces pedagogical, ethical, infrastructural, and cultural challenges that institutions must address to ensure sustainable and equitable implementation. Failure to confront these challenges risks reinforcing inequalities, undermining educational quality, and diluting the human-centered ethos of hospitality education. This section critically examines the principal challenges associated with AI integration and outlines considerations for responsible adoption.

6.1 Faculty Readiness and Professional Development

Faculty readiness remains one of the most significant barriers to effective AI integration in hospitality education. Many hospitality educators come from industry-oriented or traditionally structured academic backgrounds, where pedagogical practices emphasized face-to-face instruction, experiential training, and manual assessment. As a result, educators may lack confidence, technical expertise, or pedagogical frameworks for integrating AI tools into teaching and assessment (Zawacki-Richter et al., 2019).

Without adequate training, AI technologies risk being underutilized, misapplied, or perceived as threats rather than enablers of learning. In some cases, faculty resistance stems from fears of deskilling, loss of academic autonomy, or replacement by automated systems. Such concerns are particularly salient in hospitality education, where instructors often identify strongly with mentoring, coaching, and human interaction.

To address this challenge, institutions must invest in sustained and structured professional development initiatives. These should include hands-on workshops on AI-enabled teaching tools, interdisciplinary collaboration with instructional designers and data scientists, and opportunities for faculty to experiment with AI in low-risk pilot projects. Communities of practice can further support peer learning, enabling educators to share experiences, challenges, and best practices. Importantly, professional development should emphasize pedagogical integration rather than technical mastery alone, ensuring that AI enhances learning outcomes rather than becoming an end in itself.

6.2 Ethical and Privacy Concerns

AI's reliance on large-scale data collection and algorithmic decision-making raises profound ethical concerns in educational contexts. AI-enabled learning systems often track student behavior, performance, engagement patterns, and even emotional indicators. If not governed responsibly, such practices can compromise student privacy, autonomy, and trust (Williamson & Eynon, 2020).

Algorithmic bias represents another critical issue. AI systems trained on incomplete or biased datasets may inadvertently disadvantage students based on language proficiency, cultural background, gender, or socioeconomic status. In hospitality education—where diversity and inclusion are core professional values—such biases can undermine both educational equity and ethical credibility.

Institutions must therefore implement transparent data governance frameworks that clearly define data ownership, consent mechanisms, storage practices, and access rights. Compliance with data protection regulations and regular algorithmic audits are essential to mitigating risks. Beyond institutional policy, ethical AI literacy should be embedded within hospitality curricula to prepare students for responsible AI use in professional environments. Ethical decision-making, transparency, and accountability must be treated as core competencies alongside technical proficiency.

6.3 Access, Equity, and the Digital Divide

The adoption of AI in hospitality education requires substantial investment in digital infrastructure, including reliable internet connectivity, learning platforms, licensed software, and compatible devices. While well-resourced institutions in developed regions may readily adopt advanced AI systems, underfunded institutions—particularly in developing economies—risk exclusion from AI-enhanced education.

This digital divide threatens to exacerbate existing inequalities in hospitality education, contradicting the inclusive ethos of competency-based education. Students from rural areas or low-income backgrounds may face barriers to participation, limiting their access to personalized learning and advanced skill development.

Addressing this challenge requires coordinated policy interventions, including public–private partnerships, subsidized access to digital tools, and the adoption of open-source AI platforms. International organizations emphasize that equitable AI integration must prioritize accessibility and inclusivity to ensure that technological innovation does not deepen structural inequities (UNESCO, 2021).

6.4 Balancing Technology and Human-Centered Learning

Hospitality is fundamentally a human-centered industry grounded in empathy, emotional intelligence, and interpersonal connection. Excessive reliance on AI risks marginalizing these essential competencies by prioritizing efficiency and automation over relational skills. If AI becomes the dominant mode of instruction and assessment, students may have fewer opportunities to develop nuanced communication, cultural sensitivity, and ethical judgment (Lashley, 2018).

Therefore, AI should be positioned as a complement rather than a substitute for human interaction. Blended pedagogical models that integrate AI tools with internships, role-play, teamwork, mentoring, and face-to-face engagement are essential. Such approaches preserve hospitality's human essence while leveraging AI to enhance learning depth and relevance. Educators must consciously design learning experiences that balance technological innovation with social and emotional development.

7. CASE STUDIES: AI IN ACTION IN LEADING HOSPITALITY INSTITUTIONS

As hospitality education responds to the growing influence of AI, several globally recognized institutions have emerged as pioneers in integrating AI within Competency-Based Education (CBE) frameworks. These institutions demonstrate how AI can be leveraged strategically to enhance experiential learning, personalize education, and improve graduate employability. This section examines three leading hospitality schools—**Les Roches Global Hospitality Education, Glion Institute of Higher Education**, and **Cornell University School of Hotel Administration**—to illustrate best practices, implementation strategies, and measurable outcomes.

7.1 Les Roches Global Hospitality Education: AI Simulations for Real-World Skill Development

Institutional Overview

Founded in Switzerland, Les Roches Global Hospitality Education is internationally recognized for its experiential, industry-aligned pedagogical model. The institution adopts a competency-based approach emphasizing practical learning, project-based assessment, and immersive industry exposure.

AI Integration Strategy

Since 2020, Les Roches has incorporated AI-powered simulation platforms into core modules covering hotel operations, revenue management, guest services, and event management. These simulations replicate real-world operational constraints, including fluctuating demand, staffing shortages, budget limitations, and diverse guest profiles.

Pedagogical Application

Students participate in role-based simulations where they act as members of a virtual hotel management team. Scenarios may involve managing overbooking during peak seasons, resolving guest complaints, or optimizing pricing strategies during high-demand events. AI systems evaluate student decisions in real time, providing immediate feedback aligned with predefined competency rubrics encompassing communication, emotional intelligence, teamwork, problem-solving, and digital fluency.

Outcomes and Educational Impact

Institution-reported evaluations indicate that students exposed to AI simulations demonstrate higher engagement, accelerated skill acquisition, and increased confidence in operational decision-making. Industry partners have highlighted the digital readiness and operational competence of graduates, suggesting that AI-enhanced experiential learning effectively bridges the gap between education and professional practice (Les Roches, 2023; Ivanov & Webster, 2019).

7.2 Glion Institute of Higher Education: AI-Driven Personalized Learning

Institutional Overview

The Glion Institute of Higher Education is known for its academically rigorous hospitality programs and strong graduate placement outcomes. Glion emphasizes leadership development, adaptability, and lifelong learning within a global hospitality context.

AI Integration Strategy

In 2022, Glion adopted AI-based learning analytics platforms to personalize student learning pathways. These systems analyze performance metrics, engagement data, and assessment outcomes to identify individual strengths and learning gaps.

Pedagogical Application

For example, students excelling in guest interaction but struggling with financial modeling receive targeted learning resources, adaptive exercises, and mentoring focused on revenue forecasting and financial analysis. AI-supported reflective journals further encourage self-assessment by providing feedback on leadership, teamwork, and emotional intelligence.

Outcomes and Educational Impact

Institutional evaluations reveal improved mastery rates, higher student satisfaction, and enhanced learner confidence. Employers report that Glion graduates exhibit a rare combination of analytical rigor and service empathy—an essential competency blend for leadership in AI-enabled hospitality environments (Gervais, 2016; Holmes et al., 2022).

7.3 Cornell University School of Hotel Administration: AI-Aligned Curriculum Design

Institutional Overview

Cornell University's School of Hotel Administration (SHA), part of the SC Johnson College of Business, is a global leader in hospitality education and research. The institution adopts a data-driven, research-informed approach to curriculum design.

AI Integration Strategy

Cornell SHA uses AI analytics to monitor hospitality industry trends, graduate career trajectories, and student performance data. These insights inform continuous curriculum updates, particularly in areas such as AI-driven personalization, dynamic pricing, service automation, and digital marketing.

Pedagogical Application

Students engage with AI-enhanced case studies addressing real-world challenges, such as pricing strategies during global disruptions or the design of digital concierge systems for boutique hotels. Partnerships with hospitality technology firms provide access to enterprise-level tools, ensuring authentic industry exposure.

Outcomes and Educational Impact

Cornell reports strong graduate placement across global hotel chains, hospitality technology startups, and consulting firms. Graduates are widely recognized for their ability to operate effectively in data-intensive, technology-mediated hospitality environments, reflecting the success of AI-aligned curriculum design (Buhalis & Leung, 2018; Mariani & Borghi, 2019).

8. FUTURE DIRECTIONS IN AI-ENABLED HOSPITALITY EDUCATION

The rapid diffusion of Artificial Intelligence (AI) across the hospitality industry has initiated a fundamental rethinking of how hospitality education is designed, delivered, and evaluated. As AI continues to evolve, its role in hospitality education is expected to move beyond isolated tools toward fully integrated, intelligent learning ecosystems. These future directions emphasize immersive learning, lifelong competency development, global collaboration, ethical responsibility, and systemic institutional transformation. Together, they represent a paradigm shift from static curricula to adaptive, future-ready educational models that align with competency-based education (CBE) principles and industry demands (Buhalis & Leung, 2018; Zawacki-Richter et al., 2019).

8.1 Integration of Augmented and Virtual Reality (AR/VR)

One of the most promising future developments in AI-enabled hospitality education is the integration of immersive technologies such as Augmented Reality (AR) and Virtual Reality (VR). When combined with AI, AR/VR technologies can create intelligent, responsive, and immersive learning environments that closely replicate real-world hospitality contexts.

AI-powered VR simulations can allow students to explore fully immersive hotel properties, restaurants, kitchens, cruise ships, or event venues in 360-degree environments. These simulations can dynamically adapt to learner behavior, creating real-time consequences for decisions related to guest service, staffing, pricing, or crisis management. For example, a VR-based hotel front desk simulation may respond differently depending on a student's tone of communication, service recovery strategy, or cultural sensitivity, thereby reinforcing both technical and emotional competencies.

AR applications, on the other hand, can enhance real-world training by overlaying digital information onto physical environments. Hospitality students undertaking internships or laboratory sessions could use AR glasses or mobile devices to receive contextual guidance on equipment use, safety protocols, or service standards. AI-driven AR systems could analyze learner performance and provide instant feedback, transforming traditional workplace learning into a continuous competency assessment process (Radianti et al., 2020).

From a pedagogical perspective, AR/VR aligns strongly with experiential and constructivist learning theories, which emphasize learning through active engagement and reflection. When integrated into CBE frameworks, immersive technologies enable precise assessment of competencies such as decision-making under pressure, teamwork, cultural awareness, and leadership—skills that are difficult to measure through conventional examinations (Kolb, 2015). As costs of immersive technologies decrease and accessibility improves, AR/VR is expected to become a mainstream component of hospitality education.

8.2 AI-Supported Lifelong Learning Ecosystems

The hospitality industry is characterized by rapid technological change, evolving guest expectations, and increasing competition. As a result, hospitality professionals can no longer rely solely on initial qualifications; instead, they must engage in lifelong learning to remain relevant and competitive. AI is expected to play a central role in enabling continuous, personalized, and flexible learning ecosystems.

Future AI-supported lifelong learning models will emphasize micro-credentials, digital badges, and modular certifications aligned with specific competencies. These credentials will allow professionals to acquire targeted skills—such as AI-driven revenue management, sustainability analytics, or digital guest experience design—without committing to full-degree programs. AI-curated learning pathways will analyze individual career goals, performance history, and labor market trends to recommend personalized learning opportunities (Holmes et al., 2022).

For example, a mid-career hotel manager transitioning into a data-driven role may receive AI-generated recommendations for short courses in predictive analytics, CRM systems, and ethical AI use. As competencies are demonstrated, learners can earn verifiable digital credentials that are portable across institutions and recognized by employers. This approach aligns with the principles of CBE by focusing on demonstrable outcomes rather than time-based progression (Mulder, 2017).

AI-supported lifelong learning ecosystems also benefit employers by enabling continuous workforce upskilling and reducing skills gaps. Hospitality organizations may collaborate with educational institutions to co-create AI-driven learning platforms that align employee development with organizational strategy. In this way, hospitality education extends beyond formal institutions to become an ongoing, industry-integrated process.

8.3 Global Collaborative Learning Platforms

Another significant future direction in AI-enabled hospitality education is the expansion of global collaborative learning platforms. AI-powered digital environments are expected to facilitate cross-institutional and cross-cultural collaboration, enabling students from different regions to work together on shared hospitality projects.

Through AI-enabled platforms, hospitality students in different countries may collaborate virtually to design sustainable resorts, develop culturally sensitive service models, or analyze global tourism data. AI tools can support multilingual communication through real-time translation, facilitate equitable participation by tracking engagement, and provide adaptive feedback based on team dynamics and project outcomes.

Such global collaboration fosters critical competencies required in the contemporary hospitality industry, including intercultural communication, global awareness, teamwork, and problem-solving in diverse contexts (Reisinger & Crotts, 2019). These experiences are particularly valuable in preparing students for leadership roles in multinational hospitality organizations.

Moreover, global AI-enabled platforms can support resource sharing among institutions, allowing developing regions to access high-quality learning materials, simulations, and expertise. This collaborative approach has the potential to reduce global disparities in hospitality education and promote more inclusive knowledge exchange.

8.4 Responsible and Ethical AI Education

As AI becomes increasingly embedded in both hospitality operations and education, ethical considerations will play a critical role in shaping future curricula. Issues such as data privacy, algorithmic bias, transparency, surveillance, and accountability are particularly relevant in hospitality, where guest trust and personal data are central to service delivery.

Future hospitality education must integrate ethical AI literacy as a core competency. Students should be equipped not only to use AI tools effectively but also to critically evaluate their social, cultural, and ethical implications. This includes understanding how AI algorithms influence pricing decisions, hiring practices, marketing personalization, and customer profiling, as well as recognizing potential biases and unintended consequences (Williamson & Eynon, 2020).

International organizations such as UNESCO have emphasized the importance of human-centered and rights-based approaches to AI in education. Ethical AI education should therefore include topics such as informed consent, data governance, transparency, fairness, and sustainability (UNESCO, 2021). Accrediting bodies may increasingly require evidence of ethical AI competence as part of professional readiness standards.

Embedding ethics within CBE frameworks allows for performance-based assessment of ethical decision-making. For instance, students may be evaluated on their ability to design AI-enabled service systems that balance efficiency with fairness, privacy, and inclusivity. Such an approach ensures that future hospitality leaders are prepared to deploy AI responsibly and sustainably.

8.5 Institutional and Policy Transformation

Realizing the full potential of AI in hospitality education requires transformation at the institutional and policy levels. AI adoption cannot be limited to classroom tools; it must be supported by systemic changes in governance, infrastructure, faculty development, and regulatory frameworks.

Institutions must invest in robust digital infrastructure, including secure data systems, learning management platforms, and AI-enabled analytics. Faculty development is equally critical, as educators must acquire both technological competence and pedagogical strategies for AI-enhanced teaching. Interdisciplinary collaboration among hospitality educators, data scientists, instructional designers, and industry partners will be essential for effective implementation (Zawacki-Richter et al., 2019).

At the policy level, national qualification frameworks and accreditation standards must evolve to recognize AI-driven competencies. The National Education Policy (NEP) 2020 of India, for example, emphasizes experiential learning, digital literacy, and vocational relevance, creating a favorable policy environment for AI-integrated CBE in hospitality education (Government of India, 2020). Similar reforms are needed globally to ensure alignment between education systems and labor market demands.

Public–private partnerships between educational institutions, hospitality organizations, and technology providers can further support sustainable AI integration. Such collaborations can facilitate access to industry-grade tools, real-world data, and applied research opportunities, enhancing both educational quality and industry relevance.

9. CONCLUSION

The integration of Artificial Intelligence into competency-based hospitality education represents a transformative evolution in preparing future professionals for an increasingly complex, technology-driven service industry. AI-enabled simulations, intelligent tutoring systems, immersive learning environments, and data-driven curriculum design offer unprecedented opportunities to enhance skill development, personalize learning, and align education with real-world industry needs.

By embedding AI within CBE frameworks, hospitality education can move beyond traditional, time-based instructional models toward outcome-oriented, adaptive, and learner-centered approaches. Graduates emerging from such systems are equipped not only with foundational hospitality competencies—such as communication, empathy, and operational knowledge—but also with emerging digital skills including data literacy, technological fluency, and ethical decision-making.

However, the successful realization of AI's potential depends on addressing critical challenges related to faculty readiness, ethical governance, digital equity, and the preservation of hospitality's human-centered values. Case studies from leading institutions such as Les Roches, Glion Institute of Higher Education, and Cornell University demonstrate that a balanced and strategic approach—combining technological innovation with experiential learning and emotional intelligence—can produce graduates who are both operationally competent and empathetically grounded.

Looking ahead, hospitality education must embrace AI not as a replacement for human interaction but as a powerful enabler of deeper learning, global collaboration, and sustainable professional development. Through responsible adoption, inclusive policies, and continuous innovation, AI-enabled competency-based hospitality education can prepare graduates to lead confidently, ethically, and creatively in an AI-transformed global service economy.

REFERENCES

Banta, T. W., & Blaich, C. (2018). Closing the assessment loop. *Change: The Magazine of Higher Learning*, *50*(5), 8–15.

Buhalis, D., & Leung, R. (2018). Smart hospitality—Interconnectivity and interoperability towards an ecosystem. *International Journal of Hospitality Management*, *71*, 41–50. DOI: 10.1016/j.ijhm.2017.11.011

Buhalis, D., & Moldavska, I. (2022). In-room voice-based AI digital assistants: Implications for hospitality management and operations. *International Journal of Hospitality Management*, *101*, 103129. DOI: 10.1016/j.ijhm.2021.103129

Buhalis, D., & Sinarta, Y. (2019). Real-time co-creation and nowness service: Lessons from tourism and hospitality. *Journal of Travel & Tourism Marketing*, *36*(5), 563–582. DOI: 10.1080/10548408.2019.1592059

Gervais, J. (2016). The operational definition of competency-based education. *The Journal of Competency-Based Education*, *1*(2), 98–106. DOI: 10.1002/cbe2.1011

Government of India. (2020). National [Ministry of Education.]. *Educational Policy*, •••, 2020.

Holmes, W., Bialik, M., & Fadel, C. (2022). *Artificial intelligence in education: Promise and implications for teaching and learning*. Center for Curriculum Redesign.

Ivanov, S., & Webster, C. (2019). Conceptual framework of the use of robots, artificial intelligence and service automation in tourism. *International Journal of Contemporary Hospitality Management*, *31*(2), 739–755. DOI: 10.1108/IJCHM-04-2018-0340

Ivanov, S., Webster, C., & Berezina, K. (2020). Adoption of robots and service automation by tourism and hospitality companies. *Revista Turismo & Desenvolvimento (Aveiro)*, *33*, 13–28.

Kolb, D. A. (2015). *Experiential learning: Experience as the source of learning and development* (2nd ed.). Pearson.

Lashley, C. (2018). *Hospitality management: A brief introduction*. Routledge.

Les Roches. (2023). *Institutional teaching and learning report*. Les Roches Global Hospitality Education.

Mariani, M., & Borghi, M. (2019). Industry 4.0: A bibliometric review of its managerial intellectual structure and potential evolution. *Technological Forecasting and Social Change*, *149*, 119752. DOI: 10.1016/j.techfore.2019.119752

Marques, C. S., & Ferreira, J. J. (2020). Digital transformation in the area of health: A systematic review of 45 years of evolution. *Health Technology (Hong Kong)*, *10*, 575–586.

Mulder, M. (2017). Competence-based education and training: Review and perspectives. *Journal of Agricultural Education and Extension*, *23*(1), 1–19. DOI: 10.1080/1389224X.2016.1274076

Noone, B. M., & McGuire, K. A. (2014). Pricing in hotel revenue management: A review of the theoretical and empirical literature. *Journal of Revenue and Pricing Management*, *13*(6), 482–501.

Radianti, J., Majchrzak, T. A., Fromm, J., & Wohlgenannt, I. (2020). A systematic review of immersive virtual reality applications for higher education. *Computers & Education*, *147*, 103778. DOI: 10.1016/j.compedu.2019.103778

Reisinger, Y., & Crotts, J. C. (2019). Cultural differences in tourism. *Journal of Travel & Tourism Marketing*, *36*(2), 155–168.

Tussyadiah, I. P. (2020). A review of research into automation in tourism: Launching the *Annals of Tourism Research* curated collection on artificial intelligence and robotics in tourism. *Annals of Tourism Research*, *81*, 102883. DOI: 10.1016/j.annals.2020.102883

UNESCO. (2021). *AI and education: Guidance for policymakers*. UNESCO Publishing.

Williamson, B., & Eynon, R. (2020). Historical threads, missing links, and future directions in AI in education. *Learning, Media and Technology*, *45*(3), 223–235. DOI: 10.1080/17439884.2020.1798995

Zawacki-Richter, O., Marín, V. I., Bond, M., & Gouverneur, F. (2019). Systematic review of research on artificial intelligence applications in higher education. *International Journal of Educational Technology in Higher Education*, *16*(39), 1–27. DOI: 10.1186/s41239-019-0171-0

Compilation of References

Abilkaiyrkyzy, A., Elhagry, A., Laamarti, F., & El Saddik, A. (2023). Metaverse key requirements and platforms survey. *IEEE Access : Practical Innovations, Open Solutions, 11*, 117765–117787. DOI: 10.1109/ACCESS.2023.3325844

Abou-Shouk, M. A., Gad, H. E. S., & Abdelhakim, A. S. (2021). Exploring customers' attitudes to the adoption of robots in tourism and hospitality. *Journal of Hospitality and Tourism Technology, 12*(4), 762–776. DOI: 10.1108/JHTT-09-2020-0215

Acemoglu, D., & Restrepo, P. (2020). Robots and jobs: Evidence from US labor markets. *Journal of Political Economy, 128*(6), 2188–2244. DOI: 10.1086/705716

Adepoju, O., Akinyomi, O., & Esan, O. (2023). Integrating human-computer interactions in Nigerian energy system: A skills requirement analysis. *Journal of Digital Food, Energy & Water Systems, 4*(2).

Adidas. (2021). *Into the metaverse: How we got here and where we are headed.* Adidas News. https://news.adidas.com/originals/into-the-metaverse--how-we-got-here-and-where-we-are-headed/s/6ccb61cb-2135-453e-8626-ac3d56faab30

Adomavicius, G., & Tuzhilin, A. (2005). Personalization technologies: A process-oriented perspective. *Communications of the ACM, 48*(10), 83–90. DOI: 10.1145/1089107.1089109

Agnihotri, R., De Keyser, A., Noble, S. M., Schillewaert, N., & Thakur, R. (2024). Purchase spillovers from the metaverse to the real world: The roles of social presence, trialability and customer experience. *Journal of Business Research, 170*, 114073. https://www.researchgate.net/publication/380600660_Purchase_spillovers_from_the_metaverse_to_the_real_world_The_roles_of_social_presence_trialability_and_customer_experience

Agrawal, N., Cohen, M. A., Deshpande, R., & Deshpande, V. (2024). How machine learning will transform supply chain management. *Harvard Business Review, 103*(3-4), 128–137.

Ahmed, R., Ahmed, E., Elbarbary, A., Darwish, A., & Hassanien, A. E. (2025). Fashion industry in the age of generative artificial intelligence and metaverse: A systematic review. *arXiv*. https://arxiv.org/abs/2505.17141

Akinsola, O. K., & Liang, W. (2025). Warehouse Automation and Robotics: Revolutionizing Order Fulfillment in E-Commerce.

Al-Adwan, A. S. (2024). The meta-commerce paradox: Exploring consumer non-adoption intentions. *Online Information Review*, *48*(6), 1270–1289. DOI: 10.1108/OIR-01-2024-0017

Alexander, B., Blazquez, M., Chrimes, C., & Boardman, R. (2025). The role of immersive spaces on the customer experience: An exploration of fashion metaverses. *Psychology and Marketing*, *42*(2), 539–553. DOI: 10.1002/mar.22140

AlHogail, A. A., & Alshahrani, M. (2019). Building consumer trust to improve Internet of Things (IoT) technology adoption. *Advances in Intelligent Systems and Computing*, *775*, 325–334. DOI: 10.1007/978-3-319-94866-9_33

Ali, S. S. (2023). The future of work in the age of automation: Exploring new roles and opportunities. *Al-Behishat Research Archive*, *1*(01), 55–62.

Al-Kfairy, M., Alfandi, O., & Khaddaj, S. (2025). Psychological barriers to metaverse-based education: Examining the impact of technophobia and digital fatigue. *Journal of Information. Communication and Ethics in Society*, *23*(4), 550–566. DOI: 10.1108/JICES-04-2025-0080

Al-Romeedy, B. S., Khairy, H. A., Jasim, T. A., & Baquero, A. (2025). *Customer ability, readiness, and intention to purchase using AI in tourism and hospitality businesses: The mediating roles of trust in AI and customer engagement*. 145–166. DOI: 10.4018/979-8-3693-7447-4.ch010

Ammupriya, A., Vaishnavi, S., Ashwini, A., Kavitha, R., Paranthaman, P., & Saravanan, V. (2025, March). Cloud-Based HR Platforms for Scalable Workforce Management in Multinational Organizations. In *2025 3rd International Conference on Disruptive Technologies (ICDT)* (pp. 1607-1613). IEEE.

Andriella, A., Torras, C., Alenya, G., & Alenyà, G. (2023). Teaching robot to understand social cues: A case study in autism spectrum disorder therapy. *Frontiers in Robotics and AI*, *10*, 1092547. DOI: 10.3389/frobt.2023.1092547

Araujo, T. (2018). Living up to the chatbot hype: The influence of anthropomorphic design cues and communicative agency framing on conversational agent and company perceptions. *Computers in Human Behavior*, *85*, 183–189. DOI: 10.1016/j.chb.2018.03.051

Aryal, A. (2025). *Gucci marketing strategy: Crisis-to-comeback (2025)*. Blankboard. studio. https://www.blankboard.studio/originals/blog/gucci-marketing-strategy-crisis-to-comeback

Avolio, B. J., Walumbwa, F. O., & Weber, T. J. (2009). Leadership: Current theories, research, and future directions. *Annual Review of Psychology*, *60*(1), 421–449. DOI: 10.1146/annurev.psych.60.110707.163621 PMID: 18651820

Awad, N. F., & Krishnan, M. S. (2006). The personalization privacy paradox: An empirical evaluation of information transparency and the willingness to be profiled online for personalization. *Management Information Systems Quarterly*, *30*(1), 13–28. DOI: 10.2307/25148715

Bain, M. (2023). *How fashion NFTs fizzled*. The Business of Fashion. https://www.businessoffashion.com/articles/technology/how-fashion-nfts-fizzled/

Banta, T. W., & Blaich, C. (2018). Closing the assessment loop. *Change: The Magazine of Higher Learning*, *50*(5), 8–15.

Barone, A. M., Stagno, E., & Donato, C. (2024). Call it robot: Anthropomorphic framing and failure of self-service technologies. *Journal of Services Marketing*, *38*(3), 272–287. DOI: 10.1108/JSM-05-2023-0169

Barsom, E. Z., Graafland, M., & Schijven, M. P. (2016). Systematic review on the effectiveness of augmented reality applications in medical training. *Surgical Endoscopy*, *30*(10), 4174–4183. DOI: 10.1007/s00464-016-4800-6 PMID: 26905573

Bartlett, L., Kabir, M. A., & Han, J. (2023). A review on business process management system design: The role of virtualization and work design. *IEEE Access : Practical Innovations, Open Solutions*, *11*, 116786–116819. DOI: 10.1109/ACCESS.2023.3323445

Bartneck, C., Kulić, D., Croft, E., & Zoghbi, S. (2009). Measurement instruments for the anthropomorphism, animacy, likeability, perceived intelligence, and perceived safety of robots. *International Journal of Social Robotics*, *1*(1), 71–81. DOI: 10.1007/s12369-008-0001-3

Bass, B. M., & Riggio, R. E. (2006). *Transformational leadership* (2nd ed.). Lawrence Erlbaum Associates. DOI: 10.4324/9781410617095

Batchelor, J. (2022). *Meta's flagship metaverse Horizon Worlds struggling to attract and retain users*. GamesIndustry.biz. https://www.gamesindustry.biz/metas-flagship-metaverse-horizon-worlds-struggling-to-attract-and-retain-users

Becker, K. (2025). *NFT fashion and virtual self re-creation* [Manuscript or unpublished work].

Belanche, D., Casaló, L. V., Flavián, C., & Schepers, J. (2020). Service robot implementation: A theoretical framework and research agenda. *Service Industries Journal*, *40*(3–4), 203–225. DOI: 10.1080/02642069.2019.1672666

Bemelmans, R., Gelderblom, G. J., Jonker, P., & de Witte, L. (2021). Socially assistive robots in elderly care—A systematic review into effects and effectiveness. *Journal of the American Medical Directors Association*, *14*(2), 114–120. DOI: 10.1016/j.jamda.2010.10.002 PMID: 21450215

Berhanu, K., & Raj, S. R. (2024). The role of social media marketing in Ethiopian tourism and hospitality organizations: Applying the unified theory of acceptance and use of technology model. *Cogent Social Sciences*, *10*(1), 2318866. Advance online publication. DOI: 10.1080/23311886.2024.2318866

Bernatik, J. (2022). *Introduction to Flow blockchain*. Flow. https://flow.com/engineering-blogs/introduction-to-flow-blockchain

Bhargava, A., Bester, M., & Bolton, L. (2021). Employees' perceptions of the implementation of robotics, artificial intelligence, and automation (RAIA) on job satisfaction, job security, and employability. *Journal of Technology in Behavioral Science*, *6*(1), 106–113. DOI: 10.1007/s41347-020-00153-8

Bialkova, S. (2024). *The rise of AI user applications: Chatbots integration foundations and trends.*

Bleier, A., & Eisenbeiss, M. (2015). The importance of trust for personalized online advertising. *Journal of Retailing*, *91*(3), 390–409. DOI: 10.1016/j.jretai.2015.04.001

Blut, M., Wang, C., Wunderlich, N. V., & Brock, C. (2021). Understanding anthropomorphism in service provision: A meta-analysis of physical robots, chatbots, and other AI. *Journal of the Academy of Marketing Science*, *49*(4), 632–658. DOI: 10.1007/s11747-020-00762-y

Bommasani, R., Hudson, D. A., Adeli, E., Altman, R., Arora, S., von Arx, S., Bernstein, M. S., Bohg, J., Bosselut, A., Brunskill, E., Brynjolfsson, E., Buch, S., Card, D., Castellon, R., Chatterji, N., Chen, A., Creel, K., Davis, J., Demszky, D., . . . Liang, P. (2021). On the opportunities and risks of foundation models. arXiv. / arXiv.2108.07258DOI: 10.48550

Borghouts–van de Pas, I., Bosmans, M., & Freese, C. (2021). Unemployment prevention: The role of Human Resource Management in job-to-job transitions in the event of redundancy. *European Journal of Social Security, 23*(2), 103–119. DOI: 10.1177/1388262721995209

Bowen, J. T., & Morosan, C. (2018). Beware hospitality industry: The robots are coming. *Worldwide Hospitality and Tourism Themes, 10*(6), 726–733. DOI: 10.1108/WHATT-07-2018-0045

Breazeal, C. (2003). Toward sociable robots. *Robotics and Autonomous Systems, 42*(3–4), 167–175. DOI: 10.1016/S0921-8890(02)00373-1

British Vogue. (2021). *Dolce & Gabbana's NFT experiment is a million-dollar success story*. https://www.vogue.co.uk/news/article/fashion-nft-dolce-and-gabbana

Broadbent, E. (2017). Interactions with robots: The truths we reveal about ourselves. *Annual Review of Psychology, 68*(1), 627–652. DOI: 10.1146/annurev-psych-010416-043958 PMID: 27648986

Broadbent, E., Stafford, R., & MacDonald, B. (2009). Acceptance of healthcare robots for the older population: Review and future directions. *International Journal of Social Robotics, 1*(4), 319–330. DOI: 10.1007/s12369-009-0030-6

Brodie, R. J., Hollebeek, L. D., Jurić, B., & Ilić, A. (2011). Customer engagement: Conceptual domain, fundamental propositions, and implications for research. *Journal of Service Research, 14*(3), 252–271. DOI: 10.1177/1094670511411703

Brooks, A. (2026). *Is the metaverse fashion trend dying or evolving into something new?* Alibaba.com. https://www.alibaba.com/product-insights/is-the-metaverse-fashion-trend-dying-or-evolving-into-something-new.html

Brown, T. B., Mann, B., Ryder, N., Subbiah, M., Kaplan, J., Dhariwal, P., Neelakantan, A., Shyam, P., Sastry, G., Askell, A., Agarwal, S., Herbert-Voss, A., Krueger, G., Henighan, T., Child, R., Ramesh, A., Ziegler, D. M., Wu, J., Winter, C., & Amodei, D. (2020). Language models are few-shot learners. *Advances in Neural Information Processing Systems, 33*, 1877–1901.

Brynjolfsson, E., Li, D., & Raymond, L. (2025). Generative AI at work. *The Quarterly Journal of Economics, 140*(2), 889–942. DOI: 10.1093/qje/qjae044

Brynjolfsson, E., Rock, D., & Syverson, C. (2021). The productivity J-curve: How intangibles complement general purpose technologies. *American Economic Journal. Macroeconomics, 13*(1), 333–372. DOI: 10.1257/mac.20180386

Buhalis, D., & Leung, R. (2018). Smart hospitality: Interconnectivity and interoperability towards an ecosystem. *International Journal of Hospitality Management, 71*, 41–50. DOI: 10.1016/j.ijhm.2017.11.011

Buhalis, D., & Moldavska, I. (2022). In-room voice-based AI digital assistants: Implications for hospitality management and operations. *International Journal of Hospitality Management, 101*, 103129. DOI: 10.1016/j.ijhm.2021.103129

Buhalis, D., & Sinarta, Y. (2019). Real-time co-creation and nowness service: Lessons from tourism and hospitality. *Journal of Travel & Tourism Marketing, 36*(5), 563–582. DOI: 10.1080/10548408.2019.1592059

Burberry. (2025). *Burberry X Blankos Block Party: New NFT collection and social space*. https://www.burberryplc.com/news/brand/2022/burberry-x-blankos-block-party--new-nft-collection-and-social-sp

Business of Fashion. (2022). *Why luxury NFT collections are struggling in the secondary market*. https://www.businessoffashion.com

Butler, C. (2022). *How to play and win: Blankos Block Party*. DappRadar. https://dappradar.com/blog/how-to-play-and-win-blankos-block-party

Čaić, M., Odekerken-Schröder, G., & Mahr, D. (2018). Service robots: Value co-creation and co-destruction in elderly care networks. *Journal of Service Management, 29*(3), 519–544.

Canag, L., Loquinario, A., Pepano, R., & Peninoy, R. (2025). Democratizing fashion consumer engagement and redefining style in the digital age. *International Journal of Research, 8*(XII), 935–949. DOI: 10.47772/IJRISS.2024.8120078

Casciani, D., Chkanikova, O., & Pal, R. (2022). Exploring the nature of digital transformation in the fashion industry: Opportunities for supply chains, business models, and sustainability-oriented innovations. *Sustainability : Science, Practice, & Policy, 18*(1), 773–795. DOI: 10.1080/15487733.2022.2125640

Chain, B. N. B. (2025). *BNB Chain: Empowering the future of decentralized applications*. https://docs.bnbchain.org

Choi, S., Lim, H., & Kim, J. (2022). Service robots in the banking sector: Customer satisfaction and trust in robot-mediated service encounters. *International Journal of Bank Marketing, 40*(4), 678–697. DOI: 10.1108/IJBM-06-2021-0250

Choi, S., Mattila, A. S., & Bolton, L. E. (2021). To err is human(-oid): How do consumers react to robot service failure and recovery? *Journal of Service Research, 24*(3), 354–371. DOI: 10.1177/1094670520978798

Choi, Y., Choi, M., Oh, M., & Kim, S. (2020). Service robots in hotels: Understanding the service quality perceptions of human-robot interaction. *Journal of Hospitality Marketing & Management, 29*(6), 613–635. DOI: 10.1080/19368623.2020.1703871

Choi, Y., Oh, M., Choi, M., & Kim, S. S. (2021). Exploring the influence of culture on tourist experiences with robots in service delivery environment. *Current Issues in Tourism, 24*(5), 717–733. DOI: 10.1080/13683500.2020.1735318

Choudhuri, S. S. (2024). *AI in ERP and supply chain management.* Academic Guru Publishing House.

Chowdhury, M. M. H., Aliahmadi, S. Z., Quaddus, M. A., Ansaripoor, A. H., & Mirjalili, S. (2025). A decision support framework for resilient and sustainable service design. *Global Journal of Flexible Systems Managment, 26*(S1, Suppl 1), 25–55. DOI: 10.1007/s40171-024-00419-7

Christl, W. (2023). MONITORING WORK AND AUTOMATING TASK ALLOCATION IN RETAIL AND HOSPITALITY.

CoinDesk. (2023). *NFT market faces steep decline as trading volumes drop across major platforms.* https://www.coindesk.com

Coşkun, M. (2025). Metaverse and the future of fashion: A comparative analysis of fast-fashion and luxury fashion brands. *Sosyal Mucit Academic Review, 6*(3), 392–418. DOI: 10.54733/smar.1665701

Covington, S. M., & Soltanisehat, M. (2025). Enhancing transparency and traceability in fast fashion: The role of blockchain technology. *International Journal of Sustainable Fashion & Textiles, 4*(2), 257–280. DOI: 10.1386/sft_00060_1

Cui, Y., & van Esch, P. (2023). Artificial intelligence in customer service strategy for seamless customer experiences. In *Artificial Intelligence in Customer Service: The Next Frontier for Personalized Engagement* (pp. 73–97). Springer International Publishing. https://doi.org/DOI: 10.1007/978-3-031-33898-4_4

Dautenhahn, K. (2022). Socially intelligent robots: Dimensions of human-robot interaction. *Philosophical Transactions of the Royal Society of London. Series B, Biological Sciences, 362*(1480), 679–704. DOI: 10.1098/rstb.2006.2004 PMID: 17301026

Dauth, W., Findeisen, S., Suedekum, J., & Woessner, N. (2021). The adjustment of labor markets to robots. *Journal of the European Economic Association, 19*(6), 3104–3153. DOI: 10.1093/jeea/jvab012

Davenport, T. H., & Ronanki, R. (2018, January–February). Artificial intelligence for the real world. *Harvard Business Review*, *96*(1), 108–116.

Davis, F. D. (1989). Perceived usefulness, perceived ease of use, and user acceptance of information technology. *Management Information Systems Quarterly*, *13*(3), 319–340. DOI: 10.2307/249008

Day, D. V. (2000). Leadership development: A review in context. *The Leadership Quarterly*, *11*(4), 581–613. DOI: 10.1016/S1048-9843(00)00061-8

De Keyser, A., Köcher, S., Alkire, L., Verbeeck, C., & Kandampully, J. (2019). Frontline service technology infusion: Conceptual archetypes and future research directions. *Journal of Service Management*, *30*(1), 156–183. DOI: 10.1108/JOSM-03-2018-0082

Decentraland. (2025). *Decentraland*. https://decentraland.org/

Department for Business and Trade. (2024). Inward investment results 2023–2024 (released 13 June 2024). https://www.gov.uk/government/statistics/dbt-inward-investment-results-2023-to-2024

Department for Science. Innovation and Technology. (2023). National AI Strategy: AI Action Plan progress report. https://www.gov.uk/government/publications/national-ai-strategy

Department for Science. Innovation and Technology. (2024). AI sector study 2024 (released 24 January 2024). https://www.gov.uk/government/publications/ai-sector-study-2024

Dhakshitha, B. K. (2025). Exploring the impact of social presence cues on consumer impulse buying in E-commerce platforms. *Journal of Marketing & Social Research*, *2*, 119–125.

Dinev, T., & Hart, P. (2006). An extended privacy calculus model for e-commerce transactions. *Information Systems Research*, *17*(1), 61–80. DOI: 10.1287/isre.1060.0080

Donvito, R., Acuti, D., & Song, S. (2024). Fashion and the metaverse: Implications for consumers and firms. *Journal of Global Fashion Marketing*, *15*(1), 1–5. DOI: 10.1080/20932685.2023.2293290

Duffy, B. R. (2003). Anthropomorphism and the social robot. *Robotics and Autonomous Systems*, *42*(3–4), 177–190. DOI: 10.1016/S0921-8890(02)00374-3

Dwivedi, Y. K., Hughes, L., Wang, Y., Alalwan, A. A., Ahn, S. J., Balakrishnan, J., Barta, S., Belk, R., Buhalis, D., Dutot, V., Felix, R., Filieri, R., Flavián, C., Gustafsson, A., Hinsch, C., Hollensen, S., Jain, V., Kim, J., Krishen, A. S., & Wirtz, J. (2023). Metaverse marketing: How the metaverse will shape the future of consumer research and practice. *Psychology and Marketing*, *40*(4), 750–776. DOI: 10.1002/mar.21767

Ekman, P. (2016). What scientists who study emotion agree about. *Perspectives on Psychological Science : A Journal of the Association for Psychological Science*, *11*(1), 31–34. DOI: 10.1177/1745691615596992 PMID: 26817724

Elsayegh, O. A. E., & Mohamed, H. (2024). A proposed model for startup in the field of textile design and fashion through an analytical study in digital transformation of the global startups. *Journal of Design Sciences and Applied Arts*, *5*(2), 474–494. DOI: 10.21608/jdsaa.2023.232038.1370

Epley, N., Waytz, A., & Cacioppo, J. T. (2007). On seeing the human: A three-factor theory of anthropomorphism. *Psychological Review*, *114*(4), 864–886. DOI: 10.1037/0033-295X.114.4.864 PMID: 17907867

Fang, S., Wu, X., & Wai Lai, I. K. (2021). Factors affecting Macau undergraduate students' acceptance of hospitality English app: Applicability of UTAUT model. *International Journal of Innovation and Learning*, *29*(3), 250–266. DOI: 10.1504/IJIL.2021.114510

Fashionbi. (2025). *Fashion brands penetrating into video games*. https://www.fashionbi.com/insights/fashion-brands-penetrating-into-video-games

Fortagne, M. A. (2025). *Human-app interaction for disruptive innovations: User-centered contributions to technology acceptance and effects* [Doctoral dissertation].

Foster, W. M., Hassard, J. S., Morris, J., & Wolfram Cox, J. (2019). The changing nature of managerial work: The effects of corporate restructuring on management jobs and careers. *Human Relations; Studies Towards the Integration of the Social Sciences*, *72*(3), 473–504. DOI: 10.1177/0018726719828439

Fountaine, T., McCarthy, B., & Saleh, T. (2019, July–August). Building the AI-powered organization. *Harvard Business Review*, *97*(4), 62–73.

Frey, C. B., & Osborne, M. A. (2017). The future of employment: How susceptible are jobs to computerisation? *Technological Forecasting and Social Change*, *114*, 254–280. DOI: 10.1016/j.techfore.2016.08.019

Fuchs, H., Benkova, E., Fishbein, A., & Fuchs, A. (2023, November). The importance of psychological and cognitive flexibility in educational processes to prepare and acquire the skills required in the Twenty-First century. In *The Global Conference on Entrepreneurship and the Economy in an Era of Uncertainty* (pp. 91-114). Singapore: Springer Nature Singapore.

Fu, S., Feng, X., Sultana, A., & Zhao, L. (2023). Joint power allocation and 3D deployment for UAV-BSs: A game theory based deep reinforcement learning approach. *IEEE Transactions on Wireless Communications, 23*(1), 736–748. DOI: 10.1109/TWC.2023.3281812

Gam, A., Bahri-Ammari, N., & Soliman, M. (2023). Intelligent customer experience and behavioral brand loyalty: Empirical evidence from the commerce environment. *International Journal of Human-Computer Interaction, 39*(10), 2090–2101. DOI: 10.1080/10447318.2022.2074120

Gefen, D., Karahanna, E., & Straub, D. W. (2003). Trust and TAM in online shopping: An integrated model. *Management Information Systems Quarterly, 27*(1), 51–90. DOI: 10.2307/30036519

George, A. S. (2024). Artificial intelligence and the future of work: Job shifting not job loss. *Partners Universal Innovative Research Publication, 2*(2), 17–37.

Gervais, J. (2016). The operational definition of competency-based education. *The Journal of Competency-Based Education, 1*(2), 98–106. DOI: 10.1002/cbe2.1011

Ghazi, K. M., Kattara, H. S., Salem, I. E. B., & Shaaban, M. N. (2025). Benefit-triggered or trust-guided? Investigation of customers' perceptions towards AI-adopting hotels amid and post COVID-19 pandemic. *Tourism and Hospitality Research, 25*(1), 38–56. DOI: 10.1177/14673584231184161

Ghazizadeh, M., Lee, J. D., & Boyle, L. N. (2012). Extending the Technology Acceptance Model to assess automation. *Cognition Technology and Work, 14*(1), 39–49. DOI: 10.1007/s10111-011-0194-3

Goel, R., Singh, T. P., Sahdev, S. L., Baral, S. K., & Choudhury, A. (2022). *Impact of AI & IOT in Sustainable & Green Practices Adopted in Hotel Industry and Measuring Hotel Guests' Satisfaction.* DOI: 10.1109/ICRITO56286.2022.9965152

Gonçalves, L., Patrício, L., Grenha Teixeira, J., & Wunderlich, N. V. (2020). Understanding the customer experience with smart services. *Journal of Service Management, 31*(4), 723–744. DOI: 10.1108/JOSM-11-2019-0349

Gosling, S. D., Rentfrow, P. J., & Swann, W. B.Jr. (2003). A very brief measure of the Big-Five personality domains. *Journal of Research in Personality, 37*(6), 504–528. DOI: 10.1016/S0092-6566(03)00046-1

Government of India. (2020). National [Ministry of Education.]. *Educational Policy*, •••, 2020.

Gretzel, U., Sigala, M., Xiang, Z., & Koo, C. (2015). Smart tourism: Foundations and developments. *Electronic Markets, 25*(3), 179–188. DOI: 10.1007/s12525-015-0196-8

Grewal, D., Noble, S. M., Roggeveen, A. L., & Nordfält, J. (2020). The future of in-store technology. *Journal of the Academy of Marketing Science, 48*(1), 96–113. DOI: 10.1007/s11747-019-00697-z

GUNDA, V. N., & MUPA, M. N. (2024). A Paradox in Human Resources: Automation Vs Human Capital.

Gupta, S., Rastogi, R., & Alam, M. S. (2023). Disrupting FinTech: Revolutionizing Customer Experience With AI in Industry 5.0. In *The Impact of AI Innovation on Financial Sectors in the Era of Industry 5.0* (pp. 1-16). IGI Global.

Gupta, U., Somani, P., Behare, N., Mahajan, R., Singh, M., & Iyer, C. V. (2024). Revolutionizing Retail: IoT Applications for Enhanced Customer Experience. In *Internet of Things Applications and Technology* (pp. 60-80). Auerbach Publications.

Hahn, J. (2021). *Nike gets "metaverse-ready" with acquisition of virtual sneaker company RTFKT*. Dezeen. https://www.dezeen.com/2021/12/15/rtfkt-nike-acquisition/

Haibeh, L. A., Yagoub, M. C., & Jarray, A. (2022). A survey on mobile edge computing infrastructure: Design, resource management, and optimization approaches. *IEEE Access : Practical Innovations, Open Solutions, 10*, 27591–27610. DOI: 10.1109/ACCESS.2022.3152787

Hamamah, A., Al-Haimi, B., & Tajuri, W. (2024). Navigating the Marketing Landscape: Artificial Intelligence and Big Data Role in Digital Marketing. *International Journal of Academic Research in Business & Social Sciences, 14*(10), 2285–2299. DOI: 10.6007/IJARBSS/v14-i10/23346

Hanji, S., Navalgund, N. R., Hanji, S. S., Goud, V., Hatture, S. M., & Wadawadagi, R. S. (2024). Rise of the Machines: Investigating Service Robot Adoption Intentions in Hotel Industry. *AIP Conference Proceedings, 3131*(1), 020012. Advance online publication. DOI: 10.1063/5.0229719

Han, S. H., Lee, J., Edvardsson, B., & Verma, R. (2022). Developing a strategic viability-fit model of mobile technology adoption in hotels. *Journal of Hospitality and Tourism Technology*, *13*(1), 85–99. DOI: 10.1108/JHTT-05-2020-0109

Hardcastle, K., Vorster, L., & Brown, D. M. (2025). Understanding customer responses to AI-driven personalized journeys: Impacts on the customer experience. *Journal of Advertising*, *54*(2), 176–195. DOI: 10.1080/00913367.2025.2460985

Hayes, A. F. (2018). *Introduction to mediation, moderation, and conditional process analysis: A regression-based approach* (2nd ed.). Guilford Press.

Heerink, M., Kröse, B., Evers, V., & Wielinga, B. (2010). Assessing acceptance of assistive social agent technology by older adults: The Almere model. *International Journal of Social Robotics*, *2*(4), 361–375. DOI: 10.1007/s12369-010-0068-5

Henschel, A., Laban, G., & Cross, E. S. (2021). What makes a robot social? A review of social robots from the perspective of human-robot interaction. *International Journal of Social Robotics*, *13*(6), 1–15. DOI: 10.1007/s12369-020-00736-8

Heo, J. (2025). Marketing luxury fashion in the metaverse: The case of Gucci Town on Roblox. *Thunderbird International Business Review*, tie.70024. DOI: 10.1002/tie.70024

Ho, C. C., & MacDorman, K. F. (2010). Revisiting the uncanny valley theory: Developing and validating an alternative to the Godspeed indices. *Computers in Human Behavior*, *26*(6), 1508–1518. DOI: 10.1016/j.chb.2010.05.015

Hoch, J. E., & Kozlowski, S. W. J. (2014). Leading virtual teams: Hierarchical leadership, structural supports, and shared team leadership. *The Journal of Applied Psychology*, *99*(3), 390–403. DOI: 10.1037/a0030264 PMID: 23205494

Hochschild, A. R. (1983). *The managed heart: Commercialization of human feeling*. University of California Press.

Hoff, K. A., & Bashir, M. (2015). Trust in automation: Integrating empirical evidence on factors that influence trust. *Human Factors*, *57*(3), 407–434. DOI: 10.1177/0018720814547570 PMID: 25875432

Hollebeek, L. D. (2011). Demystifying customer brand engagement: Exploring the loyalty nexus. *Journal of Marketing Management*, *27*(7–8), 785–807. DOI: 10.1080/0267257X.2010.500132

Holmes, W., Bialik, M., & Fadel, C. (2022). *Artificial intelligence in education: Promise and implications for teaching and learning*. Center for Curriculum Redesign.

Holt, D. T., Armenakis, A. A., Feild, H. S., & Harris, S. G. (2007). Readiness for organizational change: The systematic development of a scale. *The Journal of Applied Behavioral Science*, *43*(2), 232–255. DOI: 10.1177/0021886306295295

Hompel, M., & Schmidt, T. (2006). *Warehouse management: automation and organisation of warehouse and order picking systems*. Springer Science & Business Media.

Hopf, V., Velten, L., & Rowson, B. (2018). *Roboptimism or pessimism: HR managers face a challenge*. 203–228. https://www.scopus.com/inward/record.uri?eid=2-s2.0-85206821575&partnerID=40&md5=c85876447b073baa8b4a73a8c13d0b60

Horton, D., & Wohl, R. R. (1956). Mass communication and para-social interaction: Observations on intimacy at a distance. *Psychiatry*, *19*(3), 215–229. DOI: 10.1080/00332747.1956.11023049 PMID: 13359569

Huang, M. H., & Rust, R. T. (2018). Artificial intelligence in service. *Journal of Service Research*, *21*(2), 155–172. DOI: 10.1177/1094670517752459

Huang, M. H., & Rust, R. T. (2021). A strategic framework for artificial intelligence in marketing. *Journal of the Academy of Marketing Science*, *49*(1), 30–50. DOI: 10.1007/s11747-020-00749-9

Huang, M.-H., & Rust, R. T. (2021). Engaged to a robot? The role of AI in service. *Journal of Service Research*, *24*(1), 30–41. DOI: 10.1177/1094670520902266

Hwang, B.-N., Puntha, P., & Jitanugoon, S. (2026). Human-centric AI: Transforming service ecosystems for inclusivity and excellence in resource-constrained contexts. *Journal of Service Management*. Advance online publication. DOI: 10.1177/10946705251344387

Igumenov, A., Filatovas, E., Medvedev, V., & Paulavičius, R. (2021). Estimating energy consumption of Ethereum network. In *DAMSS 2021: 12th conference on data analysis methods for software systems*.

Ikeler, P. R. (2013). *Point of consumption: Work, consciousness and organizing in the retail sector*. City University of New York.

Ikonomou, C., & Sergopoulos, K. (2025). *Artificial intelligence and tourism: Examining a relationship currently under development*. 211–224. https://www.scopus.com/inward/record.uri?eid=2-s2.0-105008487582&partnerID=40&md5=da0225886f7bf0f5579b0295a2d179c5

Independent. (2021). *Gucci digitally outfits Gen-Z in metaverse foray with Roblox*. https://www.independent.co.uk/news/gucci-digitally-outfits-genz-in-metaverse-foray-with-roblox-gucci-nfts-italian-florence-alessandro-michele-b1862331.html

Information Commissioner's Office. (2023). Guidance on AI and data protection (updated 15 March 2023). https://ico.org.uk/for-organisations/uk-gdpr-guidance-and-resources/artificial-intelligence/

Institute of Customer Service. (2024). UK Customer Satisfaction Index (UKCSI), January 2024 release; UKCSI Willingness-to-Pay Briefing 2024. London: ICS. https://www.instituteofcustomerservice.com/research-insight/ukcsi/

International Federation of Robotics. (2024). World Robotics 2024, Industrial Robots: United Kingdom country report (released 24 September 2024); World Robotics 2024, Service Robots. Frankfurt: IFR. https://ifr.org/worldrobotics/ LSEG / FTSE Russell. (2024). FTSE UK Index Series, FTSE 350 Technology sub-index (series TECH), year-end levels 2015–2024. London: LSEG. https://www.lseg.com/en/ftse-russell

ISO 9241-210. (2019). Ergonomics of human-system interaction—Human-centred design for interactive systems. International Organization for Standardization.

Ivanov, S. H., Webster, C., & Berezina, K. (2022). *Robotics in Tourism and Hospitality*. 1873–1899. DOI: 10.1007/978-3-030-48652-5_112

Ivanov, S., Seyitoğlu, F., & Markova, M. (2022). Robot human interaction in hotels: Managers' perspective. *Information Technology & Tourism*, *22*(4), 507–535. DOI: 10.1007/s40558-020-00180-2

Ivanov, S., & Webster, C. (2019). Conceptual framework of the use of robots, artificial intelligence and service automation in tourism. *International Journal of Contemporary Hospitality Management*, *31*(2), 739–755. DOI: 10.1108/IJCHM-04-2018-0340

Ivanov, S., Webster, C., & Berezina, K. (2020). Adoption of robots and service automation by tourism and hospitality companies. *Revista Turismo & Desenvolvimento (Aveiro)*, *33*, 13–28.

Jarrahi, M. H. (2018). Artificial intelligence and the future of work: Human-AI symbiosis in organizational decision making. *Business Horizons*, *61*(4), 577–586. DOI: 10.1016/j.bushor.2018.03.007

Jung, N. Y. (2025). AI-Powered Personalization Drives Loyalty in Retail. In *Leveraging AI-Powered Marketing in the Experience-Driven Economy* (pp. 105–146). IGI Global Scientific Publishing. DOI: 10.4018/979-8-3693-9561-5.ch004

Kahn, B. E. (2021). *The shopping revolution, updated and expanded edition: how retailers succeed in an era of endless disruption accelerated by COVID-19*. University of Pennsylvania Press.

Kapengut, E., & Mizrach, B. (2022). *An event study of the Ethereum transition to proof-of-stake*. arXiv. https://www.researchgate.net/publication/369670342_An_Event_Study_of_the_Ethereum_Transition_to_Proof-of-Stake

Kim, D. Y., & Lee, H. K. (2025). "I am wearing what my avatar is wearing": The effect of avatar realism on mental simulation in the metaverse. *Journal of Fashion Marketing and Management*, *29*(5), 895–914. DOI: 10.1108/JFMM-04-2024-0116

Kim, D. Y., Lee, H. K., & Chung, K. (2023). Avatar-mediated experience in the metaverse: The impact of avatar realism on user-avatar relationship. *Journal of Retailing and Consumer Services*, *73*, 103382. DOI: 10.1016/j.jretconser.2023.103382

Kim, J., & Lee, H. (2022). How virtual try-on influences purchase intention: The role of self-expressiveness, vicarious satisfaction, and stress relief. *Frontiers in Psychology*, *13*, 938060. Advance online publication. DOI: 10.3389/fpsyg.2022.938060

Kim, N. E., & Lee, J. R. (2022). The purchasing status of the avatars and digital fashion items in metaverse and consumers' purchase satisfaction and the future purchase intentions according to usage motivation. *Journal of Korean Home Economics Education Association*, *34*(3), 133–148. DOI: 10.19031/jkheea.2022.9.34.3.133

Kim, S., & Park, J. (2022). Effects of hotel service robots on customer experience and behavior. *International Journal of Contemporary Hospitality Management*, *34*(3), 1162–1180. DOI: 10.1108/IJCHM-01-2021-0024

Kim, T., Jo, H., Yhee, Y., & Koo, C. (2022). Robots, artificial intelligence, and service automation (RAISA) in hospitality: Sentiment analysis of YouTube streaming data. *Electronic Markets*, *32*(1), 259–275. DOI: 10.1007/s12525-021-00514-y

Klerk, A. D. (2021). *Louis Vuitton has launched a game as part of its founder's birthday celebrations*. Harper's Bazaar. https://www.harpersbazaar.com/uk/fashion/a37218196/louis-vuitton-game/

Kniazeva, M., Aiello, G., Dasmi, C., Mazzoli, V., Nechaeva, O., & Syed, F. U. (2024). Why fashion brands enter the metaverse: Exploring the motivations of fast fashion and luxury fashion brands. *Journal of Global Fashion Marketing*, *15*(1), 62–89. DOI: 10.1080/20932685.2023.2269952

Kolb, D. A. (2015). *Experiential learning: Experience as the source of learning and development* (2nd ed.). Pearson.

Korneeva, E., Omarova, A., & Nurova, O. (2024). Adoption of smart energy technologies in the context of sustainable development. In *E3S Web of Conferences* (Vol. 531, p. 02001). EDP Sciences. DOI: 10.1051/e3sconf/202453102001

Kotter, J. P. (1996). *Leading change*. Harvard Business School Press.

Kozinets, R. V. (2023). Immersive netnography: A novel method for service experience research in virtual reality, augmented reality and metaverse contexts. *Journal of Service Management, 34*(1), 100–125. DOI: 10.1108/JOSM-12-2021-0481

Kumari, R., & Singh, V. P. (2024). Industry 4.0 Technologies: Managing the Future of Smart Retail. In *Augmenting Retail Reality, Part A: Blockchain, AR, VR, and the Internet of Things* (pp. 59-82). Emerald Publishing Limited.

Kumar, S., Malhotra, D., & Kathuria, G. (2026). Unrestrained buying behavior in the metaverse: Exploring the moderating role of technostress and self-control. *Spanish Journal of Marketing - ESIC, 30*(2), 167–191. DOI: 10.1108/SJME-07-2024-0217

Kuo, C.-M., Chen, L.-C., & Tseng, C.-Y. (2017). Investigating an innovative service with hospitality robots. *International Journal of Contemporary Hospitality Management, 29*(5), 1305–1321. DOI: 10.1108/IJCHM-08-2015-0414

Kwon, J., Bae, S. I. S., & Blum, S. C. (2013). Mobile applications in the hospitality industry. *Journal of Hospitality and Tourism Technology, 4*(1), 81–92. DOI: 10.1108/17579881311302365

Lapatoura, I. (2023). NFTs, digital worlds and brand protection in fashion: A UK/European intellectual property perspective. *International Journal of Law in Changing World, 2*(3), 141–152. DOI: 10.54934/ijlcw.v2i3.54

Lashley, C. (2018). *Hospitality management: A brief introduction*. Routledge.

Lee, C. H., Li, Q., Lee, Y. C., & Shih, C. W. (2021). Service design for intelligent exhibition guidance service based on dynamic customer experience. *Industrial Management & Data Systems, 121*(6), 1237–1267. DOI: 10.1108/IMDS-06-2020-0356

Lee, J. D., & See, K. A. (2004). Trust in automation: Designing for appropriate reliance. *Human Factors, 46*(1), 50–80. DOI: 10.1518/hfes.46.1.50.30392 PMID: 15151155

Lee, L. Y. S. (2016). Hospitality Industry Web-Based Self-Service Technology Adoption Model: A Cross-Cultural Perspective. *Journal of Hospitality & Tourism Research (Washington, D.C.), 40*(2), 162–197. DOI: 10.1177/1096348013495695

Lemon, K. N., & Verhoef, P. C. (2016). Understanding customer experience throughout the customer journey. *Journal of Marketing, 80*(6), 69–96. DOI: 10.1509/jm.15.0420

Leong, P. Y., & Ahmad, N. S. (2024). Exploring autonomous load-carrying mobile robots in indoor settings: A comprehensive review. *IEEE Access : Practical Innovations, Open Solutions, 12*, 131395–131417. DOI: 10.1109/ACCESS.2024.3435689

Les Roches. (2023). *Institutional teaching and learning report*. Les Roches Global Hospitality Education.

Letheren, K., Jetten, J., Roberts, J., & Donovan, J. (2021). Robots should be seen and not heard… sometimes: Anthropomorphism and AI service robot interactions. *Psychology and Marketing*, *38*(12), 2393–2406. DOI: 10.1002/mar.21575

Li, Y., Cheng, X., & Yin, G. (2014). *Collaboration information system adoption in hospitality: A case study in China*. https://www.scopus.com/inward/record.uri?eid=2-s2.0-84928641058&partnerID=40&md5=fd837e8dd879b864a3f61f792c633fb5

Liang, W. (2025). Warehouse Automation and Robotics: Revolutionizing Order Fulfillment in E-Commerce.

Liang, G. (2024). Skill Shifts and Retraining Initiatives: Adapting to Post-Pandemic Labor Market Dynamics. *Law and Economy*, *3*(1), 58–61. DOI: 10.56397/LE.2024.01.06

Liberati, A., Altman, D. G., Tetzlaff, J., Mulrow, C., Gotzsche, P. C., Ioannidis, J. P. A., Clarke, M., Devereaux, P. J., Kleijnen, J., & Moher, D. (2009). The PRISMA statement for reporting systematic reviews and meta-analyses of studies that evaluate healthcare interventions: explanation and elaboration. *BMJ, 339*(jul21 1), b2700–b2700. DOI: 10.1136/bmj.b2700

Li, C., & Huang, F. (2024). The impact of virtual streamer anthropomorphism on consumer purchase intention: Cognitive trust as a mediator. *Behavioral Sciences (Basel, Switzerland)*, *14*(12), 1228. DOI: 10.3390/bs14121228 PMID: 39767370

Liew, S. Y., Rana, M. E., Hameed, V. A., & Safavi, S. (2023, December). Navigating the retail 4.0 landscape: The transformative impact of cloud computing. In *2023 IEEE 21st Student Conference on Research and Development (SCOReD)* (pp. 499-507). IEEE.

Li, J., & Liu, X. (2024). Empowering fashion design and intelligent manufacturing with digital twins in the metaverse era. *Asian Social Science*, *20*(4), 1–39. DOI: 10.5539/ass.v20n4p39

Lim, J. (2021). *In review 2021: The year that fashion truly embraced NFTs*. The Industry Fashion. https://www.theindustry.fashion/in-review-2021-the-year-that-fashion-truly-embraced-nfts/

Lin, J., Cronjé, J., Wienrich, C., Pauli, P., & Latoschik, M. E. (2023). Visual indicators representing avatars' authenticity in social virtual reality and their impacts on perceived trustworthiness. *IEEE Transactions on Visualization and Computer Graphics*, *29*(11), 4589–4599. DOI: 10.1109/TVCG.2023.3320234 PMID: 37788202

Lipton, B., Vromen, A., Rutledge-Prior, S., Good, L., Cooper, R., & Foley, M. (2022). *Technology and skills in the future of retail work: A summary of findings*. Designing Gender Equality into the Future of Work.

Li, T., Peng, G., Zhu, Q., & Başar, T. (2022). The confluence of networks, games, and learning a game-theoretic framework for multiagent decision making over networks. *IEEE Control Systems*, *42*(4), 35–67. DOI: 10.1109/MCS.2022.3171478

Liu, Y., Wang, W., & Liu, J. (2025). To NFT or not: A strategic analysis for fashion brands developing digital products in the metaverse. *Journal of Theoretical and Applied Electronic Commerce Research*, *20*(3), 155. DOI: 10.3390/jtaer20030155

Liu, Z. (2025). Human-AI co-creation: A framework for collaborative design in intelligent systems. arXiv. https://arxiv.org/abs/2507.17774 DOI: 10.54941/ahfe1007036

Long, D., & Magerko, B. (2020). What is AI literacy? Competencies and design considerations. In *Proceedings of the 2020 CHI Conference on Human Factors in Computing Systems* (CHI '20), Article 596, 1–16. Association for Computing Machinery. DOI: 10.1145/3313831.3376727

Lu, L., Zhang, P., & Zhang, T. (2023). Leveraging AI-powered service robots to deliver better customer experience: Insights from hotel guests. *International Journal of Hospitality Management*, *108*, 103328. DOI: 10.1016/j.ijhm.2022.103328

Mahmood, H. S., Abdulqader, D. M., Abdullah, R. M., Rasheed, H., Ismael, Z. N. R., & Sami, T. M. G. (2024). Conducting In-Depth Analysis of AI, IoT, Web Technology, Cloud Computing, and Enterprise Systems Integration for Enhancing Data Security and Governance to Promote Sustainable Business Practices. *Journal of Information Technology and Informatics*, *3*(2).

Mahr, D., Odekerken-Schroeder, G., & van Doorn, J. (2025). Evolution of service robots in marketing: A relational framework and future research agenda. *Journal of Business Research*, *192*, 115204. DOI: 10.1016/j.jbusres.2025.115204

Maier, S., Schneider, M., & Feuerriegel, S. (2026). Partnering with generative AI: Experimental evaluation of human-led and model-led interaction in human-AI co-creation. In Proceedings of the 2026 CHI Conference on Human Factors in Computing Systems (CHI '26) (Article 1185). ACM. DOI: 10.1145/3772318.3791185

Manyika, J., Chui, M., Miremadi, M., Bughin, J., George, K., Willmott, P., & Dewhurst, M. (2017). *A future that works: Automation, employment, and productivity*. McKinsey Global Institute.

Mariani, M., & Borghi, M. (2019). Industry 4.0: A bibliometric review of its managerial intellectual structure and potential evolution. *Technological Forecasting and Social Change, 149*, 119752. DOI: 10.1016/j.techfore.2019.119752

Marinova, D., de Ruyter, K., Huang, M.-H., Meuter, M. L., & Challagalla, G. (2017). Getting smart: Learning from technology-empowered frontline interactions. *Journal of Service Research, 20*(1), 29–42. DOI: 10.1177/1094670516679273

Markets, B. T. C. (2025). *Ethereum's role in the NFT boom: A deep dive.* https://www.btcmarkets.net/blog/ethereums-role-in-the-nft-boom-a-deep-dive

Markets.com. (2024). *The role of Polygon (MATIC) in the NFT boom.* https://www.markets.com/education-centre/the-role-of-polygon-in-the-nft-boom

Marques, C. S., & Ferreira, J. J. (2020). Digital transformation in the area of health: A systematic review of 45 years of evolution. *Health Technology (Hong Kong), 10*, 575–586.

McCrae, R. R., & Costa, P. T.Jr. (2008). The five-factor theory of personality. In John, O. P., Robins, R. W., & Pervin, L. A. (Eds.), *Handbook of personality: Theory and research* (3rd ed., pp. 159–181). Guilford Press.

McIntosh, T. R., Susnjak, T., Liu, T., Watters, P., Ng, A., & Halgamuge, M. N. (2024). A game-theoretic approach to containing artificial general intelligence: Insights from highly autonomous aggressive malware. *IEEE Transactions on Artificial Intelligence, 5*(12), 6290–6303. DOI: 10.1109/TAI.2024.3394392

McSorley, J. (2025). *Fashioning the digital self: A critical examination of identity and sustainable user practices within digital fashion ecosystems* [Doctoral dissertation or unpublished manuscript].

Mehmetoglu, M., & Engen, M. (2011). Pine and Gilmore's Concept of Experience Economy and Its Dimensions: An Empirical Examination in Tourism. *Journal of Quality Assurance in Hospitality & Tourism, 12*(4), 237–255. DOI: 10.1080/1528008X.2011.541847

Mehrabian, A., & Russell, J. A. (1974). *An approach to environmental psychology.* MIT Press.

Meirezaldi, O. (2024). The New Realm of Digital Advertising: Unveiling the Impact and Opportunities of Augmented Reality and Virtual Try-Ons Title.

Menidjel, C., Hollebeek, L. D., Leppiman, A., & Riivits-Arkonsuo, I. (2022). Role of AI in enhancing customer engagement, loyalty and loyalty programme performance. In *Handbook of research on customer loyalty* (pp. 316–331). Edward Elgar Publishing. DOI: 10.4337/9781800371637.00032

Millatunnisa, S. (2022). The 4.0 industry technology in fashion industries. In *4th International Conference on Innovation in Engineering and Vocational Education (ICIEVE 2021)* (pp. 336–340). Atlantis Press. DOI: 10.2991/assehr.k.220305.068

Misselhorn, C., Pompe, U., & Stapleton, M. (2021). Ethical considerations regarding the use of social robots in the fourth age. *GeroPsych*, *26*(2), 121–133. DOI: 10.1024/1662-9647/a000088

Mohan Banur, O., Patle, B. K., & Pawar, S. (2024). Integration of robotics and automation in supply chain: A comprehensive review. *Robotic Systems and Applications*, *4*(1), 1–19. DOI: 10.21595/rsa.2023.23349

Molina-Collado, A., Gómez-Rico, M., Sigala, M., Molina, M. V., Aranda, E., & Salinero, Y. (2022). Mapping tourism and hospitality research on information and communication technology: A bibliometric and scientific approach. *Information Technology & Tourism*, *24*(2), 299–340. DOI: 10.1007/s40558-022-00227-8

Mori, M. (1970). The uncanny valley (K. F. MacDorman & N. Kageki, Trans.). IEEE Robotics & Automation Magazine, 19(2), 98–100. (Original work published 1970)DOI: 10.1109/MRA.2012.2192811

Mori, M. (1970). The uncanny valley. Energy, 7(4), 33–35. (Trans. MacDorman, K. F., & Minato, T., 2005)

Morosan, C. (2025). Rethinking information disclosure to GenAI in hotels: An extended parallel process model. *International Journal of Hospitality Management*, *124*, 103965. Advance online publication. DOI: 10.1016/j.ijhm.2024.103965

Mubin, O., Alnajjar, F., Jishtu, N., Alsinglawi, B., & Al Mahmud, A. (2022). Service robots in healthcare: A review on current applications and future trends. *Robotics (Basel, Switzerland)*, *11*(1), 12. DOI: 10.3390/robotics11010012

Mulder, M. (2017). Competence-based education and training: Review and perspectives. *Journal of Agricultural Education and Extension*, *23*(1), 1–19. DOI: 10.1080/1389224X.2016.1274076

Mumford, M. D., Todd, E. M., Higgs, C., & McIntosh, T. (2017). Cognitive skills and leadership performance: The nine critical skills. *The Leadership Quarterly*, *28*(1), 24–39. DOI: 10.1016/j.leaqua.2016.10.012

Mystakidis, S. (2022). Metaverse. *Encyclopedia, 2*(1), 486–497. DOI: 10.3390/encyclopedia2010031

Nalbant, K. G., & Aydın, S. (2023). Development and transformation in digital marketing and branding with artificial intelligence and digital technologies dynamics in the Metaverse universe. *Journal of Metaverse, 3*(1), 9–18. DOI: 10.57019/jmv.1148015

Nass, C., & Moon, Y. (2000). Machines and mindlessness: Social responses to computers. *The Journal of Social Issues, 56*(1), 81–103. DOI: 10.1111/0022-4537.00153

Nass, C., & Reeves, B. (1996). *The media equation: How people treat computers, television, and new media like real people and places*. Cambridge University Press.

Naveen Kumar, R., Janani, M., Ramadevi, V., & Pavithra, R. (2025). *How the metaverse and virtual worlds are redefining fashion marketing and brand engagement*. Digital Transformation and Society. DOI: 10.1108/DTS-09-2024-0177

Nike. (2022). *Nike launches. SWOOSH, a new digital community and experience*. https://about.nike.com/en/newsroom/releases/nike-launches-swoosh-a-new-digital-community-and-experience

NonFungible.com. (2022). *Market trends report: Luxury NFT collections*. https://nonfungible.com

Noone, B. M., & McGuire, K. A. (2014). Pricing in hotel revenue management: A review of the theoretical and empirical literature. *Journal of Revenue and Pricing Management, 13*(6), 482–501.

Nunnally, J. C. (1978). *Psychometric theory* (2nd ed.). McGraw-Hill.

Nuryyev, G., Wang, C. Y. P., Achyldurdyyeva, J., Jaw, B., Yeh, Y. S., Lin, H., & Wu, L. (2020). Blockchain technology adoption behavior and sustainability of the business in tourism and hospitality SMEs: An empirical study. *Sustainability (Basel), 12*(3), 1256. Advance online publication. DOI: 10.3390/su12031256

OECD AI Policy Observatory. (2024). AI investment as a share of GDP, United Kingdom country profile. Paris: OECD. https://oecd.ai/en/dashboards/countries/UnitedKingdom

Office for National Statistics. (2024). Consumer Price Inflation, CPI annual rate (series D7G7), monthly release; Average Weekly Earnings (series KAB9), monthly release. London: ONS. https://www.ons.gov.uk/economy/inflationandpriceindices

Office for National Statistics. (2024). Labour Force Survey, EMP01 (employment by sector), JOBS02 (workforce jobs by industry); Regional GVA (Industry) release, December 2024. London: ONS. https://www.ons.gov.uk/employmentandlabourmarket

Office for National Statistics. (2024). Quarterly National Accounts (series ABMI/YBHA, release 28 March 2024); Index of Services (series S2KU, release 12 February 2024); Labour Productivity statistical bulletins (series LZVB, release 5 April 2024). London: ONS. https://www.ons.gov.uk/economy/economicoutputandproductivity

OKX. (2025). *XRP: The most energy efficient cryptocurrencies*. OKX Learn. https://www.okx.com/en-gb/learn/is-xrp-the-most-energy-efficient-cryptocurrency

Olaosebikan, O., Strickland, L., & Mubin, O. (2023). Exploring emotional engagement in public-space robot deployments. *Computers in Human Behavior*, *139*, 107530. DOI: 10.1016/j.chb.2022.107530

Orazi, D. C., & Van Laer, T. (2023). There and back again: Bleed from extraordinary experiences. *The Journal of Consumer Research*, *49*(5), 904–925. DOI: 10.1093/jcr/ucac022

Ouyang, F., Xu, W., & Cukurova, M. (2023). An artificial intelligence-driven learning analytics method to examine the collaborative problem-solving process from the complex adaptive systems perspective. *International Journal of Computer-Supported Collaborative Learning*, *18*(1), 39–66. DOI: 10.1007/s11412-023-09387-z

Pai, C., Wang, T., Chen, S., & Cai, K. (2018). Empirical study on Chinese tourists' perceived trust and intention to use biometric technology. *Asia Pacific Journal of Tourism Research*, *23*(9), 880–895. DOI: 10.1080/10941665.2018.1499544

Papagiannidis, E., Mikalef, P., & Conboy, K. (2025). Responsible artificial intelligence governance: A review and research framework. *The Journal of Strategic Information Systems*, *34*(2), 101885. Advance online publication. DOI: 10.1016/j.jsis.2024.101885

Parasuraman, A., Zeithaml, V. A., & Berry, L. L. (1988). SERVQUAL: A multiple-item scale for measuring consumer perceptions of service quality. *Journal of Retailing*, *64*(1), 12–40.

Parise, S., Guinan, P. J., & Kafka, R. (2016). Solving the crisis of immediacy: How digital technology can transform the customer experience. *Business Horizons*, *59*(4), 411–420. DOI: 10.1016/j.bushor.2016.03.004

Park, H. (2025). AI-assisted designers or designer-assisted AI? Human-AI collaboration in service design. In Mahamuni, R., & Onkar, P. (Eds.), *ServDes 2025: Empowering diversity, nurturing lasting impact.* Design Research Society., DOI: 10.21606/servdes2025.122

Patel, S. R. (2024). *Role of Computer Vision in Retail Stores* (Doctoral dissertation, Lamar University-Beaumont).

Pavlou, P. A. (2003). Consumer acceptance of electronic commerce: Integrating trust and risk with the technology acceptance model. *International Journal of Electronic Commerce*, *7*(3), 101–134. DOI: 10.1080/10864415.2003.11044275

Pereira, C., Marto, A., Ribeiro, R., Gonçalves, A., Rodrigues, N., Rabadão, C., Costa, R. L. D. C., & Santos, L. (2025). Security and privacy in physical–digital environments: Trends and opportunities. *Future Internet*, *17*(2), 83. DOI: 10.3390/fi17020083

Periyasami, S., & Periyasamy, A. P. (2022). Metaverse as future promising platform business model: Case study on fashion value chain. *Businesses*, *2*(4), 527–545. DOI: 10.3390/businesses2040033

Periyasamy, A. P., & Periyasami, S. (2023). Rise of digital fashion and metaverse: Influence on sustainability. *Digital Economy and Sustainable Development, 1*(1), Article 16.

Picard, R. W. (2000). *Affective computing*. MIT Press. DOI: 10.7551/mitpress/1140.003.0008

Poirier, S. M., Cosby, S., Sénécal, S., Coursaris, C. K., Fredette, M., & Léger, P. M. (2024). The impact of social presence cues in social media product photos on consumers' purchase intentions. *Journal of Business Research*, *185*, 114932. DOI: 10.1016/j.jbusres.2024.114932

Qiu, L., & Benbasat, I. (2009). Evaluating anthropomorphic product recommendation agents: A social relationship perspective to designing information systems. *Journal of Management Information Systems*, *25*(4), 145–182. DOI: 10.2753/MIS0742-1222250405

Radianti, J., Majchrzak, T. A., Fromm, J., & Wohlgenannt, I. (2020). A systematic review of immersive virtual reality applications for higher education. *Computers & Education*, *147*, 103778. DOI: 10.1016/j.compedu.2019.103778

Rai, J. S., & Goyal, M. (2024). Role of the Digital India Movement to Incorporate the Digital Technologies in the Indian System: Adoption of Artificial Intelligence. In *Omnichannel Approach to Co-Creating Customer Experiences through Metaverse Platforms* (pp. 142–155). IGI Global. DOI: 10.4018/979-8-3693-1866-9.ch011

Ranerup, A., & Henriksen, H. Z. (2022). Digital discretion: Unpacking human and technological agency in automated decision making in Sweden's social services. *Social Science Computer Review*, *40*(2), 445–461. DOI: 10.1177/0894439320980434

Ransbotham, S., Kiron, D., Gerbert, P., & Reeves, M. (2017). *Reshaping business with artificial intelligence: Closing the gap between ambition and action*. MIT Sloan Management Review and The Boston Consulting Group.

Rastogi, R., & Saha, S. (2025). Bibliometric Analysis of Omnichannel Retailing Trends: Insights from the Last 19 Years. *Indian Journal of Marketing*, 45-65.

Rastogi, R., & Gupta, S. (2025). The Role of Education in Indigenous Empowerment. In *Indigenous Empowerment through Human-Machine Interactions* (pp. 167–177). Emerald Publishing Limited. DOI: 10.1108/978-1-83608-068-820251010

Rastogi, R., & Iqbal, M. I. (2024). Data security and privacy in RegTech. In *Integrating RegTech Solutions for Industry 4.0* (pp. 185–198). IGI Global.

Recio-Román, A., Recio-Menéndez, M., & Román-González, M. V. (2024). Examining the attitude–behavior gap in EU ecolabel adoption: A mediation path analysis. *Sustainability (Basel)*, *16*(16), 7214. DOI: 10.3390/su16167214

Reisinger, Y., & Crotts, J. C. (2019). Cultural differences in tourism. *Journal of Travel & Tourism Marketing*, *36*(2), 155–168.

Robinson, H., MacDonald, B., & Broadbent, E. (2022). The role of healthcare robots for older people at home: A systematic review. *International Journal of Social Robotics*, *6*(4), 575–591. DOI: 10.1007/s12369-014-0242-2

Robinson, H., MacDonald, B., Kerse, N., & Broadbent, E. (2013). The psychosocial effects of a companion robot: A randomized controlled trial. *Journal of the American Medical Directors Association*, *14*(9), 661–667. DOI: 10.1016/j.jamda.2013.02.007 PMID: 23545466

Roblox. (2022). *Fashion metaverse trends*. https://corp.roblox.com/pdf/2022-metaverse-fashion-trends-report

Rodgers, W., Murray, J. M., Stefanidis, A., Degbey, W. Y., & Tarba, S. Y. (2023). An artificial intelligence algorithmic approach to ethical decision-making in human resource management processes. *Human Resource Management Review*, *33*(1), 100925. DOI: 10.1016/j.hrmr.2022.100925

Rogers, E. M. (2003). *Diffusion of innovations* (5th ed.). Free Press.

Romano, B. (2023). *Retail technology and the customer experience: The use of augmented reality as a shopping tool* [Doctoral dissertation, Swinburne University].

Rosenbaum, M. S., & Massiah, C. (2011). An expanded servicescape perspective. *Journal of Service Management*, *22*(4), 471–490. DOI: 10.1108/09564231111155088

Roy, S. K., Balaji, M. S., Sadeque, S., Nguyen, B., & Melewar, T. C. (2017). Constituents and consequences of smart customer experience in retailing. *Technological Forecasting and Social Change*, *124*, 257–270. DOI: 10.1016/j.techfore.2016.09.022

RTFKT. (2025). *About RTFKT Studios*. https://grokipedia.com/page/RTFKT_Studios

Rust, R. T., & Huang, M.-H. (2014). The service revolution and the transformation of marketing science. *Marketing Science*, *33*(2), 206–221. DOI: 10.1287/mksc.2013.0836

Saini, R. (2023). *Energy consumption in different blockchains*. Medium. https://medium.com/@rahuleum/energy-consumption-in-different-blockchains-9edfe4458609

Salehi, N., Martínez-García, H., Velasco-Quesada, G., & Guerrero, J. M. (2022). A comprehensive review of control strategies and optimization methods for individual and community microgrids. *IEEE Access : Practical Innovations, Open Solutions*, *10*, 15935–15955. DOI: 10.1109/ACCESS.2022.3142810

SaravanaKumar, S., & Badrinarayanan, M. K. (2018). The service automation and robotics in hospitality industry: A study on business implications. *International Journal of Mechanical and Production Engineering Research and Development*, *8*(6), 91–100. DOI: 10.24247/ijmperddec201810

Savela, N., Turja, T., & Oksanen, A. (2018). Social acceptance of robots in different occupational fields: A systematic literature review. *International Journal of Social Robotics*, *10*(4), 493–502. DOI: 10.1007/s12369-017-0452-5

Saviano, M., Thomas, A., Del Prete, M., Verderese, D., & Sasso, P. (2025). The impact of new generative AI chatbots on the switch point (SP): Toward an artificial emotional awareness (AEA). *European Journal of Innovation Management*, *28*(5), 2029–2053. DOI: 10.1108/EJIM-05-2024-0520

Sayem, A. S. M. (2022). Digital fashion innovations for the real world and metaverse. *International Journal of Fashion Design, Technology and Education, 15*(2), 139–141. DOI: 10.1080/17543266.2022.2071139

Schaub, M. (2025). *NFTs as the interface of cryptocurrency: Utility values and the act of wasteful spending.*

Schmager, S., Pappas, I. O., & Vassilakopoulou, P. (2025). Understanding human-centred AI: A review of its defining elements and a research agenda. *Behaviour & Information Technology, 44*(15), 3771–3810. Advance online publication. DOI: 10.1080/0144929X.2024.2448719

Schrier, T., Erdem, M., & Brewer, P. (2010). Merging task-technology fit and technology acceptance models to assess guest empowerment technology usage in hotels. *Journal of Hospitality and Tourism Technology, 1*(3), 201–217. DOI: 10.1108/17579881011078340

Schuller, B. W., Batliner, A., Bergler, C., Messner, E. M., Hamilton, A., Amiriparian, S., Baird, A., Rizos, G., Schmitt, M., Stappen, L., Pérez-Espinosa, H., Frick, M., Doyle, P., Triantafyllopoulos, A., & Müller, R. (2021). The INTERSPEECH 2021 computational paralinguistics challenge: COVID-19 cough, COVID-19 speech, escalation & primates. *Interspeech, 2192–2196*, 431–435. Advance online publication. DOI: 10.21437/Interspeech.2021-19

Seo, S. (2022). When female (male) robot is talking to me: Effect of service robots' gender and anthropomorphism on customer satisfaction. *International Journal of Hospitality Management, 102*, 103166. DOI: 10.1016/j.ijhm.2022.103166

Seufert, J., & Ju, C. (2025). Optimizing Store Cleaning With Robotics for Efficiency, Effectiveness, and Cost Control at Z-Group. In *Exploring Practice-Led Research for Professional Development* (pp. 151–178). IGI Global Scientific Publishing.

Seyitoğlu, F., & Ivanov, S. (2020). A conceptual framework of the service delivery system design for hospitality firms in the (post-) viral world: The role of service robots. *International Journal of Hospitality Management, 91*, 102661. DOI: 10.1016/j.ijhm.2020.102661 PMID: 32952262

Shah, T. R., Purohit, S., Das, M., & Arulsivakumar, T. (2025). Do I look real? Impact of digital human avatar influencer realism on consumer engagement and attachment. *Journal of Consumer Marketing, 42*(4), 416–430. DOI: 10.1108/JCM-05-2024-6853

Shalini, S., & Kavipriya, T. (2024). Perceived risk of disabled's consumption and behavior on clothing under digital fashion-metaverse. In *2024 IEEE 4th International Conference on ICT in Business Industry & Government (ICTBIG)* (pp. 1–11). IEEE.

Shankar, A., Gupta, M., Tiwari, A. K., & Behl, A. (2025). How does convenience impact showrooming intention? Omnichannel retail strategies to manage global retail apocalypse. *Journal of Strategic Marketing*, *33*(3), 394–415. DOI: 10.1080/0965254X.2021.1965189

Shankar, V., Kalyanam, K., Setia, P., Golmohammadi, A., Tirunillai, S., Douglass, T., Hennessey, J., Bull, J. S., & Waddoups, R. (2021). How technology is changing retail. *Journal of Retailing*, *97*(1), 13–27. DOI: 10.1016/j.jretai.2020.10.006

Shih, M. T., Lee, Y. C., Huang, C. M., & Chan, L. (2023). "A feeling of déjà vu": The effects of avatar appearance-similarity on persuasiveness in social virtual reality. *Proceedings of the ACM on Human-Computer Interaction, 7*(CSCW2), Article 1–31.

Shi, X., Luo, H., & Li, J. (2023). Service robots in the hotel industry: Role of AI features in shaping guest acceptance and trust. *Journal of Hospitality Marketing & Management*, *32*(2), 195–215. DOI: 10.1080/19368623.2022.2115279

Shumanov, M., & Johnson, L. (2021). Making conversations with chatbots more personalized. *Computers in Human Behavior*, *117*, 106627. Advance online publication. DOI: 10.1016/j.chb.2020.106627

Siengthai, S., & Pila-Ngarm, P. (2016, August). The interaction effect of job redesign and job satisfaction on employee performance. In *Evidence-based HRM: a Global Forum for Empirical Scholarship* (Vol. 4, No. 2, pp. 162-180). Emerald Group Publishing Limited. DOI: 10.1108/EBHRM-01-2015-0001

Singh, H. (2025). Artificial Intelligence and Robotics Transforming Industries with Intelligent Automation Solutions. *Available at SSRN* 5267868. DOI: 10.2139/ssrn.5267868

Singh, S., Olson, E. D., & Tsai, C. H. (2021). Use of service robots in an event setting: Understanding the role of social presence, eeriness, and identity threat. *Journal of Hospitality and Tourism Management*, *49*, 528–537. DOI: 10.1016/j.jhtm.2021.10.014

So, K. K. F., Kim, H., Liu, S. Q., Fang, X., & Wirtz, J. (2024). Service robots: The dynamic effects of anthropomorphism and functional perceptions on consumers' responses. *European Journal of Marketing*, *58*(1), 1–32. DOI: 10.1108/EJM-03-2022-0176

So, K. K. F., King, C., Sparks, B. A., & Wang, Y. (2016). The role of customer engagement in building consumer loyalty to tourism brands. *Journal of Travel Research*, *55*(1), 64–78. DOI: 10.1177/0047287514541008

Solana. (2025). *Solana blockchain fundamentals*. Oboe. https://oboe.com/learn/solana-blockchain-fundamentals-1qeokyv

Song, C. S., & Kim, Y. K. (2022). The role of the human-robot interaction in consumers' acceptance of humanoid retail service robots. *Journal of Business Research, 146*, 489–503. DOI: 10.1016/j.jbusres.2022.03.087

Soori, M., Arezoo, B., & Dastres, R. (2023). Artificial intelligence, machine learning and deep learning in advanced robotics, a review. *Cognitive Robotics*, *3*, 54–70. DOI: 10.1016/j.cogr.2023.04.001

Stanford, H. A. I. (2026). Responsible AI. In 2026 AI Index Report. Stanford University Human-Centered Artificial Intelligence. https://hai.stanford.edu/ai-index/2026-ai-index-report/responsible-ai

Stewart, J., Honig, P., AlJuburi, L., Autor, D., Berger, S., Brady, P., Fitton, H., Garner, C., Garvin, M., Hukkelhoven, M., Kowalski, R., Milligan, S., O'Dowd, L., Reilly, E., Roberts, K., Robertson, A. S., Taisey, M., Thakkar, R., Van Baelen, K., & Wegner, M. (2020). COVID-19: A catalyst to accelerate global regulatory transformation. *Clinical Pharmacology and Therapeutics*, *109*(6), 1390–1392. DOI: 10.1002/cpt.2046 PMID: 32990986

Sung, E., Kwon, O., & Sohn, K. (2023). NFT luxury brand marketing in the metaverse: Leveraging blockchain-certified NFTs to drive consumer behavior. *Psychology and Marketing*, *40*(11), 2306–2325. DOI: 10.1002/mar.21854

Sun, S., Lee, P. C., Law, R., & Zhong, L. (2020). The impact of cultural values on the acceptance of hotel technology adoption from the perspective of hotel employees. *Journal of Hospitality and Tourism Management*, *44*, 61–69. DOI: 10.1016/j.jhtm.2020.04.012

Tam, F. Y. (2025). Blockchain technology adoption in the fashion supply chain: A review on the practices of fashion brands. *Journal of Global Operations and Strategic Sourcing*.

Tapus, A., Mataric, M. J., & Scassellati, B. (2007). Socially assistive robotics [grand challenges of robotics]. *IEEE Robotics & Automation Magazine*, *14*(1), 35–42. DOI: 10.1109/MRA.2007.339605

Tarek Amer, N., Rahman, S., Baker, J., & Ashill, N. J. (2025). From the internet dilemma to the metaverse dilemma: Exploring luxury brands' adoption of metaverse technologies. *Communications of the Association for Information Systems*, *57*(1), 66. Advance online publication. DOI: 10.17705/1CAIS.05754

Taylor, J. B. (2022). *The Effect of the COVID-19 Pandemic on Online Shopping at a Traditional Ground Retailer: Walmart, Inc from the Customer Perspective* (Doctoral dissertation, Trident University International).

The Interline. (2025). *In the aftershock of the NFT collapse, fashion has a lot to discover about digital ownership.* https://www.theinterline.com/2025/04/29/in-the-aftershock-of-the-nft-collapse-fashion-has-a-lot-to-discover-about-digital-ownership/

The Sandbox. (2024). *Unbox: How top brands push fashion forward and thrive in the Sandbox.* https://www.sandbox.game/en/blog/unbox-how-top-brands-push-fashion-forward-and-thrive-in-the-sandbox/3418/

Tojib, D., Sujan, R., Ma, J., & Tsarenko, Y. (2023). How does service robot anthropomorphism affect human co-workers? *Journal of Service Management, 34*(4), 750–769. DOI: 10.1108/JOSM-03-2022-0090

Torchio, F. (2023). *Survey on automated systems for smart warehouses* (Doctoral dissertation, Politecnico di Torino).

Tornatzky, L. G., & Fleischer, M. (1990). *The processes of technological innovation.* Lexington Books.

Trujillo, A., & Bacciu, C. (2023). Toward blockchain-based fashion wearables in the metaverse: The case of Decentraland. In *2023 IEEE International Conference on Metaverse Computing, Networking and Applications (MetaCom)* (pp. 653–657). IEEE. DOI: 10.1109/MetaCom57706.2023.00115

Trunfio, M., & Rossi, S. (2022). Advances in metaverse investigation: Streams of research and future agenda. *Virtual Worlds, 1*(2), 103–129. DOI: 10.3390/virtualworlds1020007

Tung, V. W. S., & Au, N. (2018). Exploring customer experiences with robotics in hospitality. *International Journal of Contemporary Hospitality Management, 30*(7), 2680–2697. DOI: 10.1108/IJCHM-06-2017-0322

Tushar, W., Yuen, C., Saha, T. K., Nizami, S., Alam, M. R., Smith, D. B., & Poor, H. V. (2023). A survey of cyber-physical systems from a game-theoretic perspective. *IEEE Access : Practical Innovations, Open Solutions, 11*, 9799–9834. DOI: 10.1109/ACCESS.2023.3239834

Tussyadiah, I. P. (2020). A review of research into automation in tourism: Launching the *Annals of Tourism Research* curated collection on artificial intelligence and robotics in tourism. *Annals of Tourism Research, 81*, 102883. DOI: 10.1016/j.annals.2020.102883

Tussyadiah, I., & Park, S. (2018). Consumer evaluation of hotel service robots. In Stangl, B., & Pesonen, J. (Eds.), *Information and communication technologies in tourism 2018* (pp. 308–320). Springer.

UK Intellectual Property Office. (2024). Patent facts and figures, robotics-related filings, year-end 2024 tables. Newport: UKIPO. https://www.gov.uk/government/organisations/intellectual-property-office

UK Statistics Authority. (2018). Code of Practice for Statistics. London: UK Statistics Authority. https://code.statisticsauthority.gov.uk/

UNESCO. (2021). *AI and education: Guidance for policymakers*. UNESCO Publishing.

Upadhyay, N., & Upadhyay, S. (2025). The dark side of non-fungible tokens: Understanding risks in the NFT marketplace from a fraud triangle perspective. *Financial Innovation*, *11*(1), 62. DOI: 10.1186/s40854-024-00684-6

Van Doorn, J., Mende, M., Noble, S. M., Hulland, J., Ostrom, A. L., Grewal, D., & Petersen, J. A. (2017). Domo arigato Mr. Roboto: Emergence of automated social presence in organizational frontlines and customers' service experiences. *Journal of Service Research*, *20*(1), 43–58. DOI: 10.1177/1094670516679272

Vangi, F., Ricci, M., Dastan, M., & Fiorentino, M. (2024). The mixed tangible catalog: Toward tangible and sustainable B2B metaverse fashion showrooms. *IEEE Computer Graphics and Applications*, *44*(4), 40–52. DOI: 10.1109/MCG.2024.3424247 PMID: 38968010

Vega-Carillo, T., & Poikela, M. (2025). Designing co-creative systems: Five paradoxes in human–AI collaboration. *Information (Basel)*, *16*(10), 909. Advance online publication. DOI: 10.3390/info16100909

Venkatesh, V., & Davis, F. D. (2000). A theoretical extension of the Technology Acceptance Model: Four longitudinal field studies. *Management Science*, *46*(2), 186–204. DOI: 10.1287/mnsc.46.2.186.11926

Venturini, A., & Columbano, M. (2024). 'Fashioning' the metaverse: A qualitative study on consumers' value and perceptions of digital fashion in virtual worlds. *Journal of Global Fashion Marketing*, *15*(1), 6–22. DOI: 10.1080/20932685.2023.2234918

Vogue Business. (2022). *Luxury NFTs lose momentum as secondary market demand falls*. https://www.voguebusiness.com

Vogue Business. (2025). *Gucci's evolving metaverse strategy: From Roblox to digital ecosystems*. https://www.voguebusiness.com

Vogue. (2025). *What is fashion metaverse: A complete guide*. Vogue Fashion Institute. https://www.voguefashioninstitute.com/what-is-fashion-metaverse-a-complete-guide/

Vo, T. T., Phan, T. N. N., & Pham, H. H. (2026). AI-driven service design framework in e-commerce application experience. In *Proceedings of the 2025 International Conference on Human-Computer Interaction*. https://doi.org/DOI: 10.21606/servdes2025

Walport, H., Bader, V., & Schmidt, J. (2025). Between transparency and trust: Identifying key factors in AI system perception. *Behaviour & Information Technology*, *44*(6), 840–854. DOI: 10.1080/0144929X.2025.2533358

Wang, B., Han, Y., Xie, F., Kandampully, J., & Duan, Q. (2026). To whom does service robot anthropomorphism appeal? The roles of customer social context, power, and perceived social connectedness. *Service Industries Journal*, *46*(3–4), 183–216. DOI: 10.1080/02642069.2025.2460597

Weiner, B. J. (2009). A theory of organizational readiness for change. *Implementation Science : IS*, *4*(1), 67. DOI: 10.1186/1748-5908-4-67 PMID: 19840381

Williamson, B., & Eynon, R. (2020). Historical threads, missing links, and future directions in AI in education. *Learning, Media and Technology*, *45*(3), 223–235. DOI: 10.1080/17439884.2020.1798995

Wilson, H. J., & Daugherty, P. R. (2018, July–August). Collaborative intelligence: Humans and AI are joining forces. *Harvard Business Review*, *96*(4), 114–123.

Wirtz, J., Patterson, P. G., Kunz, W. H., Gruber, T., Lu, V. N., Paluch, S., & Martins, A. (2018). Brave new world: Service robots in the frontline. *Journal of Service Management*, *29*(5), 907–931. DOI: 10.1108/JOSM-04-2018-0119

Wu, C. H., Liu, C. Y., & Weng, T. S. (2023). Critical factors and trends in NFT technology innovations. *Sustainability (Basel)*, *15*(9), 7573. DOI: 10.3390/su15097573

Wüst, K., & Bremser, K. (2025). Artificial Intelligence in Tourism Through Chatbot Support in the Booking Process—An Experimental Investigation. *Tourism and Hospitality*, *6*(1), 36. Advance online publication. DOI: 10.3390/tourhosp6010036

Xiao, L., & Kumar, V. (2021). Robotics for customer service: A useful complement or an unwelcome threat? *Journal of Service Research*, *24*(1), 138–155. DOI: 10.1177/1094670519878881

Yadav, R., Kumari, S., & Singh, P. (2025). AI ethics: Integrating transparency, fairness, and privacy in AI development. *Applied Artificial Intelligence*, *39*(1), 2463722. Advance online publication. DOI: 10.1080/08839514.2025.2463722

Yang, Y., Chi, M., Bi, X., & Xu, Y. (2024). How does the anthropomorphism of service robots impact employees' role service behavior in the workplace? *International Journal of Hospitality Management*, *122*, 103857. DOI: 10.1016/j.ijhm.2024.103857

Yan, R. (2022). A sustainable fashion industry business model revolution based on the metaverse: Practices and reciprocal processes. *Highlights in Business. Economics and Management*, *4*, 363–369.

Yoganathan, V., Osburg, V. S., Kunz, W. H., & Toporowski, W. (2021). Check-in at the Robo-desk: Effects of automated social presence on social cognition and service implications. *Tourism Management*, *85*, 104309. Advance online publication. DOI: 10.1016/j.tourman.2021.104309

Yukl, G. (2012). Effective leadership behavior: What we know and what questions need more attention. *The Academy of Management Perspectives*, *26*(4), 66–85. DOI: 10.5465/amp.2012.0088

Zalan, T., & Barbesino, P. (2023). Making the metaverse real. *Digital Business, 3*(2), Article 100059.

Zawacki-Richter, O., Marín, V. I., Bond, M., & Gouverneur, F. (2019). Systematic review of research on artificial intelligence applications in higher education. *International Journal of Educational Technology in Higher Education*, *16*(39), 1–27. DOI: 10.1186/s41239-019-0171-0

Zeithaml, V. A., Berry, L. L., & Parasuraman, A. (1996). The behavioral consequences of service quality. *Journal of Marketing*, *60*(2), 31–46. DOI: 10.1177/002224299606000203

Zeng, N., Ryding, D., Vignali, G., & Pantano, E. (2025). AR atmospherics and virtual social presence impacts on customer experience and customer engagement behaviours. *International Journal of Retail & Distribution Management*, *53*(1), 58–73. DOI: 10.1108/IJRDM-08-2023-0520

Zeng, Z., Pantic, M., Roisman, G. I., & Huang, T. S. (2022). A survey of affect recognition methods: Audio, visual, and spontaneous expressions. *IEEE Transactions on Pattern Analysis and Machine Intelligence*, *31*(1), 39–58. DOI: 10.1109/TPAMI.2008.52 PMID: 19029545

Zhang, R. (2022). *Prada launches new Timecapsule NFT collection.* Medium. https://medium.com/marketing-in-the-age-of-digital/prada-launches-new-timecapsule-nft-collection-60787fa268f

Zhang, H., Li, Y., & Wang, X. (2024). Emotionally intelligent service robots powered by large language models: A framework for affective adaptation in customer-facing interactions. *International Journal of Human-Computer Studies, 181*, 103155. DOI: 10.1016/j.ijhcs.2023.103155

Zhang, X., & Bartol, K. M. (2010). Linking empowering leadership and employee creativity: The influence of psychological empowerment, intrinsic motivation, and creative process engagement. *Academy of Management Journal, 53*(1), 107–128. DOI: 10.5465/amj.2010.48037118

Zhao, X., Wei, L., & Huang, Z. (2025). AI-enhanced customer journeys: A framework for dynamic personalisation. *Journal of Information Systems Engineering & Management, 10*(2), 1–18. DOI: 10.55267/iadt/07794

Zheng, C., Ling, S., Huang, Y., & Li, X. (2025). The impact of social presence on purchase intentions of knowledge products among knowledge-based short video users: A moderated mediation model. *Behavioral Sciences (Basel, Switzerland), 15*(8), 1140. DOI: 10.3390/bs15081140 PMID: 40867496

Zhu, K., Kraemer, K. L., & Xu, S. (2006). The process of innovation assimilation by firms in different countries: A technology diffusion perspective on e-business. *Management Science, 52*(10), 1557–1576. DOI: 10.1287/mnsc.1050.0487

Zhu, R., Feng, C., & Liu, Z. (2025). Exploring creativity in human–AI co-creation: A comparative study across design experience. *Frontiers of Computer Science, 7*, 1672735. Advance online publication. DOI: 10.3389/fcomp.2025.1672735

Zulfakar, Z. A., Rahim, F. A., Ching Yat, D. N., Mun, L. H., & Cham, T. H. (2023). Say Aye to AI: Customer Acceptance and Intention to Use Service Robots in the Hospitality Industry. *Lecture Notes in Networks and Systems, 550 LNNS*, 83–92. DOI: 10.1007/978-3-031-16865-9_7

About the Contributors

Mohsin Altaf earned his PhD in Marketing from Universiti Utara Malaysia in 2018, following his MBA from the University of Central Punjab, Lahore. He is currently a Lecturer of Business and Tourism at the Global Banking School, which is affiliated with Canterbury Christ Church University, UK, where he teaches Business and Tourism modules, supervises dissertations, and contributes to curriculum development. Dr. Altaf's academic career spans more than 15 years, with prior appointments as Assistant Professor and Head of Department at The University of Lahore and leadership roles at the University of Central Punjab. His research encompasses marketing, branding, consumer behaviour, Artificial Intelligence in Consumer Behaviour, Robotics in Consumer Behaviour and sustainability, with over 40 peer-reviewed publications in international journals and book chapters. He has successfully supervised more than 50 dissertations, including master's and doctoral theses, and has actively contributed to research collaborations, conferences, and editorial work. He has also served as Editor of the UCP Management Review and as a reviewer for numerous reputable journals. In addition, he is currently an editorial board member for several academic journals, further contributing to the advancement of research and scholarship in marketing and tourism.

***

Duggirala Aravind is currently associated with the Department of Aviation, ASET College of Science and Technology, Chennai, India. He previously completed his Bachelor's degree in Electrical and Electronics Engineering at MITS University. His research interests include aviation operations, renewable energy technologies in air transport, Artificial Intelligence, Computer Science, Logistics and Business management. (aravindaset.edu@gmail.com).

Dhananjay Bhagat is a dedicated academician and researcher currently serving as an Assistant Professor in the Computer Engineering and Technology Department at Dr.Vishwanath Karad MIT World Peace University, Pune. He is pursuing his PhD in Computer Science and Engineering at the prestigious Visvesvaraya National Institute of Technology (VNIT), Nagpur. With a solid educational foundation with a Master's degree in Computer Science and Engineering from Sant Gadge Baba Amravati University, Dhananjay has developed a deep expertise in Artificial Intelligence, Machine Learning, Natural Language Processing, and Computer Vision. Over the years, Dhananjay has contributed significantly to the field through his research and publications. His work has been featured in several Scopus-indexed journals and conferences, covering diverse topics ranging from AI-powered resume optimization to deep learning applications in healthcare. He has also authored books on critical subjects like Business Intelligence and Cyber-Trolling Detection, reflecting his commitment to exploring the intersection of technology and society.

Swati Gupta is an esteemed academician with nearly a decade of international and domestic experience in education, curriculum design, and SME financial consulting. Currently serving as an Associate Professor at Universal AI University, India, she is also the Convener of the IIC Council and a Mentor of Change under the NITI Aayog's Atal Innovation Mission. Dr. Gupta has authored 14 papers, 3 case studies, 6 books, and 5 book chapters. An active researcher and author, she has published extensively and holds a patent in blockchain technology.

Sadia Idrees is a Lecturer in Digital Fashion Marketing and an interdisciplinary researcher specialising in digital transformation, consumer behaviour, and sustainable innovation within the global fashion industry. She holds a PhD in Textile Design, Fashion and Management from the University of Manchester, where her doctoral research focused on enhancing apparel size and fit prediction in fashion e-commerce through the application of 3D body scanning and mixed-methods research. She is a Fellow of the Higher Education Academy (FHEA) and brings over seventeen years of academic, research, and industry experience across the UK and international higher education sectors. Dr Idrees is currently a Lecturer in Business (Marketing) and Module Leader at Global Banking School (University of Suffolk), where she leads the design, delivery, and quality assurance of undergraduate modules in Marketing Fundamentals and Principles of Management. Her teaching practice spans digital and strategic marketing, fashion branding, consumer behaviour, marketing communications, business management, and research methods. She is particularly committed to embedding digital literacy across curricula, integrating artificial intelligence, social media strategy, immersive technologies, user experience design, and data-driven marketing analytics to prepare students for digitally transformed fashion and retail environments. Her pedagogical approach is student-centred, inclusive, and research-led, combining creative practice with analytical and strategic insight. Dr Idrees actively integrates Adobe Creative Suite tools, virtual learning platforms, and interactive digital technologies to enhance accessibility, engagement, and employability. She has extensive experience in assessment design, formative and summative feedback, curriculum review, and academic quality assurance, and is deeply committed to supporting diverse student cohorts through inclusive teaching and pastoral support. Dr Idrees' research focuses on the intersection of digital innovation, sustainability, ethics, and consumer experience in fashion marketing and management. She has published seven peer-reviewed research outputs, including journal articles in The Journal of the Textile Institute and book chapters with Springer Nature. Her work explores AI-enabled sizing systems, virtual fitting rooms, 3D body scanning, and immersive commerce, demonstrating how these technologies can reduce product returns, textile waste, and carbon footprints while enhancing inclusivity, trust, and personalisation in fashion e-commerce. Her research has received international recognition, including the Research Publication Award (Taylor & Francis), Best Researcher Award, and Best Presentation Award at international conferences. She is an active reviewer for leading journals, including Journal of Retailing and Consumer Services and Computers in Human Behavior, and regularly presents her work at global conferences such as 3DBODY.TECH and RARCS. Dr Idrees' current research agenda centres on AI-driven fashion retail systems, metaverse commerce, and responsible digital innovation, with a strong emphasis on sustainability, inclusivity, and digital ethics. She is actively developing UKRI and AHRC funding proposals focused on immersive fashion technologies and net-zero creative industries. Through her teaching, research, and academic leadership, Dr Idrees aims to contribute to globally responsible, digitally innovative, and socially impactful fashion marketing education.

Anchit Jhamb is an Assistant Professor in Management with extensive academic and administrative experience in the fields of finance, management, and business education. He is currently associated with Chandigarh Group of Colleges, Jhanjeri, and is pursuing a PhD from Maharishi Markandeshwar University, Mullana. He has served in multiple academic leadership roles including MBA Program Coordinator, NBA and NAAC Coordinator, Academic Coordinator, and Training and Placement Officer. His teaching and research interests include Financial Management, Corporate Finance, Banking and Insurance, Financial Inclusion, Digital Transformation, and Emerging Business Trends. He has published several research papers and book chapters in reputed journals and edited volumes indexed in Google Scholar and UGC-listed platforms. He is also the author and editor of books in the areas of financial management, e-commerce, and innovative management practices. In addition to academics, he has actively contributed to curriculum development, research mentoring, event management, and quality assurance initiatives in higher education. His research work primarily focuses on workforce transformation, financial systems, digitalization, and sustainable economic development.

Shweta Kaur Khalsa is an Associate Professor in the MBA Department at the School of Commerce and Business Studies, Dayananda Sagar University. She demonstrates rich consulting experience in the fields of marketing and human resource. Her research interests lies in the landscape of marketing and human resource. She is also a successful founder of Amitiel Consultants which is a 5 + years old start up into internet marketing services along with her experience in academia.

Yashwant Rawal has 15 years of experience in both industry and academics. He has worked with the Oberoi Group of Hotels and Carnival Cruise Line USA. His Doctoral Research work was in the area of Staff Turnover Intention in the Hotel industry. He has published more than 57 research papers in Hospitality, Tourism, and Management and presented more than 20 papers at International and National conferences. He was awarded the best role model teacher at the National level in 2019. He also has been awarded with high work engagement award four times at the university level

Shivansu Sachan is Pursuing Ph.D. in Management, a Master's holder from Integral University, Lucknow, Bachelor's from IHM, Lucknow. Academic and professional accomplishments of 6 years with renowned organizations and institutes. His Teaching Career has been very satisfying and have continuously upgraded the skills and knowledge by attending many courses and training programs, seminars, conferences organized by different organizations and also delivered many lectures as a Hospitality sector. He is working as an Assistant Professor at Chhatrapati Shahu ji Maharaj University Kanpur .

Siddharth Singh is an Assistant Professor at the School of Hotel Management, CSJM University, Kanpur. With a strong academic background in hospitality and food sciences, he specializes in culinary innovation, food production management, and sustainable gastronomy. His research interests include food and beverage operations, nutritional analysis of traditional cuisines, and the integration of technology in hospitality education. Siddharth Singh has contributed to several academic initiatives aimed at bridging the gap between culinary practice and scientific understanding, fostering interdisciplinary collaboration between hotel management, food technology, and nutrition. He is passionate about advancing experiential learning in hospitality studies and promoting research-driven approaches to enhance food quality, safety, and sustainability.

N.V. Suresh, renowned academician from Chennai, Tamil Nadu. He completed his schooling in Chennai. He obtained his BA degree (2012) from D.G.Vaishnav College, Madras University. He joined the Bharathiar University received an MBA degree and PGDM. He Awarded Ph.D. in Management from University of Madras. Dr.N.V.Suresh joined then IIKM in June 2014 as an Assistant Professor (2014 to 2016). He moved to Remo International College in June, 2019 as an Assistant Professor and rose through the ranks to the position of Head of the Department in Aug 2020 to July 2021 then Shifted to Working Academy of Maritime Education and Training (Deemed to be University) Presently Working as Associate Professor and Vice-Principal in ASET College of Science and Technology from March 2022 to Till Date. He is the Visiting Faculty for Institute of Distance Education, University of Madras, Tamil Nadu Open University, Anna University and Directorate of Online Education SRM University for MBA from 2015 to till date. Dr.N.V. Suresh has carried out all his research work entirely in India and has published 30 research papers in international journals.

Pradeep Yadav is a distinguished academician with over 20 years of teaching experience and expertise in Computer Science and Engineering. He holds a Ph.D. in Computer Engineering and has an extensive research portfolio, including many international journal papers, multiple conference proceedings, and several books. His research interests span machine learning, artificial intelligence, and data mining. Currently, Dr. Yadav serves as an Associate Professor and Head of Department of CSE at ITM Gwalior, MP, India. He has also contributed significantly to the field through patents and copyrights and actively engages as a reviewer, conference chair and editor.

Index

A

B

C

D

E

www.ingramcontent.com/pod-product-compliance
Lightning Source LLC
LaVergne TN
LVHW080904300926
849162LV00003B/553
9798337366401